Building Habitats on the Moon

Engineering Approaches to Lunar Settlements

More information about this series at http://www.springer.com/series/4097

Haym Benaroya

Building Habitats on the Moon

Engineering Approaches to Lunar Settlements

 Springer

Published in association with
Praxis Publishing
Chichester, UK

Haym Benaroya
Professor of Mechanical & Aerospace Engineering
Rutgers University
New Brunswick, New Jersey, USA

SPRINGER-PRAXIS BOOKS IN SPACE EXPLORATION

Springer Praxis Books
ISBN 978-3-319-68242-6   ISBN 978-3-319-68244-0   (eBook)
https://doi.org/10.1007/978-3-319-68244-0

Library of Congress Control Number: 2017958446

Back cover image courtesy of Berok Khoshnevis, Professor of Engineering, USC and CEO, Contour Crafting Corporation

Cover design: Jim Wilkie
Project Editor: Michael D. Shayler

Printed on acid-free paper

This Springer imprint is published by Springer Nature
The registered company is Springer International Publishing AG
The registered company address is: Gewerbestrasse 11, 6330 Cham, Switzerland

Contents

Acknowledgements viii

Dedication ix

About the Author.. x

Preface.. xi

1   Thoughts on the Moon... **1**
   1.1   J.F.K. at Rice University 1
   1.2   Edward Teller: Thoughts on a lunar base............................ 6

2   Overview and context **12**
   2.1   Why the Moon, and how... 12
   2.2   Other thoughts on a lunar base... 18
   2.3   The historical dimension... 21
   2.4   The economic justifications 22
   2.5   Agreements and ethics 28
   2.6   Pioneering visions... 29
   2.7   Future histories.. 33
   2.8   Interview with David Livingston 34

3   The lunar environment.. **42**
   3.1   Partial gravity... 47
   3.2   Interview with James Logan 48
   3.3   Radiation... 60
   3.4   Regolith... 71
   3.5   Soil mechanics 73
   3.6   Abrasion.. 79

4   Structures **85**
  4.1   Classes 85
  4.2   Concepts and designs 89
  4.3   Lunar architecture and engineering 92
  4.4   Interview with Marc Cohen 93
  4.5   Rigid structures 105
  4.6   Inflatables 114
  4.7   Interview with David Cadogan 130

5   Habitat studies **142**
  5.1   Frameworks 152
  5.2   Structures and living on the Moon 157

6   Lunar-based astronomy **172**

7   Materials and ISRU **178**
  7.1   Interview with Donald Rapp 183
  7.2   Lava Tubes 187

8   Structural design of a lunar habitat **197**
  8.1   Habitat geometry and loads 197
  8.2   Habitat details 201
  8.3   Habitat erection and appearance 203

9   Thermal design **208**
  9.1   A static analysis 210
  9.2   A thermal analysis 216

10 Probability theory and seismic design **224**
  10.1   A definition for probability 225
  10.2   Random variables 226
  10.3   Mathematical expectation 230
  10.4   Useful probability densities 231
  10.5   Random processes 232
  10.6   Power spectrum 235
  10.7   Random vibration 236
  10.8   Seismic activity 238
  10.9   A seismic design 239

11 Reliability and damage **249**
  11.1   Risk 250
  11.2   Risk at a lunar base 252
  11.3   Reliability engineering 262
  11.4   Failure types 266
  11.5   Fatigue life prediction 271
  11.6   Micrometeoroids 275
  11.7   Concluding comments 283

12 Airplanes, redundancy and lunar habitats **286**
   12.1 Aircraft certification 287
   12.2 Redundancy 288
   12.3 Accidents 290
   12.4 Regulation and certification 292
   12.5 Lessons learned 294
   12.6 Lunar structures 296

13 Advanced methodologies **299**
   13.1 Dempster-Shafer Theory 299
   13.2 Performance-based engineering 300

14 Concluding thoughts **307**

Index **310**

Acknowledgements

I humbly acknowledge all those who have dedicated their lives, some literally, to the exploration of space, and the acquisition of the knowledge necessary to do so. I also thank my students, who have contributed greatly to our understanding: Stephen Indyk, Sohrob Mottaghi, Zach Porter, Srikanth Reddy, Florian Ruess, Jackelynne Silva, and Jordan Smart, whose work with me has been published in archival journals and books. I am also glad to thank Dr. Yuriy Gulak for his assistance with this project, as well as many other research efforts over the decades.

I take this opportunity to thank the Department of Mechanical and Aerospace Engineering at Rutgers University for providing me with an exceptional setting to do my research, as I see fit. I thank Mike Shayler for his expert and gracious project editing of this book, making it better organized, grammatically correct, and easier to read. I am very appreciative of Maury Solomon for supporting my publishing efforts with Springer, and of Clive Horwood, who recruited me to Praxis Publishing's space book series. Gratitude goes to Jim Wilkie for the cover design.

Lastly, but not in the least, I am grateful to my family for providing me with the foundation that is needed to go forward and think about things that may not happen for decades.

To
Esther and Alfred,
my parents

About the Author

Haym Benaroya is a Professor of Mechanical and Aerospace Engineering at Rutgers University. His research interests are focused on the conceptualization and analysis of structures placed in challenging environments. These include offshore drilling structures and lunar surface structures for manned habitation.

Professor Benaroya earned his BE degree from The Cooper Union in New York, and his MS and PhD from the University of Pennsylvania in Philadelphia. Prior to joining Rutgers University in 1989, Professor Benaroya was a senior research engineer at Weidlinger Associates in New York for eight years.

While at Rutgers, Professor Benaroya has mentored twelve students to their PhDs and a similar number to their MS degrees. He is the author of over 80 refereed journal publications, two text books – one in Vibration and the other in Probabilistic Modeling – and two research monographs with two former PhD students on structural dynamics in the ocean.

His book, **From Dust to Gold - Building a Future on the Moon and Mars**, presents a vision of settling the Solar System. It was awarded the 2012 Best Engineering Sciences Book by the International Academy of Astronautics. Professor Benaroya is an elected member of the International Academy of Astronautics.

Preface

"The greatest adventure of all time."

When most of us think about the Moon, we don't really have a good sense about its size, its constituent elements, and the harshness of its environment. The Moon is truly a small body, quite a bit smaller than the Earth. Earth's total surface area is 510.1×10^6 km^2, with a land area of 149.8×10^6 km^2. This is almost four times the total lunar surface area of 37.9×10^6 km^2. Mars, for comparison, is about 144.8×10^6 km^2. Coming from Earth, humans have a certain instinct regarding how far the horizon should be, and astronauts who walked on the Moon remarked about their surprise at how close the lunar horizon seemed to them.

This is a book about the structural engineering of settlements on the lunar surface, and a bit on sublunar sites. It is about the Moon, rather than Mars, for two reasons. The first is that we know the most about the Moon. Humans have been there six times. The second reason is the Moon's proximity to Earth. Consider this: if the Earth is 100 units in diameter, the Moon is 3000 units away and is 27 units in diameter, and Mars is 428,000 units away and 53 units in diameter. The surface of Mars receives about 44 percent of the sunlight intensity that the surfaces of the Earth and Moon experience. Getting people safely to the Moon for a very long stay is barely within the abilities of our highly technological society. Some aspects are a bit beyond our abilities, but much is within that horizon of our talents. Mars, however, is another story. Much of what we need to be able to do to get people to Mars is beyond that horizon. It is not a matter of money, but rather a matter of experience in space, of surviving in space. More specifically, experience and survival on the Moon. If someone gave me as much money as I wanted to make humanity a spacefaring civilization, I would still choose the Moon as our first goal. This book will provide some insights into that choice.

We'll discuss all the issues in detail: engineering, as well as human physiology and psychology. And that is the critical difference between most engineering projects on Earth and those involving space. Human survival is part of the equation – physical survival and psychological survival. These considerations are a part of the very difficult challenge of designing habitats for humans on the Moon, or elsewhere outside of Earth's protective atmosphere.

We will approach the structural engineering in such a way that an intelligent and generally well-educated reader can follow most of the ideas that are presented. Of course, there will be points in our development where an understanding of physics and mathematics is required. While we are eager to make the material accessible to a large group of readers, we are also devoted to the creation of a book that can be used to inform the engineers that will eventually, hopefully soon, be called upon to prepare preliminary structural designs for the lunar surface.

Figure 0.1. *Earthrise* is the name given to a photograph of the Earth taken by astronaut William Anders in 1968 during the Apollo 8 mission. (Courtesy: NASA)

The material in this book could have been developed in a variety of sequences, all of which would have met their mission. That is, to provide an overview with some details about the engineering, scientific, medical, psychological, political, and economic aspects that have a role in the creation of a significant and permanent human presence on the Moon. In our effort, we first discuss the lunar environment, and how it affects humans in so many ways, because the structures designed to house them must also safeguard them against that environment. Additionally, given the indoor life awaiting the inhabitants of the Moon, we need to provide them with a psychologically pleasing and nurturing haven within which to thrive.

Of course, we discuss in great detail the design of structures primarily for the lunar surface, using ideas from the mechanics of solids, the idea of stress, strain and displacement. We detail how engineers couple material properties, structural geometry and engineering reliability to create structures that perform as desired, effectively, safely, for a prescribed amount of time. A significant list of references to the literature is provided as follow-on for the interested reader. We also include a large set of references to our own work, since we are most familiar with these. Sometimes, we are able to walk through all this a step at a time, descriptively with words and their shorthand, mathematics, while at other times the presentation is in a highlighted way, but still mentioning the key ideas. By the end, the devoted reader will have a good understanding of structural mechanics for the lunar surface, and more.

With this book, we offer the reader an appreciation of the tremendous breadth of both the intellectual and financial resources needed to embark on a manned space program to the Moon, Earth's closest body, but our focus is on the various designs proposed for safe habitats for living on the lunar surface. These designs must eventually accommodate the physiological and psychological challenges that humans will face in such environments.

A search of the literature for lunar settlements and bases will lead in the majority to the consideration of longer-term facilities that are only possible after the creation of a lunar industrial infrastructure. The key is, and has always been, how to get there (the expansive lunar bases that our mind's eye imagines will evolve over the next century or two) from here (designing and placing the first habitats, of small volume, that can house a few pioneers). At this time (late 2017), no nation or industrial entity has a program or plan to send astronauts to the Moon for permanent or semi-permanent habitation. This is not because we cannot, but because the process is long and the United States, still the preeminent space-exploring nation, does not yet see this as a high priority. Industries are making major strides, but primarily with transportation systems, with Bigelow the exception by creating a habitat. But it appears that the Moon may again be of interest to the national space programs of China, Japan, Russia, and slowly, the United States. It will happen, the question is when.

The goal of this book then is to provide an overview of lunar structural concepts, understand how they can be designed, and acknowledge the severe environmental concerns. We know that the human body and mind that has evolved for the 1 g radiation-free surface of the Earth over millions of years, both physically and psychologically, is not naturally suited to space without advances in medical science. Humans on the lunar surface can be protected against the radiation, the micrometeoroids and the vacuum, but the low gravity effects are, as yet, only partially understood. As structural designers working with architects and a host of other professionals, we recognize that there are numerous design constraints for a lunar structure, as compared to a typical Earth structure. Our experience base needs broadening, but we know that the first long-term manned facilities will be works in progress from which we learn much while they protect our settlers.

It is worth noting that while the actual analysis and design of lunar surface structures are, for the most part, understood from the perspectives of structural engineering, the challenges arise from the need to characterize the lunar environment, the uncertainties associated with that environment, and the environmental effects on biological entities, materials, electronics, machines and structures. There are also the serious challenges to our abilities when we try to estimate structural and system risk and reliability, which are partially related to the lunar environment and partially related to the complexity of the structural system. In a few words, once the environment is defined we can go forward with a structural analysis and design, but the uncertainties across the relevant disciplines preclude a straightforward path. Other engineering disciplines that are part of the overall design of a lunar habitat, from chemical to electrical, are not mentioned here. And they face similar challenges.

The question we ask ourselves is: What is the design problem? Can we define the problem in a way that helps us begin to solve it? The short answer is: To design a structure for the lunar surface that can safely be inhabited by humans, where they can live, work, and thrive as individuals. That goal may not initially be met in all its dimensions, but that is our goal in time.

This book is an effort to gather relevant ideas on the design of a lunar habitat. It is not, and cannot be, conclusive, or all inclusive. It is representative. Too much yet needs to be discovered and figured out for any study to be conclusive. Of the many thousands of published reports, papers and books, only a very small fraction have been referenced here. Many high-quality studies have not been referenced, with regret. The topic of this book could have led to an encyclopedia of many volumes that would still be incomplete. The reader, however, will come away with some general – and some specific – insights into the challenges that face us when we eventually decide to return to the Moon, with people who will reside there for the long term, and eventually, for some, the remainder of their lives.

As President Kennedy said in his speech at Rice University: "… we have given this program a high national priority – even though I realize that this is in some measure an act of faith and vision, for we do not now know what benefits await us …" We have some ideas of the benefits, but there are many benefits that will flow from discoveries that we cannot even imagine, so how can we imagine the second and third generation discoveries that will emanate from those that we cannot yet imagine?

The story is still in its infancy.

1

Thoughts on the Moon

"We choose to go to the moon. We choose to go to the moon in this decade and do the other things, not because they are easy, but because they are hard, because that goal will serve to organize and measure the best of our energies and skills, because that challenge is one that we are willing to accept, one we are unwilling to postpone, and one which we intend to win, and the others, too."

1.1 J.F.K. AT RICE UNIVERSITY

Address at Rice University on the Nation's Space Effort by President John F. Kennedy, Houston, Texas, September 12, 1962.

President Pitzer, Mr. Vice President, Governor, Congressman Thomas, Senator Wiley, and Congressman Miller, Mr. Webb, Mr. Bell, scientists, distinguished guests, and ladies and gentlemen:

I appreciate your president having made me an honorary visiting professor, and I will assure you that my first lecture will be very brief.

I am delighted to be here and I'm particularly delighted to be here on this occasion.

We meet at a college noted for knowledge, in a city noted for progress, in a State noted for strength, and we stand in need of all three, for we meet in an hour of change and challenge, in a decade of hope and fear, in an age of both knowledge and ignorance. The greater our knowledge increases, the greater our ignorance unfolds.

Despite the striking fact that most of the scientists that the world has ever known are alive and working today, despite the fact that this Nation's own scientific manpower is doubling every 12 years in a rate of growth more than three times that of our population as a whole, despite that, the vast stretches of the unknown and the unanswered and the unfinished still far outstrip our collective comprehension.

No man can fully grasp how far and how fast we have come, but condense, if you will, the 50,000 years of man's recorded history in a time span of but a half a century. Stated in these terms, we know very little about the first 40 years, except at the end of

© Springer International Publishing AG 2018
H. Benaroya, *Building Habitats on the Moon*, Springer Praxis Books,
https://doi.org/10.1007/978-3-319-68244-0_1

them advanced man had learned to use the skins of animals to cover them. Then about 10 years ago, under this standard, man emerged from his caves to construct other kinds of shelter. Only five years ago man learned to write and use a cart with wheels. Christianity began less than two years ago. The printing press came this year, and then less than two months ago, during this whole 50-year span of human history, the steam engine provided a new source of power.

Newton explored the meaning of gravity. Last month electric lights and telephones and automobiles and airplanes became available. Only last week did we develop penicillin and television and nuclear power, and now if America's new spacecraft succeeds in reaching Venus, we will have literally reached the stars before midnight tonight.

Figure 1.1. Attorney General Kennedy, McGeorge Bundy, Vice President Johnson, Arthur Schlesinger, Admiral Arleigh Burke, President Kennedy, Mrs. Kennedy watching the 15-minute historic flight of astronaut Alan Shepard on television, May 5, 1961, the first American in space. (Cecil Stoughton, photographer. Courtesy John Fitzgerald Kennedy Library, Boston, MA)

This is a breathtaking pace, and such a pace cannot help but create new ills as it dispels old, new ignorance, new problems, new dangers. Surely the opening vistas of space promise high costs and hardships, as well as high reward.

So it is not surprising that some would have us stay where we are a little longer to rest, to wait. But this city of Houston, this State of Texas, this country of the United States was not built by those who waited and rested and wished to look behind them. This country was conquered by those who moved forward – and so will space.

William Bradford, speaking in 1630 of the founding of the Plymouth Bay Colony, said that all great and honorable actions are accompanied with great difficulties, and both must be enterprised and overcome with answerable courage.

If this capsule history of our progress teaches us anything, it is that man, in his quest for knowledge and progress, is determined and cannot be deterred. The exploration of space will go ahead, whether we join in it or not, and it is one of the great adventures of all time, and no nation which expects to be the leader of other nations can expect to stay behind in this race for space.

Those who came before us made certain that this country rode the first waves of the industrial revolutions, the first waves of modern invention, and the first wave of nuclear power, and this generation does not intend to founder in the backwash of the coming age of space. We mean to be a part of it – we mean to lead it. For the eyes of the world now look into space, to the moon and to the planets beyond, and we have vowed that we shall not see it governed by a hostile flag of conquest, but by a banner of freedom and peace. We have vowed that we shall not see space filled with weapons of mass destruction, but with instruments of knowledge and understanding.

Yet the vows of this Nation can only be fulfilled if we in this Nation are first, and therefore, we intend to be first. In short, our leadership in science and in industry, our hopes for peace and security, our obligations to ourselves as well as others, all require us to make this effort, to solve these mysteries, to solve them for the good of all men, and to become the world's leading space-faring nation.

We set sail on this new sea because there is new knowledge to be gained, and new rights to be won, and they must be won and used for the progress of all people. For space science, like nuclear science and all technology, has no conscience of its own. Whether it will become a force for good or ill depends on man, and only if the United States occupies a position of pre-eminence can we help decide whether this new ocean will be a sea of peace or a new terrifying theater of war. I do not say that we should or will go unprotected against the hostile misuse of space any more than we go unprotected against the hostile use of land or sea, but I do say that space can be explored and mastered without feeding the fires of war, without repeating the mistakes that man has made in extending his writ around this globe of ours.

There is no strife, no prejudice, no national conflict in outer space as yet. Its hazards are hostile to us all. Its conquest deserves the best of all mankind, and its opportunity for peaceful cooperation many never come again. But why, some say, the moon? Why choose this as our goal? And they may well ask why climb the highest mountain? Why, 35 years ago, fly the Atlantic? Why does Rice play Texas?

We choose to go to the moon. We choose to go to the moon in this decade and do the other things, not because they are easy, but because they are hard, because that goal will serve to organize and measure the best of our energies and skills, because that challenge is one that we are willing to accept, one we are unwilling to postpone, and one which we intend to win, and the others, too.

It is for these reasons that I regard the decision last year to shift our efforts in space from low to high gear as among the most important decisions that will be made during my incumbency in the office of the Presidency.

In the last 24 hours we have seen facilities now being created for the greatest and most complex exploration in man's history. We have felt the ground shake and the air shattered by the testing of a Saturn C-1 booster rocket, many times as powerful as the Atlas which launched John Glenn, generating power equivalent to 10,000 automobiles with their accelerators on the floor. We have seen the site where five F-1 rocket engines, each one as powerful as all eight engines of the Saturn combined, will be clustered together to make the advanced Saturn missile, assembled in a new building to be built at Cape Canaveral as tall as a 48 story structure, as wide as a city block, and as long as two lengths of this field.

Within these last 19 months at least 45 satellites have circled the Earth. Some 40 of them were made in the United States of America and they were far more sophisticated and supplied far more knowledge to the people of the world than those of the Soviet Union.

The Mariner spacecraft now on its way to Venus is the most intricate instrument in the history of space science. The accuracy of that shot is comparable to firing a missile from Cape Canaveral and dropping it in this stadium between the 40-yard lines.

Transit satellites are helping our ships at sea to steer a safer course. Tiros satellites have given us unprecedented warnings of hurricanes and storms, and will do the same for forest fires and icebergs.

We have had our failures, but so have others, even if they do not admit them. And they may be less public.

To be sure, we are behind, and will be behind for some time in manned flight. But we do not intend to stay behind, and in this decade, we shall make up and move ahead.

The growth of our science and education will be enriched by new knowledge of our universe and environment, by new techniques of learning and mapping and observation, by new tools and computers for industry, medicine, the home as well as the school. Technical institutions, such as Rice, will reap the harvest of these gains.

And finally, the space effort itself, while still in its infancy, has already created a great number of new companies, and tens of thousands of new jobs. Space and related industries are generating new demands in investment and skilled personnel, and this city and this State, and this region, will share greatly in this growth. What was once the furthest outpost on the old frontier of the West will be the furthest outpost on the new frontier of science and space. Houston, your City of Houston, with its Manned Spacecraft Center, will become the heart of a large scientific and engineering community. During the next 5 years, the National Aeronautics and Space Administration expects to double the number of scientists and engineers in this area; to increase its outlays for salaries and expenses to $60 million a year; to invest some $200 million in plant and laboratory facilities; and to direct or contract for new space efforts over $1 billion from this Center in this City.

To be sure, all this costs us all a good deal of money. This year's space budget is three times what it was in January 1961, and it is greater than the space budget of the

Figure 1.2.   *From the Earth to the Moon* (French: *De La Terre à la Lune,* 1865) is a humorous science fantasy novel by Jules Verne and is one of the earliest entries in that genre. It tells the story of a Frenchman and two well-to-do members of a post-American Civil War gun club who build an enormous sky-facing cannon, the *Columbiad*, and launch themselves in a projectile/spaceship from it to a Moon landing.

previous eight years combined. That budget now stands at $5400 million a year – a staggering sum, though somewhat less than we pay for cigarettes and cigars every year. Space expenditures will soon rise some more, from 40 cents per person per week to more than 50 cents a week for every man, woman and child in the United States, for we have given this program a high national priority – even though I realize that this is in some measure an act of faith and vision, for we do not now know what benefits await us. But if I were to say, my fellow citizens, that we shall send to the moon, 240,000 miles away from the control station in Houston, a giant rocket more than 300 feet tall, the length of this football field, made of new metal alloys, some of which have not yet been invented, capable of standing heat and stresses several times more than have ever been experienced, fitted together with a precision better than the finest watch, carrying all the equipment needed for propulsion, guidance, control, communications, food and survival, on an untried mission, to an unknown celestial body, and then return it safely to Earth, re-entering the atmosphere at speeds of over 25,000 miles per hour, causing heat about half that of the temperature of the sun – almost as hot as it is here today – and do all this, and do it right, and do it first before this decade is out – then we must be bold.

I'm the one who is doing all the work, so we just want you to stay cool for a minute. [laughter].

However, I think we're going to do it, and I think that we must pay what needs to be paid. I don't think we ought to waste any money, but I think we ought to do the job. And this will be done in the decade of the sixties. It may be done while some of you are still here at school at this college and university. It will be done during the terms of office of some of the people who sit here on this platform. But it will be done. And it will be done before the end of this decade.

And I am delighted that this university is playing a part in putting a man on the moon as part of a great national effort of the United States of America.

Many years ago, the great British explorer George Mallory, who was to die on Mount Everest, was asked why did he want to climb it. He said, "Because it is there."

Well, space is there, and we're going to climb it, and the moon and the planets are there, and new hopes for knowledge and peace are there. And, therefore, as we set sail we ask God's blessing on the most hazardous and dangerous and greatest adventure on which man has ever embarked.

Thank you.

1.2 EDWARD TELLER: THOUGHTS ON A LUNAR BASE

"The Moon can be a nascent civilization."

I would like to start with a statement that I expect, and even hope, may be controversial.[1] I believe there is a very great difference between the space station now being planned and any activity on the Moon now under discussion. I believe that in the space station we should do as much as possible with robots for two simple reasons. There is nothing in space—practically nothing—except what we put there. Therefore, we can foresee the conditions under which we are going to work, and, in general, I think robots are less trouble than people.

The other reason is that, apart from experiments and special missions that we have in space, we do not want to proceed to change anything in space, whereas on the Moon we will want to change things. Likewise, on the Moon, we will find many things that we do not expect. Adapting robots to all the various tasks that may come up, and that we do not even foresee, is not possible.

The space station is obviously extremely interesting for many reasons. However, that is not what I want to talk about except to state that, of course, the space station is apt to develop into a transfer station to the Moon. Therefore, its establishment is not independent of what we are discussing here.

I would like to look forward to an early lunar colony. I do not want to spend time in making estimates but simply want to say that it would be nice to have a dozen people on the Moon as soon as possible. I think we could have it in ten years or so. When I say 12 people, I do not mean 12 people to stay there but to have 12 people at all times, to serve as long as it seems reasonable. To me, three months is the kind of period from which you could expect a good payoff for having made the trip. Longer rotations than that might be a little hard, and efficiency might come down. But all this is, of course, a wild estimate on my part.

What kind of people should be there? It will be necessary to have all of them highly capable in a technical manner, and I believe that they should perform all kinds of work. Probably at least half of them, after coming back to Earth, should get the Nobel Prize. The result will be that we will soon run out of Nobel Prizes because I believe there will be very considerable discoveries.

Also, if you have 12 people you probably ought to have a Governor. I have already picked out the Governor to be, of course, Jack Schmitt. Furthermore, I would like to tell you that when I first testified about space, and was asked whether there should be women astronauts, I proposed that all astronauts should be women. The packaging of intelligence in women is more effective in terms of intelligence per unit weight. However, in view of the strong sentiment for ERA, I think I might compromise with an equal number of women and men. That arrangement has all kinds of advantages.

I believe that the discussion here has had plenty of emphasis on what I know will be the main practical result of a lunar base — use as a refueling station. It will supply both portable energy in a concentrated form and portable fuel for refueling rockets, primarily in the form of oxygen extracted out of lunar rocks. The only question is how to do it. My first idea was, of course, we should do it with nuclear reactors. Perhaps the environmental movement, the Sierra Club, may not have an arm that extends beyond one light second. On the other hand, we will have some problems, problems of cooling. However, most of the energy might be needed to squeeze oxygen out of iron oxide, and that simply means a high temperature. You may not need a lot of machinery, and some of the energy can be, in this way, usefully absorbed right inside the reactor. What remains probably should be converted to electricity.

The other possibility is solar energy. I am strongly inclined to believe that solar energy will be quite useful for two reasons. First, great advances have been made in solar cells, particularly with regard to Ovshinsk's idea of utilizing amorphous semi-conductors. The point is that they are not very good conductors of electricity and therefore must be thin, but, on the other hand, amorphous materials are very good absorbers of light and therefore can be thin. Methods to fabricate them have indicated that you can, with practical certainty, get down to one dollar per peak watt.

Figure 1.3. (Left) Launching of the Mercury-Redstone 3 rocket from Cape Canaveral on astronaut Alan B. Shepard's Freedom 7 suborbital mission. NASA research mathematician Katherine Johnson did the trajectory analysis for the mission, America's first human spaceflight. (Right) Shepard, in his silver pressure suit with the helmet visor closed, prepares for his launch on May 5, 1961. Shepard's capsule lifted off at 9:34 a.m. from Launch Complex 5 at Cape Canaveral Air Force Station, and flew a suborbital trajectory lasting 15 minutes and 22 seconds. During the rocket's acceleration, Shepard was subjected to 6.3 g just before shutdown of the Redstone engine, two minutes and 22 seconds after liftoff. Soon after, America's first space traveler got his first view of the Earth. "What a beautiful view," Shepard said. His spacecraft splashed down in the Atlantic Ocean, 302 miles from Cape Canaveral, where he and Freedom 7 were recovered by helicopter and transported to the awaiting aircraft carrier USS Lake Champlain. (Courtesy: NASA)

There is, however, another advantage to solar power and that is if you do not want power, but just want high temperature for driving oxygen out of oxides, you may not need mirrors that have to be moved. It might be sufficient to have the right kind of surface that absorbs and emits ultraviolet but is highly reflective in the visible and the infrared. In equilibrium with solar radiation, this will give high temperatures; the farther you go in the ultraviolet the more you can approach the maximum temperature obtainable, the surface temperature of the Sun. If you try to approach this limit, then the energy content — the power — will be small because it utilizes a smaller portion of the solar spectrum. But the temperature you can get is high. What the optimum is where you will want to compromise, I do not know.

Let me extend this idea one step further. I would not only like to get very high temperatures; I also want to get very low temperatures as cheaply as possible. You can achieve the latter during the 14-day lunar night. If you isolate yourself from the surface of the Moon, put your apparatus on legs and put some space in between — all very cheap arrangements — you can approach temperatures in the neighborhood of 2.7 degrees Absolute. In this way, you can get low temperature regions of large volume and high temperature regions of large volume.

Now, I would like to talk about one practical point that may not have been discussed, namely, the question of where on the Moon the colony should be. I would like to go to one of the poles because I would like to have the choice between sunlight and shade with little movement. Furthermore, it would be a real advantage to establish the colony in and around a crater where you might have even permanent shade in some places and where moving away from the rim on one side or the other you can vary conditions quite fast. Of course, it is of importance not only to position yourself in regard to the Sun but also in regard to the Earth. For many purposes, you want to see the Earth in order to observe it. For other purposes, for instance astronomy, you want to be shielded from the Earth, not to be disturbed by all the terrestrial radio emission. All these conditions will be best satisfied in a crater near a pole.

I have a little difficulty in reading the lunar maps. There seem to be three good craters in the immediate vicinity of the south pole but no good craters near the north pole, or vice versa I am not quite sure. At any rate, I want to go to the pole that has the craters.

The purpose of all this is obviously what I have said to begin with and what you all realize — refueling and energy. Oxygen is the main point, but it would be nice also to have hydrogen. Hydrogen we could get from the Earth much more cheaply than the oxygen, but still it is one-ninth the cost of oxygen plus the considerable weight of the tank. Hydrogen has been deposited in the lunar dust by the solar wind over geologic time, and the mass of hydrogen in that lunar dust, as far as I know, is not much less than one part in ten thousand. Without having made a decent analysis, my hunch is that it is easier to move the lunar dust a few miles on the Moon than to come all the way from the Earth even though you have to move ten thousand times the mass. If you can distill oxygen out of iron oxide, you certainly can distill hydrogen out of the lunar dust. Furthermore, Jack Schmitt tells me that there is a possibility of finding hydrogen, perhaps even hydrogen that is four and one-half billion years old, in other parts of the Moon in greater abundance than what we see in the average lunar dust.

All of this is, of course, of great importance and perhaps serves as a little illustration of what kind of constructions we are discussing. Obviously, we will have to try to make these constructions with tools as light as can be transported from the Earth. In planning the lunar colony, special tools and special apparatus have to be fabricated on the Earth, specifically adapted to the tasks already described as well as others.

I would like to make a special proposal. I believe that surveillance of the Earth — permanent, continuous surveillance that is hard to interfere with — is an extremely important question; important to us, important for the international community, important for peace keeping. There have been proposals, and I am for them, to guarantee present observation of facilities by treaties. On the other hand, treaties not only can be broken; treaties have been broken. It is in everyone's best interest to have observation stations that are not easy to interfere with.

I would like to take the biggest chunks that I could get off the Moon and put them into a lunar orbit, perhaps 120 and 240 degrees away from the Moon. Of course, they will be very small compared to the Moon but maybe quite big compared to other objects that we put into space. If the Moon and these two additional satellites are available for observation, then we can have a continuous watch on all of the Earth with somewhat lesser information around the pole. The latter also can be obtained with additional expenditure, but to have 95 percent of the most interesting part of the Earth covered continuously would be already a great advantage. I would be very happy if, on these observation stations, we would do what we should have done with our satellites and are still not doing, namely, make the information of just the photographs obtained from the satellites universally available. I believe that would be a great step forward in international cooperation, international relations, and peace keeping.

Traveling to these artificial satellites from the Moon is a much smaller job than reaching them from the Earth. Since you stay on the same orbit, you just have to have a very small additional velocity after leaving the Moon, wait until you are in the right position, and then use a retrorocket. The total energy for that is small, and if you produce the rocket fuel on the Moon, then I think you have optimal conditions.

I also would like to have a satellite with a special property. It should have as big a mass as possible, built up from a small mass in the course of time. But, furthermore, I want it to rotate in such a manner that instead of turning the same face all the time to the Earth it should turn the same face all the time to the Sun. If you can do that, then half of the surface will be in permanent night, half in permanent illumination, and whatever we can do on the Moon, for instance setting up a permanent low-temperature establishment, you can do that very much better on these satellites.

Now, I would like to finish up by making a very few remarks on purely scientific work that will become possible. In the vacuum of the Moon we can work with clean surfaces. It is obvious that surface chemistry could make big strides. This can be done equally well in the space station, and, in this respect, the Moon does not have an obvious advantage.

Where you do get an obvious advantage is in astronomical observations where you want the possibility to collimate in a realty effective manner. When you want to look at X-rays or gamma rays from certain directions, all you need to do is to drill a deep hole that acts as a collimator and have the detectors at its bottom. You would have to have a considerable number of these holes, but I believe that it will be much cheaper than to have a considerable number of observation apparatus shot out from the Earth, particularly because the mass for collimation will be not available in space stations except at a considerable cost. The same holes may be used for high energy cosmic rays.

Another obvious application is in high-energy physics. As the size of accelerators kept going up, many years ago our very good friend Enrico Fermi, at a Physical Society meeting as far as I know, made the proposal in completely serious Italian style that sooner or later we will make an accelerator around the equator of the Earth. Well, we are approaching that — at least we are planning an accelerator that takes in a good part of Texas. I am not quite sure that we should do that. Let us wait until we get to the Moon. (That might happen almost as soon as a giant accelerator can be constructed.) We actually could have an accelerator around the equator of the Moon. Taking advantage of the vacuum available, you only need the deflecting magnets and the accelerating stations, and these can be put point for point rather than continuously.

I have been interested for many years in the remarkable discovery of Klebesadel at Los Alamos of gamma ray bursts that last for longer than 15 milliseconds and less than 100 seconds, have their main energy emission between 100 and 200 kilovolts, but seem to have components far above a million volts, too. I believe everybody is in agreement that these come from something hitting neutron stars and converting the energy into gamma rays. But most people believe that they come from nearby regions of our galaxy and are, therefore, isotropic. Actually, the number of observations depends on the intensity in such way as though from more distant places we do not get as many as expected. The usual explanation is that we get these from farther places and we get them only from the galactic disc rather than a sphere. Unfortunately, these bursts are so weak that the directional determinations cannot be made. On the Moon, you could deploy acres of gamma ray detectors of various kinds and leave them exposed to the gamma rays or cover them up with one gram per square centimeter, five grams per square centimeter, or ten grams per square centimeter so that with some spectral discrimination you will get a greater intensity from perpendicular incidence than from oblique incidence. As this apparatus will look into the plane of the galaxy, into the main extension of the galaxy, or toward the galactic pole, you should see a difference, a deviation from spherical distribution, for these weakest bursts, essentially bursts of 10^{-5} to 10^{-7} ergs/cm^2/s.

A very good friend, Montgomery Johnson (who unfortunately died a few months ago) and I had made an assumption that these radiations really do not come from the galaxy but from outer space, from regions where the stars are dense and where collisions between neutron stars and dense stars like the white dwarfs may occur. Good candidates are the globular clusters, but there may be other dense regions in the universe as well. If this hypothesis turns out to be correct, then the reason you find fewer events at great distances are cosmological reasons—curvature of space, a greater red-shift, lesser numbers of neutron stars and white dwarfs in the distant past, which was closer to the beginning of the universe. Actually, if this hypothesis is correct, then the gamma-ray bursts would, in the end, give us information about early stages of the universe. No matter which way it goes, the gamma-ray bursts are interesting phenomena, and the Moon is one of the places where they could be investigated with real success.

I am sure that in these ways and many others an early lunar colony would be of great advantage.

Reference

1. Teller, E., *Thoughts on a Lunar Base*, **1985 Lunar Base Conference**, Lunar and Planetary Institute, reproduced courtesy LPI. Teller was at the Lawrence Livermore Laboratories, University of California, Livermore, CA at the time of this keynote lecture.

2

Overview and context

"So what did we get in return?
… So much!"

2.1 WHY THE MOON, AND HOW

The case for the permanent manned return to the Moon – as a destination in its own right, and as a platform for the human and robotic exploration of the Solar System – is clear.[1]

Great societies and civilizations advance in evolutionary ways, as well as revolutionary ways. Positive revolutionary progress can be categorized as social and technological. Social progress advances personal freedoms and opportunities. Technological progress advances the power of the individual and groups of individuals, and potentially also personal freedoms and opportunities, although such advances can also be used to repress and intimidate.

The Moon is our closest planetary body, roughly three days' flying time away, with almost instantaneous communication with Earth. The rival Mars is essentially as hostile to human life as the Moon, but also requires about a year of travel time from Earth, with a significant communications delay. A strategic view of space exploration and settlement places the Moon and Mars in their proper order, based on their proximity to Earth.

While space activities during the Apollo program of the 1960s were purely a government-led effort supported by American industrial might, today we see the beginnings of a transition, where commercial interests are staking claims to the space economic sector beyond the needs of the government. This is evident in the emerging space tourism market, commercial launch systems that service the government and private sectors, resource recovery plans via asteroid mining and sample return from the Moon, and privately financed space-based science.

Without a doubt, governments are still the largest customers. This will change as launch costs decrease, a space/lunar infrastructure is created, space resources become more valuable, and the space/lunar environment becomes critical for certain types of manufacturing and processing.

© Springer International Publishing AG 2018
H. Benaroya, *Building Habitats on the Moon*, Springer Praxis Books,
https://doi.org/10.1007/978-3-319-68244-0_2

Far from being a barren wasteland, the Moon has a regolith composed of many of the elements needed to build an infrastructure for human activity. Hydrogen, oxygen, silicon, magnesium and, it is strongly believed, water in ice form are found in the lunar regolith. Solar power can be viable on the Moon with its two weeks of daylight per month, and the solar panels could be manufactured on site using local resource silicon. Ideas for vast solar farms embedded on the surface of the Moon have been suggested. Significant quantities of helium-3 can be tapped as nonradioactive nuclear fusion reactors become feasible as a source of power.

With the current revolution in 3D manufacturing technologies, we can envision sending robots to the Moon, in advance of people, to begin to build fully functional structures for habitation and to mine the regolith for the above-mentioned elements. Even today, such advanced manufacturing can create objects of significant complexity using multiple materials. In-situ resource utilization (ISRU), coupled with advanced robotic manufacturing capabilities, implies that our lunar facilities will be almost autonomous, with full self-repair capabilities.

Figure 2.1. The Arabian Peninsula can be seen at the northeastern edge of Africa. The large island off the coast of Africa is Madagascar. The Asian mainland is on the horizon toward the northeast. This photograph is known as *The Blue Marble*, and was taken on December 7, 1972 at a distance of about 29,000 km (18,000 miles) as Apollo 17 was heading to the Moon. NASA officially credits this photo to all three astronauts, Eugene Cernan, Ronald Evans and Harrison Schmitt. Some credit Harrison Schmitt as the photographer. (Courtesy NASA)

The challenges and risks are significant, however. There are gaps in our knowledge of how to keep humans alive and robust in the space environment in general and on the Moon in particular. Engineering reliable hardware and software for long lives in the harsh space and lunar environments also requires the solution of a number of difficult technological problems.

But we need to keep in mind that the health and engineering issues that existed on the day President John F. Kennedy gave his speech challenging the United States to send man to the Moon before the end of the decade were even more difficult than those we face today. We did not know what many of the problems were, much less how to solve them. We had the faith, though, that with a sizable, sustained effort we would be able to match the challenges. And we did.

What did we get in return? We landed men on the Moon. On a political level, the nation demonstrated its engineering and scientific superpower status. To paraphrase Kennedy, the United States was able to marshal tremendous intellectual and material resources in a short period of time to solve a problem that only a few years before was deemed beyond humanity's reach. The space race born of the Cold War gave birth to a very long list of technologies, resulting in numerous industries that gave impetus to our economy and from which we benefit to this day. Included in this bounty are the medical sciences and technologies that we depend on, as well as the rarely mentioned ability to manage super-large projects of tremendous intricacies and logistical challenges.

What do we need in order to return people to the Moon with an eye toward permanence? We need to be able to send mass to the Moon, of course, but the strategic vision requires us very quickly to be able to use local lunar resources to cover most of our needs. In-situ resource utilization will allow us to use lunar materials to build structures, manufacture very large solar panels for energy, and extract valuable elements from the lunar regolith that can be used to create an industrial infrastructure. This capability, and advanced manufacturing techniques – also known as 3D printing technologies – are the keys to a viable manned exploration and settlement effort.

In order to advance the mission outlined above, we will need the following: access to orbit; low Earth orbital operations; human-rated transportation to the Moon along with all the technologies for descent and landing; lunar habitats; solar, battery and nuclear power systems; life-support and shielding systems to safeguard against radiation, micrometeorites, and zero- and one-sixth gravity; the ability to perform surface missions; in-situ resource utilization in conjunction with necessary logistics and technologies; and fuel to ascend into lunar orbit for a return to Earth. We will need to be able to ameliorate the adverse psychological effects of close-quarters cohabitation and isolation from Earth and family. Supporting human life requires a number of additional basic capabilities – in particular, plant growth in a closed and reduced gravity environment, waste processing and nutrient recovery, atmosphere revitalization and water management. Engineering challenges include propulsion, power, structures, optics, instrumentation, environmental controls, guidance and control, data management and storage, and communications. And more.

These are all difficult challenges. Research and development will solve these problems and, as a bonus, lead to tremendous advances in engineering, medical sciences and technologies. It is not possible to predict all of these spinoffs to the Earth economies, but the history of Apollo, which contributed to the advancement of many sectors of the U.S. economy and gave birth to many more, gives us reasonable credibility when we say that we expect many advances that will feed into the Earth economy.

Clearly, the human settlement of the Moon implies the support of robots and automated systems. Research is progressing rapidly in these disciplines, but even the most advanced robots today cannot autonomously explore and build on the Moon. They require human guidance and participation.

A lunar base will first be an engineering and medical laboratory, for the study of extra-terrestrial infrastructure development and for the creation of a safe environment for human habitation. Access to lunar resources will drive industrial activity. Public interest in space travel will also develop as it has today for tours of low Earth orbit. Second, it will be a site for the scientific study of the Moon and the Solar System.

In conjunction with these, the Moon will become an economic nodal point that will support space transportation in cislunar space, and outward to Mars and the asteroids and outer planets. Resources recovered on the Moon will be used to support the manufacture of items needed locally, as well as of use beyond the Moon.

Culturally, humans will evolve in other ways as well. Some predict that in a matter of a few generations, the human species will bifurcate as a result of exposure to the lunar environment. We can be certain that there is no turning back from spacefaring, and the positive feedback to life on Earth.

Why the Moon, and how – addendum

We periodically return to the onset of humanity's return to the Moon. United States administrations change every four to eight years, and with new administrations, space policies and goals go through discontinuities that have been generally painful and costly. Looking back at 1989, 1993, 2001, 2009, and now 2017, we are again at the beginning of choices to be made about space generally, and the Moon particularly. What were some of the views?

A 1992 study known as the Exploration Task Force Study concluded the following about President G.H.W. Bush's Space Exploration Initiative:[2]

"NASA's research in strong, light-weight metal alloys and plastics will support the efforts of such industries as car manufacturing, residential and commercial construction, and aircraft manufacturing. Miniaturization of electronic components will allow future designers to downsize many electronic systems. Sophisticated, portable, light-weight, health-monitoring and medical care equipment ... will become available in emergencies, at remote locations, and for local first aid squads. Long duration space missions will require partially closed life support systems in which air, water, and food must be conserved and recirculated, which could lead to advanced air purification devices for industry and medicine, water purification equipment for homes and industry, and vital new dietary information. Software ... can help manage terrestrial hazardous materials and waste. Automation and robotics ... can be applied to automobile assembly, undersea research, robotic manipulators, and vision systems. [The requirement for] high output, low weight, portable power supplies ... may yield a portable energy source for scientific, industrial, and military outposts at remote sites on Earth. Advances in the energy industry [may result], reducing our dependency on fossil fuels. In short, ... our lives [can be made] more comfortable and our environment more secure."

Others also agreed the likelihood of many benefits from space exploration. Livingston, the host of a popular space show, expresses the following thoughts.[3] "The money spent on space exploration is spent on Earth and in the most productive sectors of the economy. The money that is spent goes to manufacturing, research and development, salaries, benefits, insurance companies, doctors, teachers, scientists, students, blue- and white-collar workers, and corporations and businesses both large and small. The money disperses throughout the economy in the same way as money spent on medical research, building houses, or any other activity we engage in with government or even private spending."

The debate regarding space exploration has often centered on potentially better uses for the 'space money' that the government would spend. Of course, there are many worthy causes for government expenditures that can be considered to be investments to benefit many people over a long period of time (although some say – with some justification – that the private sector is better suited to make such choices).

But such expenditures "inspire others to do hard work, to go the next step, to push the envelope for the next level of advancement for all our benefit ... manned space is able to do it all." Thirty-four years after Apollo, 80 percent of people who were involved in science, engineering and space-related fields and businesses said they were inspired and motivated because of our having gone to the Moon. "Thirty-four years after all funding had stopped for the Apollo program, investment and wealth building ... was still going on as a result of our manned space exploration years earlier."

In the U.S., the return to the Moon was seen as a way to "replace the retiring generation of scientists and engineers inspired by Apollo."[4] Some say that the money spent on the space program is better spent directly on the education of scientists and engineers. But this is a common misconception: that more money spent on education leads to a better education, even though the evidence does not show such a correlation. Also, students need something to inspire their efforts. Figure 2.2 demonstrates the correlation between the vigorous Apollo program and the interest in advanced engineering, physical sciences, and mathematics degrees.

Safeguarding the species should Earth be destroyed – be it by a wayward asteroid or as a result of war – has always been a reason to place people extra-terrestrially. Another related reason is to have a safe site for the scientific, technical and cultural information that has been created and upon which the survival of our civilization rests.[5] Such a secure facility on the Moon, tended by a permanent staff, could provide humans with the intellectual and other resources needed to rebuild a shattered people. "Unlike ancient manuscripts which have survived for centuries in unattended storage, the data collections of the future will require continual attention from trained staffs. Skilled individuals will be required not only to update the software and hardware, but to control the environment."

In a related effort to safeguard our Earth heritage, the *Svalbard Global Seed Vault* was created. It is a secure seedbank located on the Norwegian island of Spitsbergen, near the town of Longyearbyen, in the remote Arctic Svalbard archipelago. The island of Spitsbergen is about 1120 kilometers (700 miles) from the North Pole. The facility was established to preserve a wide variety of plant seeds, from locations worldwide, in an underground cavern. The Seed Vault holds duplicate samples of seeds held in gene banks worldwide and provides insurance against the loss of seeds in those gene banks, as well as a refuge for seeds in the case of large scale regional or global crises. The Seed Vault is managed under

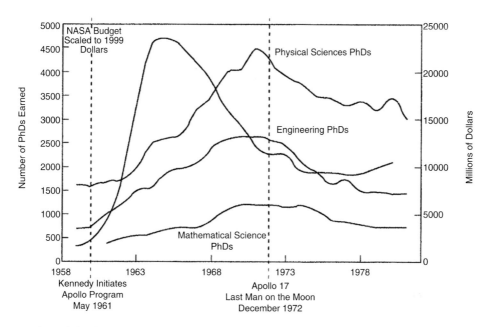

Figure 2.2.   Correlation between the NASA budget in 1999 dollars and the number of PhDs that were granted in the physical sciences, engineering and mathematics. The curve labeled 'NASA Budget Scaled to 1999 Dollars' peaks at about 1965, at the height of the Apollo funding. (Courtesy William Siegfried)

terms spelled out in a tripartite agreement between the Norwegian government, the Global Crop Diversity Trust (GCDT), and the Nordic Genetic Resource Center.

There was much debate regarding the multiplier[1] attached to NASA spending. An early study utilized input-output analysis to investigate the impact of the U.S. space program on the economy.[6] An extensive detailing of *all* of the industries that were involved in creating the hardware for NASA included *all* the interrelationships between *all* of the companies. Direct requirements generated by NASA expenditures resulted in indirect requirements, in a cascading way, as each contractor placed its orders with its vendors. "Specifically, in 1987, the NASA procurement budget generated $17.8 billion in total industry sales, had a 'multiplier effect' of 2.1 on the economy, created 209,000 private-sector jobs and $2.9 billion in business profits, and generated $5.6 billion in federal, state and local government tax revenues." These benefits cascaded throughout the United States, even though the initial NASA expenditures were in only a few states.

---

[1] A multiplier is the net economic effect of a dollar's expenditure. As an example, suppose $100 is spent to procure an item. That $100 spent at one company results in the company also spending part of that $100 to buy the parts that it needs to create the requisitioned item from a number of other companies. Those companies, similarly, spend part of their share to purchase their raw materials. So that $100 can generate additional economic activity in a trickle-down fashion. The multiplier is the factor that multiplies the $100 to yield the total dollar expenditures across all companies downstream.

While the overall multiplier was 2.1, several states had multipliers greater than 10 – "for every dollar Indiana receives directly in space program funds, it also receives $12 indirectly in business arising from the program" – and certain industries also had high multipliers; for example, electronic components had a multiplier of 5.9.

"Our findings are significant because we have for the first time estimated the benefits flowing from the second, third, and fourth rounds of industry purchases generated by NASA procurement expenditures. ... Many workers, industries and regions benefit substantially, and these benefits are much more widespread throughout the United States than has heretofore been realized. We believe our results imply that the economic benefits and costs of space exploration need to be reassessed." Indeed, that conclusion needs repeating as a reminder that the nation benefits in tremendous ways from the epic and sustained journey of humans into space.

2.2   OTHER THOUGHTS ON A LUNAR BASE

Summarizing the thoughts of Edward Teller in 1985, the great nuclear physicist believed that we could be back on the Moon within ten years, with a rotation of 12 people who would stay there in three-month periods. He viewed that the lunar facility would have a role as a refueling station, utilizing oxygen extracted from the lunar rocks. Solar and nuclear power would be the likely source of power to the lunar inhabitants. He liked the poles, especially the southern pole, or at a crater rim, for possible sites for the colony, because at those locations small distances result in very large changes, from daylight to nighttime, from very hot to very cold. The southern pole has many craters, much more than the northern, and so it was a likely choice.

The Moon could be a place from which to observe the Earth, for peace-keeping purposes. Astronomical observatories of all kinds on the Moon were of interest to Teller. Very large accelerators set up on the Moon would have the advantage of the vacuum when designing and placing the magnets needed to keep the particles on track.

Mendell pointed to the historical interest by humans in discovery, citing space as the ultimate opportunity for discoveries of all kinds.[7] He noted the rapidity of change in today's world; change that has come to signify progress. Generally, progress has been tied either to technology and science, or to abilities derived from their expansion. The scientific method and process of inquiry is seen to be the best approach we have to understand nature, and to create the technology upon which humanity depends for its survival and satisfaction. According to Bacon, knowledge is power.

Mendell mentioned the close connection between the science fiction literature and the imaginations of the great scientists and inventors, in particular with regard to space exploration. "Only one thing is sure: humans will continue to dream, and space is the destination." That dream continues today and broadens its appeal, drawing in entrepreneurs with a passion for space and its promise, along with the resources to develop the very large and very expensive technologies requisite for humanity to become a spacefaring civilization.

As part of the general discourse by those who are interested in manned space exploration and settlement, there are two topics for discussion. One is the perpetual hope that

Figure 2.3. This painting was used as a visual at an April 1988 conference in Houston titled 'Lunar Bases and Space Strategies of the 21st Century'. A deep drill team (lower center) obtains cores for petrological studies of the floor units of the young, 30 km, 4200 m crater, Aristarchus. The pea of Aristarchus is a few kilometers to the south of the drill rig. This perspective from the crater floor shows the prominent slump terraces of the crater walls and the solidified impact melt rivulets which flowed down the steep inner wall immediately after the crater was formed. Because the crater is very 'young', the rivulets and the volcanic-like features and cooling cracks of the impact melt floor unit are only slightly muted by meteorite erosion and ejecta blanketing. The drilling activities are taking place at 23.7 degrees north and 47.5 degrees west. The painting was a joint effort between Pat Rawlings and Doug McLeod of Eagle Engineering, one of a series of paintings done on subcontract to, and under the technical and scientific direction of, Lockheed Engineering and Management Services Company. The work was sponsored by the NASA Johnson Space Center. (S88–3127, April 7, 1988. Courtesy NASA)

those in political power will have an epiphany about the importance of space and its settlement, and will then commit the necessary funds to propel us back to the Moon, and on to Mars. Every decade or so since the Apollo era, such hopes have been raised and then summarily dashed.

The other is the debate about the target for our renewed exploration of space. Should it be the Moon, or should we focus first on Mars? People of great talent and achievements have taken both sides of this debate. Some have changed sides. The view here is that the Moon makes the most sense, for numerous reasons that will become apparent as we proceed through our discussions in these chapters.

Launius addressed both topics.[8] Regarding the confusion about why we did not move rapidly after the Apollo successes to begin to settle the Moon, he concluded that a unique confluence of political necessity, personal commitment and activism, scientific and technological ability, economic prosperity, and public mood made possible the 1961 decision to carry out Apollo. Kennedy's political objectives behind the decision to land Americans on the Moon allowed expenditures that were approximately six times those of NASA today, as a percentage of the Federal budget. The second question was answered in the following way.

"Using the space station as a base camp, humans may well be able to return to the Moon and establish a permanent human presence there. It is no longer hard to get there. All of the technology is available to land and return. Such an endeavor requires only a modest investment, and the results may well be astounding. Why return to the Moon? This is a critical question, especially since humans have already 'been there, done that.' There are six compelling reasons:

- It is only three days travel time from Earth, as opposed to the distance to Mars, allowing greater safety for those involved.
- It offers an ideal test bed for technologies and systems required for more extensive space exploration.
- It provides an excellent base for astronomy, geology, and other sciences, enabling the creation of critical building blocks in the knowledge necessary to go farther.
- It extends the knowledge gained with the space station in peaceful international cooperation in space and fosters stimulation of high-technology capabilities for all nations involved.
- It furthers development of low-cost energy and other technologies that will have use not only on the Moon but also on Earth.
- It provides a base for nuclear weapons that could be used to destroy near-Earth asteroids and other threats to Earth."

An interesting and sobering statistic provided by Launius is that half of the eight probes sent to Mars in the 1990s ended in failure, confirming that manned travel to Mars will be at least an order of magnitude more complex, riskier and costlier than a similar trip to the Moon.

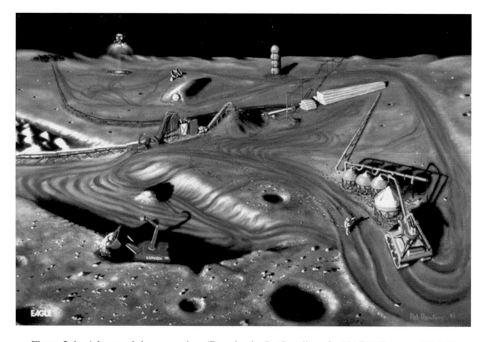

Figure 2.4. A lunar mining operation. (Drawing by Pat Rawlings for NASA. Courtesy NASA)

Rovetto emphasized that space, as a human activity, is visionary and forward looking.[9] The population can look at this adventure and be inspired by it, and also benefit from the job creation that results from a robust space program. "The nature of the enterprises involved – their scale, novelty, and complexity – requires a steady and continuous upward progression toward greater societal, scientific and technological development."

Bainbridge discussed many reasons for, and 18 broad practical goals to be achieved by, space flight.[10] Yarymovych discussed the challenges, both programmatic and technical, to manned space travel and settlement, as well as the benefits that accrue from them.[11] A brief summary was provided of primarily the American space program. Crawford entered the debate between the costs associated with manned exploration versus robotic exploration, and whether the savings from a robotic program can accelerate the scientific exploration of the Solar System.[12] He presented two arguments against a robotic-centered program. The first is that science benefits by including humans in the missions, because the exploration becomes more dynamic and difficulties can be overcome more often than by a robot. The second argument is that it is unlikely that savings garnered by switching from a manned space exploration program to a robotic one would go back into more space exploration. More likely, it would revert to other government programs on Earth.

2.3   THE HISTORICAL DIMENSION

Launius wrote about the need to approach the history of space, the program, its people and its artifacts, from the scholarly perspective.[13] An important point was made "that scientific and technological issues are simultaneously organizational, social, economic, and political. Various interests often clash in the decision-making process as difficult calculations have to be made. These interests could come together to make it possible to develop a project that would satisfy the majority of priorities brought into the political process, but at the same time many others would undoubtedly be left unsatisfied." Critical to a successful process of space exploration is a long-term plan for human exploration and settlement. With the necessary priority given to human safety and survival in space, requiring very large expenditures in addition to a long-term plan, there must be agreed-upon goals that transcend budgetary cycles and, in the United States, administrations. The changes in priorities every four to eight years have led to substantial costs but little progress of the type mentioned here.

Part of the historical dimension is a recognition that life on Earth today would be significantly different had it not been for the technologies that originated with the space program and were spun off into commercial products, though it is a complicated proposition to paint the spinoffs as completely positive. It is a mixed picture, and some of the technologies may have evolved similarly without Apollo, but at different paces. Clearly, though, technologies that depend on the microchip, such as medical diagnostics, communications and software for example, have become foundational in our society, and can trace their origins to the Apollo program.

2.4   THE ECONOMIC JUSTIFICATIONS

There have been numerous studies on the spinoffs, as well as the development of tools to assess these in a quantitative way.[14] Top-down and bottom-up approaches have been used for such assessments and while most assessments of such technology transfers have been qualitative and difficult to trace, there are quantitative measures.[15] Studies have concluded that approximately "90 percent of the long-term increase in output per capita in the U.S. has been attributable to technological change, increasing educational achievement. ... The Department of Defense typically accounted for 50 percent or more of all federal R&D expenditures, while NASA and the Atomic Energy Commission ranked as the second and third largest spenders. ... From 1960 to 1974, Chase [Econometric Associates] concluded that society's rate of return on NASA R&D was 43 percent," yielding significant positive impact on national productivity and employment levels.

Based on such studies of the Apollo program, proponents of a manned and permanent return to the Moon have expressed the belief that over time, such activity can propel the Earth economy to a higher level of activity due to federal and industrial expenditures in high-tech research and development. An expanding lunar economy will also supply products and generate demand that benefits the Earth economy.[16] [17]

Williamson emphasized that a sustainable model for lunar exploration and development is one that protects the space environment.[18] Sadeh *et al.* argued that an important factor in advancing missions such as the exploration and settlement of the Moon lies in partnerships between the private sector and the public government sector.[19] SpaceX is a public-private partnership, while Blue Origin has developed launch capabilities wholly supported by private capital. Hickman has pointed to the difficulty of capitalizing the very large investments for space.[20] Investors and lenders compare the choices they have for their funds and rates of return, usually in less than five years, govern investment and lending decisions. Lack of private funding is the reality for the private space community. The crucial difference between governments and private firms is not that governments are better at *managing* very large projects, but that they are better at *financing* very large projects. It is noted that the very large endeavors used as examples for motivating space development, for example the European colonization of America, Asia and Africa, the development of the U.S. Western Frontier, and the building of the Panama Canal, were all results of massive government expenditures and involvement.

The idea of public-private partnerships has also been suggested in the form of a Lunar Development Corporation that supports long-term efforts aimed at achieving lunar settlements by creating a hybrid finance model that integrates spinoffs strategically.[21] Tables 2.1, 2.2, and 2.3 provide a number of technologies that have both terrestrial and lunar markets. By spinning off technologies created for a trip to the Moon, a funding stream can be developed. This additional source of funds may need to be supplemented in some way.

Crawford discussed what it would take to create and sustain a space development program that would last for centuries, eventually leading to interstellar spaceflight capability.[22] He pointed to the possibility of replacing war by space development as the adventure that humanity requires. Funds going to war fighting could be moved to space flying. Geopolitical stability could set the stage for this next generation of space activities.

Table 2.1   Partial list of dual-use technologies: Communications and Information Systems; In-Situ Resources Utilization; Surface Mobility – Suits

Dual-Use Technologies: Communications/Information Systems

Terrestrial Applications	Technology	Space/lunar/Mars applications
• Communications • High-definition TV broadcast • Business video conferencing	• Ka band or higher	• Telepresence: vision and video data • Interferometers: raw data transmission
• Entertainment industry • Commercial aviation • Powerplant operations • Manufacturing operations	• Machine-human interface	• Control stations • System management
• Communications • Archiving • Computer operating systems	• Data-compression information processing • Large-scale data management systems	• Interferometers: raw data transmission information processing • System management expert data • Archiving/neural nets

Dual-Use Technologies: In-Situ Resource Utilization

Terrestrial Applications	Technology	Space/lunar/Mars applications
• Mineral analysis, yield estimation – deep mine vein location and tracking • Wall and cell integrity	• Advanced sensors	• Mineral analysis, yield estimation of surface mineral analysis and resource location
• Deep mine robotic operations for – mining – beneficiating – removal	• Advanced robot mining	• Surface mine robotic operations for – mining – beneficiating – removal • Remote, low-maintenance processing
• Improved automated processing: increased efficiency	• Automated processing technology	
• Reliable, low-pollution personal transmission • Regenerable energy economies • Small, decentralized power systems for remote third world applications	• Alternative, regenerable energy economies – methane/O_2, H_2/O_2	• ISRU-based engines • Regenerable energies • High-density energy storage
• Environmentally safe energy production	• Space-based energy generation and transmission	• Surface power generation and beaming

(continued)

Table 2.1 (continued)

	Dual-Use Technologies: Surface Mobility – Suits	
Terrestrial Applications	**Technology**	**Space/lunar/Mars applications**
• Hazardous materials cleanup • Fire fighting protection • Underwater equipment • Homes • Aircraft	• Lightweight superinsulated materials	• Surface suits: thermal protection • Surface facilities: thermal protection
• Robotic assisted systems • Orthopedic devices for mobility of impaired persons • Human power enhancement	• Robotics • Mobility enhancement devices and manipulators	• Robotic assisted suit systems • Human power enhancement
• Hazardous materials cleanup • Fire fighting protection • Underwater equipment	• Dust protection, seals, abrasive resistant materials	• Surface suits: outer garment
• Hazardous materials cleanup • Underwater breathing gear	• Lightweight hi-reliability life support	• Portable life support for surface suits • Backup life support systems
• Remote health monitoring	• Portable biomedical sensors and health evaluation systems	• Surface EVA crew member health monitoring
• Hypo-hyper thermal treatments • Fire fighting protection and underwater equipment • Arctic/Antarctic undergarments	• Small, efficient, portable cooling and heating systems	• Surface suits: thermal control systems • Rovers: thermal control systems

Table 2.2 Partial list of dual-use technologies: Surface Mobility – Vehicles, Human Support, and Power.

Dual-Use Technologies: Surface Mobility – Vehicles		
Terrestrial Applications	**Technology**	**Space/lunar/Mars applications**
• All-terrain vehicles for research (volcanoes), oil exploration • Automobiles	• Mobility	• Surface transportation for humans, science equipment, maintenance and inspection
• Reactor servicing/hazardous applications • Military	• Robotics and vision systems	• Teleoperated robotic systems
• Earth observation, weather, research	• Super-pressure balloons (110,000 ft. – Earth equivalent)	• Mars global explorations
• Efficient, long-term operations with low maintenance machines in Arctic/Antarctic environments	• Tribology	• Surface vehicles: drive mechanisms, robotic arms, mechanisms
• Helicopters, autos	• Variable speed transmissions	• Surface vehicles
• Automated, efficient construction equipment • Military	• Multipurpose construction vehicle systems and mechanisms	• Robotic construction and set-up equipment

Dual-Use Technologies: Human Support		
Terrestrial Applications	**Technology**	**Space/lunar/Mars applications**
• Stored food: US Army, NSF polar programs, isolated construction sites	• Long-life food systems with high nutrition and efficient packaging	• Efficient logistics for planetary bases, long spaceflights, space stations
• Improved health care • Sports medicine – cardiovascular safety	• Physiological understanding of the human chronobiology	• Countermeasures for long-duration and/or micro-g space missions
• Osteoporosis – immune systems • Isolated confined environments/polar operations	• Understanding of psychological issues	• Health management and care
• Noninvasive health assessments • Military	• Instrumentation miniaturization	• Systems/structural monitoring and self-repair

(continued)

Table 2.2 (continued)

Terrestrial Applications	Technology	Space/lunar/Mars applications
• Health care • Disaster response • Military	• Long-term blood storage	• Health care for long-duration space missions
• Office buildings ('sick building" syndrome) • Manufacturing plants • Homes	• Environmental monitoring and management	• Environmental control for spacecraft cabins, planetary habitats, pressurized Rovers
• Contamination cleanup • Waste processing • Homes	• Waste processing • Water purification	• Closed water cycles for spacecraft cabins, planetary habitats, pressurized Rovers
• Long-life clothes • Work clothes in hazardous environments • Military	• Advanced materials/fabrics	• Reduced logistics through long-life, easy-care clothes, etc. • Fire-proof/low-outgassing clothes • Building material for inflatable structures
• Efficient food production	• Advanced understanding of food production/hydroponics	• Reduced logistics through local food production for spacecraft cabins, planetary habitats

Dual-Use Technologies: Power

Terrestrial Applications	**Technology**	**Space/lunar/Mars applications**
• Batteries for: autos – remote operations for DOD, NSF polar programs	• High-density energy storage • Alternate energy storage (flywheels)	• Reduced logistics for planetary bases • High reliability, low-maintenance power systems • Spaceship power storage
• Clean energy from space	• Beamed power transmission	• Orbital power to surface base • Surface power transmission to remote assets
• Remote operations for: DOD, NSF polar programs	• Small nuclear power systems	• Surface base power • Pressurized surface Rover • Interplanetary transfer vehicle
• Remote operations for: DOD, NSF polar programs • High efficiency auto engines	• High efficiency, high reliability low-maintenance heat-to-electric conversion engines	• Energy conversion for planetary bases: low servicing hours – little or no logistics

Table 2.3   Partial list of dual-use technologies: Structures and Materials, Science and Materials, Science and Science Equipment, and Operations and Maintenance

Dual-Use Technologies: Structures and Materials

Terrestrial Applications	Technology	Space/lunar/Mars applications
• Vehicles • Fuel-efficient aircraft • Modular construction, homes	• Composite materials: hard – soft • Advanced alloys, high-temperature	• Cryogenic tanks • Habitat enclosures • Pressurized Rover enclosures • Space transit vehicle structures
• Aircraft fuel tanks • Home insulation • Large structures, high-rise buildings, bridges • Commercial aircraft: improved safety – lower maintenance	• Superinsulation • Coatings • Smart structures • Imbedded sensors/actuators	• Cryogenic tanks • Habitable volumes • Space transit vehicle structures • Planetary habitat enclosures • Surface power systems • Rover suspensions

Dual-use Technologies: Science and Science Equipment

Terrestrial Applications	Technology	Space/lunar/Mars applications
• Energy resource exploration • Environmental monitoring, policing	• Spectroscopy: gamma ray – laser – other	• Geo-chem mapping • Resource yield estimating • Planetary mining operation planning
• Undersea exploration • Hazardous environment assessments and remediation	• Telescience	• Remote planetary exploration
• Environmental monitoring • Medicine • Improved health care • Sports medicine – cardiovascular • Osteoporosis – immune systems • Isolated confined environments/Polar operations • Noninvasive health assessments	• Image processing: compression technique – storage – transmission – image enhancements • Physiological understanding of humans • Instrumentation miniaturization	• Communication of science data • Correlation of interferometer data • Countermeasures for long-duration and/or micro-g space missions • Health management and care

Dual-use Technologies: Operations and Maintenance

Terrestrial Applications	Technology	Space/lunar/Mars applications
• Military • Systems and structures health monitoring • Inventory management	• Task partitioning • Reliability & quality assurance in long-term hazardous environments • System health management and failure prevention through AI and expert systems, neural nets	• Systems and structures health monitoring • Inventory management • Self-repairing technologies • Logistics improvement

Beyond these hopes, also identified were "the following elements of a space-based infrastructure that can ultimately be capable of supporting the human colonization of the Solar System:

- efficient transportation between the Earth's surface and low Earth orbit, for example single-stage-to-orbit spaceplanes, and a new generation of heavy-lift launch vehicles
- the ability to build large structures in space, for example space stations, space factories, interplanetary vehicles, and lunar and planetary outposts
- the ability to tap and utilize large quantities of solar energy, and
- the ability to extract and process extraterrestrial materials, especially from near-Earth and main-belt asteroids, but possibly also from the Moon and Mars."

Clearly, this list of capabilities are the core requirements. We can add a fifth, which is implicit in the list, namely the ability to survive, and flourish, as humans in the radioactive, low, and micro-gravity space environment.

Figure 2.5.   A bond issued by the Dutch East India Company, dating from November 7, 1623, for the amount of 2400 florins. While a proposed Lunar Development Corporation is not envisioned as monopolistic – as were the East India Companies of Holland and England – it could operate in a similar way.

2.5   AGREEMENTS AND ETHICS

Once activity begins on the Moon, Mars, the asteroids and beyond, we expect to see a rapid development of the rules that govern such activities. Laws and ethics need to be written that will be viewed as legitimate by the participants, as well as the observers on the side.

Livingston proposed a code of ethics with the aim of facilitating an increase in commercial activity in space.[23] The code of ethics "covers areas such as environmental stewardship of space, the promotion of honest dealings, making safety an important concern, ensuring a free-market economy, and disclosure of conflicts of interest or political contributions." In addition, the unique nature of outer space requires that its development be done with care, with thoughtful long-term planning, and with consideration of future generations.

The rules by which we will operate in space are not yet set, but there have been efforts to begin to outline them. There is a United Nations report that provides the text of treaties and principles that govern the activities of countries in the exploration and use of outer space.[24] While the U.N. General Assembly has ratified these treaties, there is little legal weight to them without further ratification by all the individual members of the U.N. There are five space treaties:

- Treaty on Principles Governing the Activities of States in the Exploration and Use of Outer Space, including the Moon and Other Celestial Bodies (Outer Space Treaty, 1967)
- Agreement on the Rescue of Astronauts, the Return of Astronauts and the Return of Objects Launched into Outer Space (Rescue Agreement, 1968)
- Convention on International Liability for Damage Caused by Space Objects (Liability Convention, 1972)
- Convention on Registration of Objects Launched into Outer Space (Registration Convention, 1975)
- Agreement Governing the Activities of States on the Moon and Other Celestial Bodies (Moon Agreement, 1979).

We do not enter into a discussion of these issues here, except to agree that the nature, details and extent of these laws and agreements will have serious impacts on the rate and breadth at which activities in space can move forward. Some of these aspects have been discussed in an extensive literature, for example by Reynolds and Bini.[25] [26] There is disagreement about what activities are allowed and what activities are not allowed, while the issues of ownership and "first possession" rights are also interpreted in various ways. Regarding property rights, Reynolds pointed out that "the mere absence of regulation is not enough to encourage investment: there must be positive legal protection for property rights."

2.6 PIONEERING VISIONS

Numerous studies have been commissioned at inflection points of the United States space program, to assess its status with an effort to extrapolate optimal trajectories going forward. In 1986, *The Report of the National Commission on Space* predicted a human outpost on the Moon by 2005, and one on Mars by 2015.[27] The Commission was created by Congress and appointed by President Ronald Reagan, and was designed to formulate a space agenda for the United States into the 21st century.

After President George H.W. Bush took office and gave his July 20, 1989 speech known as the *Space Exploration Initiative*, NASA Administrator Richard H. Truly commissioned a study – to be led by Johnson Space Center Administrator Aaron Cohen – to provide a database for the National Space Council led by Vice President Dan Quayle, as a reference for strategic planning. The so-called *90-Day Study* considered infrastructure, meeting human needs in space, science opportunities and strategies, technology, and national and

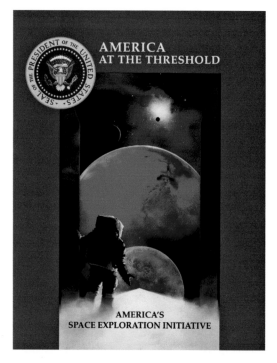

Figure 2.6.   America at the Threshold: America's Space Exploration Initiative 1989

Figure 2.7.   Pioneering the Space Frontier 1986

institutional impact, with a focus on achieving the "earliest possible landing on Mars."[28] Five reference approaches were studied, with lunar operations commencing between 2005–2014, and humans on Mars between 2011 and 2018.

A more fleshed-out study, sometimes called the *Synthesis Group*, under the direction of Lieutenant-General Thomas P. Stafford, was released in May, 1991.[29] This report provided scenarios with technical details and specific goals.

There were other commissions and reports, including one after President George W. Bush gave his *New Vision for Space Exploration* on January 14, 2004. In this vision, the president proposed extending human presence across the Solar System, beginning with a human return to the Moon by 2020, where preparations for the human exploration of Mars could begin.

Other reports have included the *Rogers Commission* report that investigated the *Challenger* explosion shortly after liftoff on January 28, 1986, issued in June 1986; the *Ride Report*, the informal name of the report titled *NASA Leadership and America's Future in Space: A Report to the Administrator*, issued in 1987; and the *Columbia Accident Investigative Board Report*, issued in August 2003, on the *Columbia* disintegration upon return to Earth on February 1, 2003.

McNutt provided a summary of the beginnings of the space program, a discussion of space science goals, exploration with probes, and cost comparisons with the United States GDP as well as with United States war costs.[30] Interesting mass comparisons were made, listing the masses of the great engineered structures as well as various space structures. For example, a fully fueled Saturn V rocket had a mass of approximately 3039 tons, whereas the Eiffel Tower has a mass of approximately 10,000 tons, and the Empire State building has a mass of approximately 365,000 tons. The Saturn V-D concept had a mass of approximately 9882 tons, and the L5 space habitat concept for 10,000 people had an estimated mass of almost 11,000,000 tons. The roles of humans were explored and tied to psychological and physiological issues of great significance. McNutt suggested a permanently staffed lunar base in 2030, as well as a human mission to Mars in 2030, with a permanent base there in 2080. Human missions were also listed to Callisto (Jupiter system) in 2050, and to Enceladus (Saturn system) in 2075, thus postulating human exploration of the Solar System before the permanent settlement of Mars.

Koelle and the International Academy of Astronautics reviewed "plans for an extensive program to survey and develop the Moon and to explore the planet Mars during the 21st century."[31] The premise was that the exploration of space reflects humanity's highest aspirations. The study placed the output of a potential lunar base into the following categories: new knowledge, services, and material goods and energy. These product categories can be of interest to buyers within the lunar economy, in-space enterprises, and Earth-based enterprises. Possible lunar programs could be developed to establish a temporary lunar outpost, a permanent lunar outpost, a lunar base, a lunar factory, or a lunar settlement. More details were provided in the paper about the technical and economic aspects. Much more detail was provided in the report by Koelle.[32]

The Rutgers 2007 *Lunar Symposium on Lunar Settlements* brought together some of the most recognized and influential researchers and scientists in various space-related disciplines. The almost five-day meeting addressed the many issues that surround the permanent human return to the Moon. Numerous national and international contributors offered their insights into how certain technological, physiological and psychological challenges must be met to make permanent lunar settlements possible.

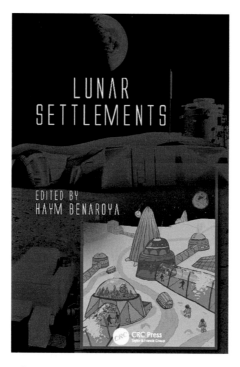

Figure 2.8.   The Symposium on Lunar Settlements was held at Rutgers University during the summer of 2007, an almost five-day event covering all the aspects of lunar settlements. This cover is from the CRC book, which includes chapters beyond those presented at the symposium. (Cover by A. Benaroya)

The Symposium book, shown in Figure 2.8, first looked to the past, covering the Apollo and Saturn legacies.[33] In addition, former astronaut and United States Senator Harrison H. Schmitt discussed how to maintain deep space exploration and settlement. The book then discussed economic aspects, such as funding for lunar commerce, managing human resources, and commercial transportation logistics. After examining how cultural elements would fit into habitat design, the text explored the physiological, psychological and ethical impacts of living in a lunar settlement. It also described the planning and technical requirements of lunar habitation, the design of both manned and modular lunar bases, and the protection of lunar habitats against meteoroids. Focusing on lunar soil mechanics, the book concluded with discussions on lunar concrete, terraforming, and using greenhouses for agricultural purposes.

Drawing from the lunar experiences of the six Apollo landing missions, as well as the many American and Soviet robotic missions, and including current space activities and research, this volume summarized the problems, prospects and practicality of enduring lunar settlements. It reflected the key disciplines, including engineering, physics, architecture, psychology, biology, and anthropology, that will play significant roles in establishing these settlements.

The Lunar Exploration Analysis Group (LEAG), established in 2004, has met annually to discuss issues and goals in preparation for the return to the Moon. It released a recent roadmap that details three themes:[34]

> "The *Science Theme* has a long heritage of study, including NRC studies, and represents community consensus. The *Feed Forward Theme* has been presented to the Mars Exploration Program Analysis Group and their comments have been incorporated. This theme is now expanded to include using the Moon to go to other airless bodies. The *Sustainability Theme* is at the lowest fidelity, representing a small (but growing) body of opinion, and will require refinements, which have begun at the LEAG Annual Meetings.
>
> "Overall, the roadmap is intended to layout an integrated and sustainable plan for lunar exploration that will allow NASA to transition from the Moon to Mars (and beyond) without abandoning the lunar assets built up using tax payer dollars. As such, the roadmap will enable commercial development, through early identification of 'commercial on-ramps' that will create wealth and jobs to offset the initial investment of the taxpayer. In addition, the roadmap will, with careful planning, enable international cooperation to expand our scientific and economic spheres of influence while enabling an expansion of human and robotic space exploration."

All three main themes focus on learning to live and work on another world, to expand Earth's economy to the Moon yielding benefits to both worlds, to strengthen and create new global partnerships, and to engage, inspire and educate the public. Within the sustainability theme: "The fundamental purpose of activity involving the Moon is to enable humanity to do there permanently what we already value doing on Earth: science, to pursue new knowledge; exploration, to discover and reach new territories; commerce, to create wealth that satisfies human needs; settlement, to enable people to live out their lives there; and security, to guarantee peace and safety, both for settlers and for the home planet." Economic activities, ethically carried out, will be the foundation for self-sufficient settlements, that will also continue exploration and science, feeding forward to the human exploration of the Solar System.

2.7   FUTURE HISTORIES

At the 1985 Lunar Base Conference at the Lunar and Planetary Institute, the 'official historian of Luna City' gave an address 'on July 4, 2076', to commemorate 'the 75th year of the permanent settlement of the Moon'.[35] Thus, in 1985, it was reasonable to estimate, perhaps hope, that by 2001 there would be the beginnings of a permanent manned settlement of the Moon. This future historian extolled the benefits accrued due to the existence of this settlement. There were people born on the Moon and there were intelligent robots, with equal rights, who were extolled as critical to the human settlement. People settled from all parts of the Earth. Great resources were developed and wealth created. It was a harmonious city, with prosperity for all. Noted was the possible discovery of extraterrestrial beings, based on faint microwave signals from the Veil Nebula in the constellation Cygnus.

In another effort at 'future history', Benaroya extrapolated the present, and its promise, into a possible future.[36] In that book, the timeline shown below was proposed for this extrapolation, from the perspective of Yerah Timoshenko, who is the protagonist and narrator.

1969 First men on the Moon
2009 Chandarayaan-1/Moon Mineralogy Mapper reveals H_2O molecules
2014 Space tourism reaches $1 billion
2024 Humans return to the Moon
2029 Permanent lunar colony
2034 Humans land on Mars
2041 Permanent Martian colony
2046 Space elevator prototype construction begins in orbit around Earth
2049 Lunar space elevator construction begins
2059 Yerah Timoshenko's (YT) great grandparents go to the Moon
2060–2061 YT's grandparents born on Earth (conceived on the Moon)
2070 Fusion reactors go online on the Moon, a year later on Mars
2084 First families settle on the Moon
2089–2090 YT's parents born on Earth
2094 First lunar Olympics held
2099 First human birth on the Moon
2115 YT's parents move to the Moon
2119 YT born on the Moon
2142–43 YT's son and daughter born on the Moon
2169 The present – 200th anniversary of the first men on the Moon
Now Terraforming on Mars started
2179 Upcoming 150th anniversary of off-Earth permanent habitation.

2.8   INTERVIEW WITH DAVID LIVINGSTON

This interview with David Livingston was conducted via email, beginning on June 25, 2017 and concluding on July 19, 2017. We begin with a brief professional biography, and then continue with the interview.

Brief Biography
Dr. David Livingston is the founder and host of The Space Show®, with the website www. thespaceshow.com. For nearly 17 years and close to 3000 interviews, it has been the nation's only talk radio program exclusively focusing on space commerce, space tourism, and facilitating our becoming a space-faring economy and society. Dr. Livingston is also the Executive Director of the One Giant Leap Foundation, Inc. (OGLF), the 501(C)3 that controls The Space Show®. OGLF strives to promote space education.

David is also an Adjunct Professor of Space Studies in the Odegard School of Aerospace Sciences at the University of North Dakota. He teaches graduate classes on commercial space. Livingston has a BA in Political Science, an MBA specializing in International Business Management, and a Doctorate in Business Administration (DBA). His dissertation was titled *Outer Space Commerce: Its History and Prospects.*

Livingston is a frequent speaker at space conferences and has published more than 50 papers, including chapters in books. For example, he authored the Space Tourism chapter in the *Space Encyclopedia* and coauthored three business and financial chapters in the New SMAD Textbook, *Space Mission Engineering: The New SMAD.* David has lectured at several universities and has been a guest on many national radio programs, including a weekly segment on the John Batchelor Show, a nationally syndicated radio program. When not teaching, broadcasting, writing, or just playing, Livingston is a business consultant working with clients on business, financial and strategic planning.

Our Interview

Lunar structures have been grouped into three categories: I – habitats brought from Earth, II – habitats with components from Earth, but also components made from lunar resources, and III – habitats primarily made from lunar resources; ISRU, perhaps 3D printed structures. The longer it takes for us to get back to the Moon to stay, the more certain technologies advance, for example 3D printing. Would you keep the above categories, or modify them in some way?

> *I would keep them because in my opinion we are still pretty far out from even starting a lunar structures project. I suggest lots and lots of flexibility as this will change and evolve once a valid project is actually being planned for real, financed, and implemented. In addition, much might depend on the launch vehicles available to carry out a lunar structures project. While today there is much talk regarding SLS, Falcon Heavy and possible entries from Blue Origin, none of these have yet flown even as much as a demo flight and the only economics we have for the vehicles, costs, and performance are projected economics. Nothing yet stems from operations. We might have surprises ahead that would definitely influence the course of action and direction.*

Given that we have been waiting almost 50 years to return to the Moon, do you have a feeling as to when we might return, with people, for an extended stay?

> *No. I don't even have a feeling when we will return to competent government, sane economic and growth policies, and much more. In my opinion, we need such policies and competence to undertake lunar development. Additionally, talk, media reports, and good intentions and plans don't equal reality.*

In your mind, what is a likely scenario and timeline for manned space?

> *As I said above, I don't really have a timeline. If we had leadership, economic sanity, and a decent growth economy, I suspect that within a decade we would be well along with a lunar structures project and the first habs would be on the Moon. But we don't have that, so I think we might spin our wheels for several more years just to be able to accomplish the most basic part of the project.*

Do you agree with those who have made a case for settling the Moon first and then using it as a base for travel to Mars?

Yes, I do. Closer, cheaper, easier, and a good learning environment. An important stepping-stone if done right.

Is ISRU oversold for near-term lunar development?

Yes, most likely for the close-in projects. I believe it will become very important and useful as advances in our TRLs [Technology Readiness Levels] and lunar habitats progress.

Key challenges to engineers and scientists who will be tasked to develop the lunar mission and structural design are the knowledge gaps that cannot be filled before being on site. For example, the lunar 1/6 *g* environmental effects on biology, and systems reliability in the lunar dust environment. Also of great concern is our understanding of the very complex systems that will not be integrated before they are placed on the Moon. Even if the behavior of all the components are fully understood, the complete integrated system will generally have unanticipated characteristics. Is there a way of dealing with these knowledge gaps before being there?

I'm not an engineer so this question is really beyond my technical level. I do think that, as progress is made and as we evolve in our lunar structures work, engineers and scientists will mitigate many of the challenges and complexities they face, but they might be creating new challenges as well.

Dust mitigation is a critical issue generally on the Moon. How far are we from getting a handle on that problem, both the biological risks and the engineered systems risk?

I am not totally up to date on this but I would hope this has been a major study and research item since the Apollo days. We should be pretty far along in containment and mitigation strategies. If we are not, then it supports my argument about incompetent government and policy makers, and the lack of quality of leadership.

The isolated environment that astronauts will face on the Moon, and even more isolated on Mars, makes critical the need for high levels of reliability. Part of that implies a self-healing capability. Such technologies are being studied and developed but are far from being usable. Do you see self-healing as a critical technology?

Yes. When fiber optic cables were being put underseas connecting the world, those cables were largely self-correcting, and this was a few decades ago. I would like to think we have advanced in these areas but I could understand needing more advancement. I do think that as we advance with lunar development, self-correcting technologies will prove very important and useful.

Manned space settlement has been promoted as a justified expenditure, financed by taxpayers, for many reasons: economic development, scientific advances, SETI, manifest destiny, building bridges between nations, limitless energy production, for example. What is your case for manned space settlement? Do you have an overarching reason for supporting such an endeavor?

I am a big fan of human spaceflight and development. I believe people want to know things, learn, and get answers, plus explore. Also, we like risk taking and excitement, so while robots can do much, they cannot really satisfy all of what humans want in our lives. I believe that science and technology development and advancement make strong cases to undertake human spaceflight. We all benefit from what we do and learn and develop. That said, since human spaceflight is risky, what we do with it needs to be worth the risk. If all we do is risk people for mundane things in space or LEO, then I think we have a right to question such action. Since people are being put at risk, let's make sure the risk is well worth it.

It appears that missions to settle the Moon or Mars have two underlying potential show stoppers: (1) assuring human survival and good health, and (2) assuring very high reliability for all engineered components. While we have shown with the ISS and Apollo that we can manage short forays to the Moon, or longer ones to LEO, satisfying (1) and (2) in limited ways, the bar is an order of magnitude higher for long-duration lunar stays, and an order of magnitude higher than that for Mars human missions. Are guaranteeing (with high probability) human and machine survival and high operability on the horizon, or beyond it?

Depending on the timeline for the horizon, I would say yes, but we are not there yet. I do believe that we will get there by doing the real thing. I don't think we can get there in the lab or just with simulations. That is why I say, if we are going to undertake these human risks, let's make sure that what we are doing is well worth the risk in the first place.

Who would you consider as your key influence in your pursuit of space studies? Who inspired you?

My key interest has been on the policy, leadership and economic side of things. I was actually inspired by early Sci-Fi writers, early science fiction movies and TV shows, the thought of others out there in the universe, and the quest to know more about our planets, the Moon, and to go to another solar system. Growing up I could not get enough of those topics. Von Braun and his vision was inspirational, as were the Mercury 7 astronauts, and the lunar astronauts, but I was also inspired by the Sci-Fi characters, Tom Corbett, Rocky Jones, the spaceship drawings and art by Chesley Bonestell.

All of us working on space recognize that much of what we do, and are interested in promoting, will not happen in our lifetimes. Of course, we have no choice but to accept this reality, but how do you view this?

I want to leave the planet a better place for my kids, their kids, and the future, than how I found it. There are several ingredients to that goal and space is one of them. I think that a valid roadmap for a better Earth, more prosperity and peace, runs through space on a rocket ride. It is a balancing act to avoid getting caught up in the fantasy and to say grounded and realistic, but I think it is possible to contribute to this vision. I hope my Space Show educational interviews help us along the way.

Is there an issue or concern of yours that I have not touched upon?

Space advocacy is both a blessing and a curse. If the advocacy is too rooted in fantasy, and la la land, and cheerleading, it can disappoint people, discredit others, including those on the financial and policy side. On the other hand, if all it is rooting for is the next big engineering success, then it can put people to sleep faster than what happens in an operating room in the hospital. Space advocacy can be, and is, an important tool for advancing multiple space agendas, such as commercial space development. I think everything in moderation is a good approach. A little of the crazy and fantasy mixed in with being grounded and steeped in at least some basic technology, engineering, and science, can go a long way to inspire, motivate, and lead us to a really great space future. In my opinion, many in the space advocacy and enthusiast community on both sides of what I am talking about have yet to learn about balance, synergy, and moderation.

Your Space Show has been broadcasting interviews with many of the people who impact space in some way. You have been broadcasting since 2001 and there have been nearly 3000 shows. Are there several overarching themes or views that come across in your interviews with all the many hundreds of guests?

Lots of programs address policy issues regarding returning to the Moon or going to Mars. One theme that always grabs attention has to do with rockets and size and the companies behind them. Topics run in cycles. Right now, we are seeing lots of discussion about changes to the Outer Space Treaty and for the Musk plans for going to Mars. At the time of the VSE, returning to the Moon and Constellation were regular topics. SLS and Orion continue to be popular topics. The search for ET life and interstellar flight and technology are long-standing popular topics. EM Drive has also risen to a top topic over the past few years. AI and robotics are also front and center right now.

There are numerous dichotomies in the space community. There are: lunar vs. Mars; robotic vs. manned; government vs. industry; science vs. commercial; and there may be other sub-group debates. Which are the most contentious?

For The Space Show, contentious might be too strong as a good descriptive word. Based on my Space Show experience, the past 16 1/2 years, I would list the topics most likely to cause a spirited debate to be the lunar and Mars issue, which can really get nasty quickly but, fortunately, most of the time it just remains spirited. For the past year or so, there has been an ongoing debate topic regarding whether capitalism in space is good, and government in space is bad, or vice versa. Sometimes, emotions and facts seem to serve to excite instead of inform and persuade, but I've not seen this topic devolve into a contentious shouting and name calling argument. The one topic that perhaps could be classified as contentious would be SLS. Passions run very high on both sides of the argument for SLS. Usually the participants in the discussion voluntarily constrain themselves, but I can think of a few people who frequent The Space Show that often go over the line, but fortunately the other side does not take the bait. Space discussions are passionate. Over the years I have found that there are multiple paths to accomplishing things in space for commerce, science, and yes, even for humanity and our nation. Being a good listener is as important as being a good proponent of the position at hand, but I think in total, we collectively come up on the short side of being good listeners to one another.

One of the goals of your show is to bring people back to reality. People find it more enjoyable to fantasize about futuristic colonies, but the most difficult part of sending humans to space and settling them on the Moon and Mars is getting to first base (the Moon?), the first steps. There are so many resources available that explain the difficulties involved in the space enterprise, and yet there is this disconnect. Why do you think so?

This is a very complex question to answer and I am not sure I have any better take on it than a very experienced space professional shrink. For the most part, I think people find it easy and very enjoyable to dream big, present bold visions for the future, talk about grandiose projects as though they are real today because they have zero accountability for making them happen in real time. Without accountability, one can propose just about anything on any timeline and then complain when others don't meet their timeline. It is as if people give themselves a pass from reality to enjoy their own fantasy world, which in this case focuses on big space projects. The truth is they know the complexities and challenges but they don't let that knowledge interfere with their passion and their dreams. Remember, they have no responsibility for making their dreams come true. They just get to talk and write about them and be the expert for the day.

I used to think all this was a negative, but I don't think that anymore. Most big projects that have happened were the big dreams and visions of someone. We need the big, the bold and the dreams, even if it is just fantasy. We can't pick who the winner will be in the dream fantasy game so let everyone play. Maybe a few will be the winners that lead us forward. Thus, I think these fantasies and unrealistic concepts play a role in keeping these issues on the table for many of us and enabling others working their way through education or something else to discover them and carry them forward to hopefully a next level. They key for many of us, myself included, is to be able to do some basic due diligence and apply discernment so we help advance the plausibility and let lie the absurd. This is why I give air time on The Space Show to the absurd and whacky, and very creative. Not only is it fun, it can be informative and inspiring, if not for me, then for someone else, and that someone else may be the one person to make a huge difference for the better in our world. The trick for me is to balance the whacky with the more grounded programming because, trust me on this, there are far [more] whackier programming options than grounded options. The key is everything in moderation.

Are there a few exceptional guests that you wish to point to whose interviews created tremors? Or changed the paradigm of the conversation? And perhaps gave you and your audience whole new perspectives?

Obviously, Dr. Benaroya as a guest has been everything and more as a change artist for the better. Other than that, with nearly 3000 guests, my approach to answering this question is to not really answer it for fear of leaving out a very important guest or for my comments being taken in a way that demeans a guest because that person was not mentioned. I believe that all The Space Show programs have contributed to our industry. I trust The Space Show audience in that they can decide for themselves who has been tops, who has been awful, etc. In addition, sometimes a person is not good with a microphone but their work and ideas are the best. That may not come out in an interview because of microphone or speaking issues. To not consider such a person a great change artist of importance because of style would be unfair and counterproductive. So I don't answer this question with names other than the obvious, which, of course, for this interview, is you, Haym. Take a bow please.

Thank you.

References

1. Originally published as a commentary in Space News (spacenews.com) by H. Benaroya, November 25, 2013. (Courtesy Space News)
2. R.McC. Adams: *Why Explore the Universe*, 30th Aerospace Sciences Meeting & Exhibit, Reno, January 6–9, 1992.
3. D. Livingston: *Is space exploration worth the cost?* The Space Review, January 21, 2008.
4. E. Sterner: *More than the Moon*, washingtontimes.com, April 11, 2008.
5. R. Shapiro: *A New Rationale for Returning to the Moon? Protecting Civilization with a Sanctuary*, Space Policy, Vol.25, 2009, pp.1–5.
6. R.H. Bezdek and R.M. Wendling: *Sharing Out NASA's Spoils*, Nature, Vol.355, January 9, 1992, pp.105–106.
7. W.W. Mendell: *The vision of human spaceflight*, Space Policy 21 (2005) pp.7–10.
8. R.D. Launius: *Why go to the Moon? The many faces of lunar policy*, Acta Astronautica 70 (2012) pp.165–175.
9. R.J. Rovetto: *The essential role of human spaceflight*, Space Policy 29 (2013) pp.225–228.
10. W.S. Bainbridge: *Motivations for space exploration*, Futures 41 (2009) pp.514–522.
11. M.I. Yarymovych: Santini memorial lecture: *Space challenges and opportunities for human benefit*, Acta Astronautica 75 (2012) pp.63–70.
12. I.A. Crawford: *The scientific case for human space exploration*, Space Policy 17 (2001) pp.155–159.
13. R.D. Launius: *The historical dimension of space exploration: reflections and possibilities*, Space Policy 16 (2000) pp.23–38.
14. J. Clark, C. Koopmans, B. Hof, P. Knee, R. Lieshout, P. Simmons, and F. Wokke: *Assessing the full effects of public investment in space*, Space Policy 30 (2014) pp.121–134.
15. J. Smart, and H. Benaroya: *An examination of nonlinear and passive technology transfer in the space sector: Consideration of the Contingent Effectiveness Model as a basis for formal modeling*, Space Policy 38 (2016) pp.39–47.
16. *Lunar Industrialization and Colonization*, Journal of the British Interplanetary Society, Editor H. Benaroya, Vol.47, No.12, December 1994.

17. *Lunar Commercial Development (Part II)*, Journal of the British Interplanetary Society, Editor H. Benaroya, Vol.51, No.2, February 1998.
18. *M. Williamson: Lunar exploration and development - A sustainable model*, Acta Astronautica 57 (2005) pp.161–166.
19. E. Sadeh, D. Livingston, T. Matula, and H. Benaroya: *Public-private models for lunar development and commerce*, Space Policy 21 (2005) pp.267–275.
20. J. Hickman: *The Political Economy of Very Large Space Projects*, Journal of Evolution and Technology, Vol.4, November 1999.
21. H. Benaroya: *Economic and Technical Issues for Lunar Development*, Journal of Aerospace Engineering, Vol.11, No.4, October 1998, pp.111–118.
22. I.A. Crawford: *Viewpoint: Space development - social and political implications*, Space Policy 1995 11(4) pp.219–225.
23. D. Livingston: *A Code of Ethics for Conducting Business in Outer Space*, Space Policy 19 (2003) pp.93–94.
24. **United Nations Treaties and Principles on Outer Space**, United Nations, New York, 2002.
25. G.H. Reynolds: *The Moon Treaty: Prospects for the Future*, Space Policy 1995 11 (2) pp.115–120.
26. A. Bini: *The Moon Agreement in the 21st Century*, Acta Astronautica 67 (2010) pp.496–501
27. **Pioneering the Space Frontier, An Exciting Vision of Our Next Fifty Years in Space**, Bantam Books, 1986.
28. **Report of the 90-Day Study on Human Exploration of the Moon and Mars**, November 1989, Internal NASA Report.
29. **America at the Threshold, America's Space Exploration Initiative**, May 3, 1991, published by the Government Printing Office.
30. R.L. McNutt Jr.: *Solar System Exploration: A Vision for the Next 100 Years*, Johns Hopkins APL Technical Digest, Vol.27, No.2 (2006).
31. H.H. Koelle, and D.G. Stephenson: *International Academy of Astronautics 5th cosmic study preparing for a 21st century program of integrated, Lunar and Martian exploration and development (executive summary)*, Acta Astronautica 52 (2003) pp.649–662.
32. H-H. Koelle: *Prospects of a Settlement on the Moon - Development, Operation, Cost, Benefits*, Technical University Berlin, Institute of Aeronautics and Astronautics, ILR 364 (2002).
33. **Lunar Settlements**, H. Benaroya, Editor, CRC Press, 2010.
34. The Lunar Exploration Roadmap: Exploring the Moon in the 21st Century: Themes, Goals, Objectives, Investigations, and Priorities, 2016.
35. B. Bova: Epilogue: Address Given at a 'Tricentennial Celebration, 4 July 2076', by Leonard Vincennes, Official Historian of Luna City, 1985 Lunar Base Conference, Lunar and Planetary Institute.
36. **Turning Dust to Gold, Building a Future on the Moon and Mars**, H. Benaroya, Springer-Praxis, 2010.

3

The lunar environment

"It's hot, toxic, and raining bullets,
… but it's a great place to live."

A primary source for the physical and chemical properties of the Moon and its environment was written and edited by Heiken *et al.*[1] The lunar physical environment is dramatically different than that of the Earth.[2] [3] Human health risks are continually studied and updated with new data.[4] A summary of lunar physical parameters is presented in Table 3.1.

The diurnal cycle on the Moon is 29.53 Earth days, almost evenly split between daylight and nighttime. The lack of atmosphere has many implications for potential lunar dwellers, such as a lack of shielding against radiation and micrometeoroids, but also that daylight is in extreme contrast, which has implications for outposts at the poles. The temperature transition from daylight to nighttime is rapid (about 5°C/hr). At the Apollo landing sites, the temperature ranged from 111°C to −171°C, resulting in major thermal expansion/contraction and thermal cycling challenges to surface structures. If a structure is to be directly exposed to these extreme temperatures it must be made of highly elastic materials, and materials with different coefficients of thermal expansion must be used carefully. Material fatigue due to thermal cycling is generally a problem that needs to be ameliorated. Even those structures that are shielded are susceptible to material fatigue and brittle fracture and since everything will be exposed during construction (because the shielding is not yet in place), designs must be careful of this phase.

Lindsey, in a paper on the use of regolith as shielding for a surface lunar structure, noted that lunar test data indicates that under a few centimeters of regolith, or in a lava tube about 10 m below the surface, the temperature is relatively constant in the range of −35°C to −20°C.[5]

The dangers to people of certain environments, and the use of certain materials, is well known. On Earth, concerns about the environment and its toxic components is a serious issue. Carrying those concerns into space, and to the Moon and Mars, adds uncertain

© Springer International Publishing AG 2018
H. Benaroya, *Building Habitats on the Moon*, Springer Praxis Books,
https://doi.org/10.1007/978-3-319-68244-0_3

Table 3.1. Comparison of Earth and lunar physical parameters.

Property	Moon	Earth
Surface area [km²]	37.9×10^6	510.1×10^6
Radius [m]	1738	6371
Gravity at Equator [m/s²]	1.62	9.81
Escape velocity at Equator [km/s]	2.38	11.2
Surface temperature range		
°C	−173 to 127	−89 to 58
°F	−279 to 261	−128 to 136
Seismic energy [J/year]	$\simeq 10^9$ to 10^{13}	10^{17} to 10^{18}
Magnetic vector field [A/m]	0	24 to 56
Surface atm pressure [kPa, psi, mbar]	0, 0, 3×10^{-12}	101.3, 14.7, 1000
Day length [Earth days]	29.5	1
Sidereal rotation time	27.332 d	23.9345 h

Seismic energy does not account for seismic activity due to meteoroid impacts. The magnetic vector field is in units of Ampere/meter. For the Moon, there is a small paleofield, that is, a very small ancient magnetic field.

dimensions to toxicity risk characterization. In particular, there is a significant lack of understanding about how the micro- and low gravity environment affects a number of issues regarding human health and wellbeing. Clearly, structures need to be designed with a thorough understanding of how their component materials behave and age in the low gravity lunar environment.

This following discussion is based on the work of Irons et al.[6] While this work focused on human exposure to substances during long-duration space flights, many of the concerns and conclusions apply to extended stays at bases on the Moon and Mars. We quote the framework:

> "Risk associated with chemical exposure is defined as the probability that a substance will produce adverse effects under a given set of exposure conditions. In characterizing risk, the toxicologist attempts to integrate basic knowledge of the harmful effects of an agent, its mechanism(s) of action and its potency, together with an assessment of the nature and likelihood of exposure. What constitutes an acceptable risk is often a matter of judgement that must take into account the nature of the adverse outcome predicted, the opportunity or likelihood of its occurrence, and the costs associated with preventing it. Risk characterization in the extended space flight environment presents the toxicologist with a unique set of challenges."

In the space environment, the uncertainties are large and therefore margins of safety must be large, while remaining practical. The forces that the human body endures, as well as the physiological changes it undergoes in the space environment, affect its ability to resist the toxicity of a given compound. There is a coupling between the environment experienced by the human body and how it responds physiologically to everything else it experiences.

The challenges of reduced gravity and ionizing radiation on the human body, and perhaps the challenges imposed by a closed recycling life support system, are appreciated, but less discussed are the issues of the body's exposure to system and materials degradation during long-duration missions. "This may result in either the outgassing or release of the products of material decomposition, or result in the failure of containment vessels for waste or hazardous materials."

Our Apollo and Shuttle missions point to the following exposure possibilities:

"… thermodegradation of synthetic materials, primarily electrical insulation; contamination from extinguishing agents; leakage or release of payload chemicals; leakage of waste chemicals or recycling products from closed systems; and contamination from extravehicular activities with potential exposure resulting from external venting. In addition, the potential for the growth of microbial flora in a confined, recycled atmosphere raises the possibility of additional secondary production of chemical contaminants."

Figure 3.1. Apollo 11 astronaut Edwin Aldrin poses for a photo beside the U.S. flag on the Moon. Mission time: 110:10:33, July 20, 1969. (Courtesy NASA/Neil A. Armstrong)

Low gravity affects all these possibilities in ways that we cannot test on Earth. For example, the way fires initiate, propagate and extinguish in a low gravity environment is not sufficiently understood and the choice of extinguishing agents for (likely) electrical fires is critical. For example, Halon cannot be used in a lunar habitat because of the way it operates and the products that result from its use (see Irons *et al*. p.245). Whether through EVA operations, or on the lunar surface, exposure to hydrazine, found in spacecraft propulsion systems, can be fatal, "with reported effects involving the central nervous system, the blood, the liver, the kidneys, and the immune system."

The enclosed spaceflight and lunar habitat environment represent serious challenges, in particular to "controlling the growth of microflora [and] biofilms ... from the standpoint of the potential for contaminant generation, and the possibility of their deliberate controlled use in bioregenerative life support systems ... may pose additional complications such as 1) a source of chemical corrosion on containment vessel surfaces, which are usually extremely thin, and 2) a secondary source for the generation of toxic chemical species. ... There is the question of whether the chemical products of biotransformation are more toxic than the parent compounds."

Irons *et al*. expressed concerns about the correlation between physiological changes and susceptibility to chemical toxicity; the influence of body fluid redistribution, and cephalic fluid shifts due to low gravity, on pharmacokinetics and the blood-brain barrier; and the influence of microgravity, stress, and radiation on chemical toxicity to the blood and immune systems. Simply put, the new environment changes the human body's response to toxicity in ways that we do not fully understand.

An interesting model, proposed in that paper for the prediction and modeling of contaminant-induced toxicity and performance degradation, is the SMAC model. The Spacecraft Maximum Allowable Concentration value is a measure of the maximum acceptable concentration of a contaminant in the closed space environment, and is based on toxicological data. The toxicity of a defined mixture is estimated by adding the ratios of measured concentrations for each compound to the appropriate SMAC for every compound in the mixture that has the same target organ and the same mechanism of action:

$$T = \frac{C_1}{SMAC_1} + \frac{C_2}{SMAC_2} + \cdots \frac{C_n}{SMAC_n},$$

where T is the toxicity value; $C_1, C_2, \ldots, C_n$ represent realized concentrations of the different chemicals, and $SMAC_1, SMAC_2, \ldots SMAC_n$ represent their respective maximal allowable concentrations. The maximum allowable concentration for such a mixture is one that results in a T value of 1.0. There are potential issues with this formulation, but it provides a simple approach to satisfying required specifications. Those who are familiar with fatigue life modeling of mechanical components will recognize this approach as identical to Miner's rule for the fatigue life of a vibrating system, where fractions of life expended are added for each oscillation/amplitude and with failure defined when the sum of life fractions equals 1.0. In the Miner model for fatigue, failure sometimes occurs for sums less than 1.0 and other times for sums greater than 1.0. Such variability occurs for SMAC estimates as well. These are only simple order of magnitude models.

Numerous factors affect living organisms on the Moon. These factors have been known for a considerable period of time.[7] On the Earth, surface radiation dosages range between 1 and 2 mSv/year (0.001 to 0.002 Sv/year) due to the protective atmosphere. At the lunar surface, the galactic cosmic radiation gives rise to a dose equivalent of about 0.3 Sv/year.[1] But at a depth of 1 m of regolith, the annual radiation dose equivalent due to cosmic-ray particles decreases to about 2 mSv/year – comparable to the Earth's surface. Shielding materials can interact with the cosmic-ray particles and release neutrons with an additional dose equivalent of 0.1 Sv/year.

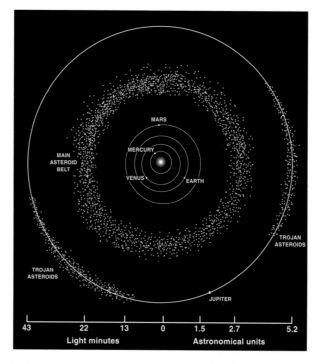

Figure 3.2. The inner Solar System, from the Sun to Jupiter, including the asteroid belt, the Jupiter Trojans, and the near-Earth asteroids. The image represents a radius of 43 light minutes, or about 5.2 astronomical units. Positions on 14 August 2006.

At the Apollo landing sites, a 282°C range of temperatures existed, between 111°C and −171°C. At a depth of 20 cm, the variation decreases to about 10°C, and at a depth of 40 cm, the variation is 1°C, where the nominal temperature is approximately −20°C. The regolith has low thermal conductivity, which is why we see such a drastic reduction in the temperature variations.

---

[1] The sievert (Sv) is the SI-derived unit of dose equivalent. It attempts to reflect the biological effects of radiation as opposed to the physical aspects, which are characterized by the absorbed dose, measured in gray. It is named after Rolf Sievert, a Swedish medical physicist famous for work on radiation dosage measurement and research into the biological effects of radiation.

Space flight has been shown to induce varied, potentially detrimental, immune responses, some of which occur immediately after arriving in space while others develop throughout the span of the mission. The major causes include microgravity and the lack of load bearing, the stress due to the high-demand activities, the confined spaces for social interactions, the uncommon diet, and radiation.[8]

We can envision that the criticality of such issues will require an ability to monitor astronauts' vital signs so that corrective actions can be regularly implemented. In-flight measuring capabilities have been available since the first manned missions, but the level of sophistication needed, beginning with the return to the Moon, will likely lead to the development of devices at the micro and nano scale. With time, genetically engineered nanodevices will be designed for each astronaut's DNA, leading to the ability to tailor medicines for each individual. DNA-customized nano-pharmaceuticals for the delivery of an exact dosage at the molecular level is not far-fetched.

From the perspective of radiation health, a regolith layer of 2–3 m depth provides radiation shielding of about 40 g/cm², whereas the Earth's atmosphere provides a shielding equivalent of 100 g/cm². Rare, large solar flares require shelters with shielding of at least 700 g/cm².

In addition to radiation effects, the lunar regolith dust is abrasive and toxic – bad for machines, deadly for people. Shielding or burial of structures will be required to solve the radiation problem almost immediately upon our settlement of the Moon, but the dust problem will linger much longer. Sheaths for machines and robots will need to be designed to keep moving parts and joints clear of the dust. Environmental suits will be built into the walls of the habitats and structures so that astronauts will not have to go through hatches to go to the lunar surface and back.

3.1 PARTIAL GRAVITY

At the lunar surface, gravitational acceleration is about 1/6 g, where $g = 9.8$ m/s² on Earth. This means that the same structure will have six times the weight-bearing capacity on the Moon as it would on the Earth. Conversely, to support a certain loading condition, one-sixth the load bearing strength is required on the Moon as on the Earth. Therefore, the concepts of dead loads and live loads[2] within the lunar gravitational environment have to be reconsidered. Mass-based rather than weight-based criteria will need to be developed for lunar structural design codes, because mass is invariant whereas weight depends on the gravitational acceleration, as per the equation:

$$weight = mass \times gravitational\ acceleration$$

Lower gravity on the Moon means that much longer spanned structures are possible, but it also means that gravity plays a much smaller role in anchoring the structure.

---

[2]A dead load is a weight that is permanently in the structure, for example, floor weight or machine weight. A live load is a weight that can move. It does not have to be alive, it just has to be capable of being at different locations at different times. For example, cars moving on a bridge are live loads.

The human physiology and medical issues are perhaps more critical, suggesting that there is much to learn from on-site experiments, first with tissue, plants and lower mammal life forms, and then with humans. Horneck summarized the concerns succinctly:

"One point of concern is human stay at 1/6 g, which will trigger a string of adaptational processes, in the field of neurophysiology, the cardiovascular system, oxygen metabolism, calcium turn-over, and in the blood forming system. These same topics have been identified as the most sensitive ones during the weightlessness of space flight. In order to study the adaptational processes during 1/6 g, it is important to identify their gravity thresholds. It is further required to establish countermeasures against the ill effects of the reduced gravitational field.

"Medical care and the knowledge of pharmacokinetics are important issues when treating a sick astronaut and for certain preventive procedures. Studies under weightlessness have shown increased blood concentrations of some pharmaceuticals, which has to be investigated also for 1/6 g conditions.

"Finally, a database should be exploited and applied for the health, safety and well-being of the population on the Moon, in order to optimize the living and working environment on the Moon. The information laid down in the database should also be made available to architects, engineers, agencies, politicians, policy and law people and other decision-makers, and constructors. Hence, a lunar base which will eventually have a permanent human presence may be considered as a first step towards living on another celestial body."

Pharmacokinetics and pharmacodynamics in micro- and low gravity have been recognized as something requiring technological or gene therapy advances.[9]

3.2   INTERVIEW WITH JAMES LOGAN

This interview with James Logan was conducted via email beginning on June 25, 2017 and concluding on July 10, 2017. We begin with a brief professional biography, and then continue with the interview.

Brief Biography
Dr. Jim Logan held numerous positions in his twenty-two-year career at NASA, including Chief of Flight Medicine and Chief of Medical Operations at Johnson Space Center in Houston. He served as Mission Control Surgeon, Deputy Crew Surgeon or Crew Surgeon for numerous Space Shuttle missions and as Project Manager for the Space Station Medical Facility, developing the initial design for the first in-flight medical delivery system for long-duration missions.

A founding board member of the American Telemedicine Association, Dr. Logan has served as a consultant to the RAND Corporation, a variety of professional organizations, international and domestic hospital-based health care systems, and the Department of Defense. He was the first Provost for the International Space University in Strasbourg, France.

Board certified in Aerospace Medicine, he completed a medical fellowship in Undersea & Hyperbaric Medicine at Duke University Medical Center in 2013. A recipient of NASA's Distinguished Speakers Award, Dr. Logan has lectured in Australia, France, Germany, the United Kingdom, Japan, Canada, Iceland, Russia, Argentina, Costa Rica, Guam, South Korea, New Zealand and the People's Republic of China. As an expert in human space operations and space settlement, he has been featured on the Public Broadcast System, CanadaAM, The History Channel, National Geographic Channel, numerous radio talk shows, and a variety of print and online publications.

Dr. Logan is now the CEO and Co-founder of the Space Enterprise Institute, a non-profit 501(C)3 educational foundation.

Our Interview
Given that we have been waiting almost 50 years to return to the Moon, do you have a feeling as to when we might return, with people, for an extended stay?

> *This is an excellent question that is difficult to answer. Despite what the space community would like to believe, going to the Moon the first time had essentially nothing to do with space, per se. It was a proxy competition (i.e., Space Race) to prove to the world what the most successful economic model was at the time: Capitalism or Communism. Once the question was settled (i.e., Capitalism 'won' the race by landing men on the Moon and returning them safely to Earth first), national interest in human space exploration for its own sake waned and has not returned to pre-Moon landing levels.*
>
> *The next 'Moon rush' won't be the result any kind of 'race'. It will be for control of resources. Unfortunately, in the last 50 years neither NASA nor the DoD has embraced the concept of in-situ resource utilization (ISRU) with the singular exception of electrical energy from sunlight. The private sector has an appreciation of the necessity of ISRU for the next phase of human expansion into the solar system but has grossly underestimated the complexities and costs of implementing it. Until some entity (private or otherwise) actually demonstrates the economic efficacy of operational ISRU, 'progress' will be limited to flashy PowerPoint presentations and speculation rather than objective experience.*

In your mind, what is a likely scenario and timeline for manned space?

> *In my personal opinion, the 'scenario', when and if it happens in the foreseeable future, will be based initially on harvesting off-Earth water but primarily if water is utilized as a 'multiuse propellant', that is as a propellant, a consumable and radiation protection (shielding) agent. The best propellant scenario is one that utilizes liquid water as propellant for nuclear thermal propulsion ($I_{sp} > 800$ s with exhaust temperatures in excess of 3000°C; an admittedly challenging but solvable materials problem) because liquid water is easy to store, transport and will likely be more compatible with the vast majority of ISRU scenarios. (Note: The prospects of finding significant concentrations of extra-terrestrial xenon or hydrazine are remote.) Propulsion using cryogens derived from water or ice are*

technically feasible but would require massive amounts of power, infrastructure and increased complexity for cooling/storage. Furthermore, cryogens would supply only half NTP [nuclear thermal propulsion] performance with I_{sp} [specific impulse] capped at no greater than 365 s. Utilizing liquid water as both propellant and radiation shielding exhibits the added synergy of being able to 'burn' part of the shield at destination capture (e.g., moons of Mars). This 'multiuse propellant' scenario has been described previously in general terms by D.R. Adamo and J.S. Logan (http://www.thespacereview.com/article/2631/1 and http://www.thespacereview.com/article/2637/1). It is also discussed in expanded technical detail.[10]

Do you agree with those who have made a case for settling the Moon first and then using it as a base for travel to Mars?

There is no case for establishing a human presence on Mars and the prospects for a reasonable case diminish with each new attribute we learn about the red planet (see answer to next question).

Regarding the Moon, it all depends on the precise definition of the word 'settling'. Unfortunately, terms like 'base', 'outpost', 'settlement', 'frontier' and 'colony' are used more or less interchangeably in the space lexicon. NASA is perhaps the worst offender. The space agency's lexicon has much more to do with public relations hype and funding rather than reality. NASA actually has significant incentives to muddle definitions because NASA's prime goal is public 'support'. If potential supporters (citizens as well as politicians) hear what they want to hear in the words of NASA press releases and publications, so much the better. Why split hairs?

However, words are important; subtle differences in meaning have significant implications. The entire 'human expansion into the solar system' discussion could benefit from general agreement on the definition of terms. Exactly what is meant by 'sortie', 'outpost', 'settlement', 'frontier' and 'civilization'?

A 'sortie' is an 'out and back' mission. All the Apollo missions to the Moon were little more than two-week sorties. The hallmark of a sortie is intermittent visitation.

An 'outpost' is place you go to repeatedly for a specific reason, whether it be scientific or economic, then come back. Antarctica is a scientific outpost. An offshore drilling platform is a resource extraction outpost. The International Space Station is an outpost. There can be so-called 'permanent human presence' but by different humans. The hallmark of outpost is 'rotating crews'. ISS has been permanently 'manned' since Halloween of 1999 but it's still only an outpost. Any lunar 'base' or Mars 'base' will also only be an outpost. Another hallmark of outposts is no one is born there. No one (with one exception) has ever been born in Antarctica even though there has been permanent human presence there for over 60 years. There are no natural born citizens of Antarctica.

'Settlements' are different. People go there with the intent to stay. And, in fact, they do. Settlements consist of men, women and children. Roanoke Island was a settlement – albeit a failed one. People went there to stay in 1585. The hallmark of settlement is babies are born there. Virginia Dare was the first child of European descent to be born in the New World even though we don't know what became of her because the colony vanished. It is now known as the Lost Colony and is located in Dare County, North Carolina.

The settlement of Jamestown in tidewater Virginia was founded in 1607. Plymouth and Boston were founded in 1620 and 1630, respectively.

Settlements are considered community. Unlike outposts, which are more or less 'resupplied' from someplace else on an ongoing basis, settlements have the added burden of having to achieve self-sustainability early on.

The next stage is 'frontier'. Like settlements, frontier consists of men, women and children, but the hallmarks of frontier are multiple generations and multiple settlements. Another hallmark of frontier is low population density overall and perhaps even economic isolation between settlements.

'Civilization' is the final stage. Civilization implies men, women, children, multiple generations with ongoing physical and economic interaction – with shared (or at least very similar) language, traditions, geography, culture, religion and perhaps shared governance (even taxes!) and what would be considered political capitals or 'seats of power'. Historical examples include the Greek, Roman and Chinese civilizations.

By these definitions it's extremely unlikely the Moon will be a true settlement destination in the foreseeable future. Although no definitive proof exists either way, the consensus of most life science experts is [that] lunar gravity (one-sixth that of Earth) is insufficient for normal human growth, development and reproduction. Furthermore, chronic radiation levels at the lunar surface would severely constrain routine EVAs as well as surface habitation. All habitats will be subsurface by necessity. It is not hyperbole to make the statement [that] if humans ever reside on the Moon, they will have to live like ants, earthworms or moles. The same is true for all round celestial bodies without a significant atmosphere or magnetic field – Mars included.

Extended lunar surface operations will primarily be done utilizing robotic techniques, overseen by humans hunkered down in radiation-protected habitats.

It is very possible the Moon will remain, like terrestrial off-shore drilling platforms and Antarctica, only an outpost for quite some time.

How do you view the push for Mars settlements at this time in our spacefaring?

The particular topic always hits a raw nerve with me.

Mars has even less going for it than the Moon. Other than rabid enthusiasm of Mars zealots, mostly due to the role Mars has played in science fiction the past hundred years, there is no rational argument for placing humans on the Martian surface and several excellent arguments against doing so.

As I have stated many times in public presentations and written comments, Mars is not, repeat not, the destiny of mankind. It's not even a viable side trip. It's a dangerous, expensive and destined-to-fail diversion. If we continue pursuing this mass delusion, Mars will be to us what Greenland was to the Vikings.

The truth is, Mars is the most tragic planet in our solar system. Initially promising with a dense atmosphere, liquid water and perhaps the possibility of primitive life, it is now a frigid (average surface temperature of minus 82 degrees Fahrenheit), poisonous (remnant atmosphere of mostly CO_2 and perchlorate-infused regolith approximately one percent concentration planet-wide), radiation-soaked, sterile (as far as we know), dead-end-destination ghost of a planet with only 50 percent the ambient sunlight of Earth. It exists at the bottom of a large gravity well that's too big for rockets and (very likely) too small for people, plants and animals.

Mars should be pitied, not coveted.

There is absolutely nothing of any value except for the singular exception of raw materials to synthesize propellant to get off Mars!

Mars colonization is a 'zombie' idea; that is, one that died previously (when Mariner 4 sent back the first close up pictures in 1965) but like 'real' zombies, 'Mars colonization' still lifelessly roams the intellectual landscape wreaking havoc among the living. It embodies the worst aspects of dust bowl Oklahoma, Antarctica and a toxic superfund site.

All the talk of 'one million people on Mars' is utter bunk, sheer fantasy and completely uncoupled from any logic or rationale whatsoever – other than Mars just happens to 'be there' whatever that means. It's the 21st century equivalent of the 1950s era flying car. Almost everyone assumed it would happen (except the people who 'ran the numbers') but it never did for very valid reasons.

Your extensive twenty-year career at NASA, including Chief of Flight Medicine, and Chief of Medical Operations at NASA's Johnson Space Center in Houston, has provided you a rare and inside view of the physiological and psychological side of the human spaceflight equation. While the engineering of manned spaceflight and the eventual lunar habitats is a serious challenge to our abilities, it appears that the other side of the equation, the biological side, is an order of magnitude more difficult. Do you agree?

This is an excellent question that would require an entire book to answer completely. But here's the short answer: There are many challenges to long-duration human spaceflight, especially interplanetary human spaceflight. But the two biggest are flight dynamics (always constrained by the cold hard realities of the Rocket Equation) and bioastronautics (the psycho-physiological adaptation, or lack thereof, of biological organisms to the deep space environment). The human exploration and expansion into space cannot, repeat cannot, violate the laws of physics and biology. Period.

NASA is an organization of engineers, by engineers and for engineers. One of the first things I learned at the agency was that if engineers could have sent men to the Moon and back without any life science input whatsoever, they would have done so. But of course, they couldn't. Nevertheless, the space life science has always been the redheaded stepchild of the space program.

I spent the last ten years of my NASA career begging Space Life Sciences management (unsuccessfully) to address the critical aspects of bioastronautics to prepare for the coming 'interplanetary phase' of human spaceflight. Instead of heeding the advice of myself (and others), NASA consciously chose not *to address the issues. The result is what we have now: No beyond Earth orbit (BEO) human operations (or viable plans for such operations) except in flashy PowerPoint slides.*

Dust mitigation is a critical issue generally on the Moon. How far are we from getting a handle on that problem, both the biological risks and the engineered systems risk?

We haven't even adequately characterized the biological risk of lunar 'fines' (extremely small highly biologically active particles of regolith that can pass through a 1 mm sieve) and 'dust' (particles that can pass through a 100-micron sieve), much less developed potential mitigation strategies. Lunar regolith is fundamentally different from Earth soil.

The concern is the possibility lunar dust exposure could produce a clinical syndrome like Silicosis. What used to be called stone-grinder's disease first came to widespread public attention during the Great Depression when hundreds of miners drilling the Hawk's Nest Tunnel through Gauley Mountain in West Virginia died within half a decade of breathing fine quartz dust kicked into the air by dry drilling – even though they had been exposed for only a few months. It was (and still is) one of the largest occupational-health disasters in U.S. history.

The problem with lunar dust isn't limited to biology. Lunar fines are very abrasive and easily creep into seals and gaskets, rendering critical equipment inoperative. Apollo 17 Commander Gene Cernan, the last man to walk on the Moon, told me personally on several occasions he and Harrison Schmitt would not have been able to do another EVA on the lunar surface because lunar dust had functionally frozen (or severely limited the range of motion of) most of the joints in the space suit.

Key challenges to engineers and scientists who will be tasked to develop the lunar mission and habitat design are the knowledge gaps that cannot be filled before being on site. For example, the lunar 1/6 *g* environmental effects on biology, and systems reliability in the lunar dust environment. Also of great concern is our lack of understanding of the very complex systems that will not be integrated before they are placed on the Moon. Even if the behaviors of all the components are fully understood, the complete integrated system will generally have unanticipated characteristics. Is there a way of dealing with these knowledge gaps before being there?

Certain knowledge gaps can be addressed prior to going back to the Moon. The role of 1/6 g on a 24/7 basis could be generally assessed in a LEO variable-g research facility (which could also facilitate certification of essential subsystems). However, certifying functionality of equipment in an integrated lunar environment, including radiation and lunar dust, may have to wait until initial deployment.

It appears that missions to settle the Moon or Mars have two underlying potential show stoppers: (1) assuring human survival and good health, and (2) assuring very high reliability for all engineered components. While we have shown with the ISS and Apollo that we can manage short forays to the Moon, or longer ones to LEO, satisfying (1) and (2) in limited ways, the bar is an order of magnitude higher for long-duration lunar stays, and an order of magnitude higher than that for Mars human missions. Are guaranteeing (with high probability) human and machine survival and high operability on the horizon, or beyond it?

Sailing ships plied the Mediterranean for thousands of years without much difficulty. However, once transoceanic voyages were attempted beginning in 15th century it didn't take captains, crews or their respective patrons long to discover long-duration voyages across vast expanses of open ocean were a whole new ball game.

Ferdinand Magellan left Spain in August of 1519 with five ships and a crew of 265 men. Three years later, only one stricken ship and 18 men, half-dead from starvation and disease, limped into Seville harbor after circumnavigating the globe. The great navigator lost fully 80 percent of his crew to Scurvy, a known but heretofore rare disease, crossing the Pacific. Even so, he and his voyage changed the course of history.

Magellan's voyage was undertaken with primitive 16th century technology. Likewise, the first forays into interplanetary space will no doubt be attempted with rather primitive legacy technologies, many of which date from the 1960s. Propulsion and re-entry techniques for human spaceflight haven't fundamentally changed since the dawn of the space age.

History may not repeat but it does recycle. The Mediterranean may be to LEO what early attempts at transoceanic voyages will be to BEO and interplanetary spaceflight.

Long-duration transoceanic voyages required breakthroughs in technology and operational medicine. Unless we are creative as well as informed (by valid data and reasonable extrapolation), we may be ushering in what in retrospect will be a completely unnecessary 'scurvy phase' (metaphorically) of human spaceflight.

The numbers illustrate what neophytes we are regarding BEO human operations. Six Apollo missions (1969-1972) cumulatively logged slightly less than 300 hours on the lunar surface, including 81 total hours of lunar EVA. Since each Lander had a two-person crew, human beings have only 600 man-hours of lunar surface experience, approximately 162 of which are lunar EVA. Although the final and longest mission (Apollo 17) spent a little more than three days on the Moon, the average lunar time per Apollo lunar astronaut was only 2.08 days and the average lunar EVA time per astronaut was only 13.5 hours. While significant, these exposure times are not compelling either from a biomedical or occupational exposure perspective.

Yet space advocates talk blithely among themselves and in the blogosphere about interplanetary spaceflight and colonies on the Moon and Mars like it will be no big obstacle when, in fact, nothing is further from the truth.

Who would you consider as your key influencer in your pursuit of space studies? Who inspired you?

I was a child of the space program. I was one of those lucky kids who knew by the time I was seven years old exactly what I wanted to do with my life. Even though I was no genius (as my friends and co-workers can attest), through a very fortuitous series of events I was able to take (and pass) a college course in astronomy when I was ten years old. The course was taught by astronomer James A. Westphal, then only 30 years old, who later became a professor at Caltech, Director of the Palomar Observatory in the 1990s, a recipient of MacArthur Foundation genius grant and lead investigator for the development of the Hubble Space Telescope's Wide-Field and Planetary Camera. Six months after completing the course, I met Werner von Braun and was invited to attend the First National Conference on the Peaceful Uses of Space.

But it was the rescue of the Apollo 13 crew in the spring of 1970 that gave me the 'compass heading' for my path. Flight Director Gene Kranz and his flight control team had to deal with the initial hours of an unfolding catastrophe, essentially a race against time to keep the spacecraft (and the crew) alive. Kranz and his team (1) Set the constraints for the consumption of spacecraft consumables (oxygen, electricity and water); (2) Made the decision to use the Lunar Module as a lifeboat; (3) Created a jury-rigged mechanism to reverse dangerously increasing levels of carbon dioxide using cardboard, paper, scissors and duct tape from available materials; (4) Controlled the three course-correction burns during the trans-Earth trajectory; as well as (5) Created in-space power-up procedures for an essentially dead Command Module (never before done or even simulated) that allowed the astronauts to jump into it at the last second, jettison the Lunar Module and parachute safely back to Earth.

The ethos of keeping your cool, not *freaking out, getting a grasp of the situation no matter how dire, then thinking it through,* doing the math, *prioritizing and sequencing the problems, then focusing on one problem at a time, always mindful the solution to one problem can inadvertently create other ones, is essentially the same philosophy embodied in the character of Mark Watney in* The Martian.

The Apollo 13 team, including the astronauts, had to deal with truth *as it was – in all of its ugly and complex manifestations – not what they wanted it to be. For their truly heroic efforts, Gene Kranz, his team and the crew subsequently received the Presidential Medal of Freedom.*

I followed the emerging crises of Apollo 13 in real time completely transfixed by the tremendous competence, tenacity and creativity of Kranz and his team. I remember consciously thinking that *was one of the things I wanted to do in my life.*

Twelve years later, through a series of seemingly random events and more than a little luck; after years of college, graduate school, the military, medical school, a surgical internship and residency training in Aerospace Medicine; as well as months of intense study covering every aspect of spaceflight operations including guidance, navigation, communications, propulsion, flight dynamics, Shuttle systems and subsystems training including Environmental Control & Life Support; as well as successful completion of scores of intense Mission Control ascent, orbit and entry simulations, I was granted admission to one of the most exclusive clubs in the world – Gene Kranz's Mission Control Flight Control Team. It was, and still is, one of the proudest accomplishments of my life.

Gerard K. O'Neill, father of the modern space colony, author of The High Frontier *and tireless proponent of the 'humanization of space', is another very significant professional influence, as is the work of Dr. John Lewis, author of* Mining the Sky *and other publications.*

However stellar the previous 'influencers' may be, the prime motivator of my intellectual life, my alpha inspiration, the 'north star' of my attitudes, beliefs and feelings about science, technology, nature and philosophy, as well as the 'mentor' I always return to, is R. Buckminster (Bucky) Fuller (1895-1983).

I met Bucky Fuller briefly twice, once in the late 1960s when I was a college student and once in 1975 while I was still in medical school. In my opinion Bucky Fuller, Gerry O'Neill and perhaps Carl Sagan, whose writings and insights continue to inspire me to this day, were the last visionary geniuses of the 20th century (unfortunately none have appeared yet in the 21st century).

Rather than a scientific revolutionary, Bucky Fuller was an evolutionary. His genius was grounded in his unique breakthrough perspectives. He championed the terms 'Spaceship Earth', 'ephemeralization' and 'synergetics'. In 1968, he even published a book entitled Operating Manual for Spaceship Earth. *Fuller would have been fifth-generation Harvard had he not been expelled, twice, the first time for skipping an exam because he was dating a showgirl and the second time for lack of ambition (he was eventually awarded an Honorary Degree by Harvard – as well as the Presidential Medal of Freedom by President Ronald Reagan). Rated number four by TIME magazine in their 2010 list of the Top Ten College Dropouts (Bill Gates, Steve Jobs and Frank Lloyd Wright were one through three respectively), he actually appeared on the cover of TIME in 1964.*

Bucky Fuller never saw himself, or anyone else, as a noun. He was described by others frequently as a philosopher, mathematician, architect, author, poet, inventor, teacher, etc. Although he was definitely all these things, he saw himself as a verb, a comprehensive anticipatory design science innovator as he put it.

Fuller's goal, after he seriously contemplated suicide at age 27 because he considered himself a complete failure, was to discern what he called Nature's Coordinate System for all structures. He surmised Nature, by its very essence, would express the most efficient way of doing everything. After thinking about it, utilizing various thought experiments (like Einstein), he concluded the simplest, most efficient unit of structure was the tetrahedron, not the square or rectangle. He then derived the math to describe it (since the technology didn't exist yet to actually see it) – which became the science of geodesics.

The geodesic sphere is mathematically the most efficient way to enclose three-dimensional space with the least amount of material. It is also the most stable and doesn't depend on gravitational compression for strength - unlike standard brick and mortar buildings. A geodesic sphere is inherently stable in any configuration and any gravity level ... or vector, including zero g. It embodies what Fuller referred to as "tensional compression."

When the technology finally got good enough (e.g., electron microscope) to actually see minute natural structures, it became obvious the atomic and/or molecular lattice upon which structures 'in nature' are built is based on the tetrahedron and geodesics. He predicted groups of carbon atoms that could form small nanoscale geodesic structures. When they were eventually discovered in Nature, physicists termed them Buckyballs or buckminsterfullerenes.

Fuller passionately believed when thinking about any issue or problem you must start with the universe (i.e., move from the general to the specific rather than the reverse). When I am working on a problem, he famously said, I never think about beauty, but when I have finished, if the solution is not beautiful, I know it is wrong.

He also stated categorically the universe never contradicts itself. If it appears to do so, check your assumptions because one of them is wrong.

Fuller's poetic side (in 1962 he was the visiting Charles Eliot Norton Professor of Poetry at Harvard University) was never far from display in his writings. He famously said, Love is metaphysical gravity.

All of us working to promote space exploration and settlement recognize that much of what we do will not happen in our lifetimes. Of course, we have no choice but to accept this reality, but how do you view this?

This is an excellent question, one that generates a very visceral personal response. I have always been a fan of good science fiction. I say 'good' Sci-Fi because at its best Sci-Fi can address aspects of technology, culture, society, justice and civilization that can be problematic for straight nonfiction, philosophy or social critique. I vividly remember how profoundly affected I was by Arthur C. Clarke's vision of the future depicted in Stanley Kubrick's stunning 1968 motion picture 2001: A Space Odyssey. *I'll come back to Sci-Fi at the end so bear with me ...*

By July of 1981, I was a NASA employee at Houston's Johnson Space Center, directly supporting human space operations. The first Space Shuttle launch had occurred in April of that year. Hyped by NASA for over a decade as the ultimate solution to cost effective launch systems, I had a ringside seat in the slow but inexorable disillusion many, including myself, felt as the many shortcomings of the Shuttle became evident during the next five years. By mid-1985, I knew beyond doubt the year 2001 would be nothing like the movie 2001.

My disillusion turned into seething frustration in January 1986 with the death of three close friends and four other colleagues on the Challenger *mission – especially after I learned there had been multiple partial as well as complete burn-throughs on the infamous primary and secondary O-Rings on several previous missions, which NASA summarily ignored even though it violated its own flight rules. When it became obvious senior NASA management had decided to 'double down' on a human launch vehicle with proven fatal design flaws, I reluctantly realized my cherished goal of playing a supporting role in the transition of mankind from a terrestrial to a celestial species, the reason I aspired to the space program in the first place, would not happen in my lifetime.*

I spent a year at NASA Headquarters in Washington, D.C. beginning in May of 1988, during which I became completely demoralized with the agency, and eventually resigned in May of 1989 to become the first Provost of International Space University.

During the next ten years I had no direct interaction with the space agency. In January of 1999 I was recruited back to NASA to provide medical support for the space station program (I was a founding board member of the American Telemedicine Association and had worked the previous decade on the design and implementation of telemedicine systems across the globe). However, before I agreed to put the NASA badge back on, I had to come to terms with my rather low expectations. Coming back to NASA was like getting together with your 'one true love' of years past. You are elated to be back with your first love but it doesn't take long to realize why you broke up in the first place.

Ironically, one of my best friends at the time, a fellow flight surgeon, lost his wife (another flight surgeon) when Space Shuttle Columbia *disintegrated over central Texas during entry in early 2003 dooming the entire crew – almost 17 years to the day of the* Challenger *disaster. Again, NASA ignored its own flight rules by discounting multiple documented instances of debris coming off the external tank during ascent because they couldn't bring themselves to believe a 1.3 pound 'piece of foam' could bring down a Space Shuttle. It was, however, another fatal design flaw.*

Now back to Sci-Fi …

In the movie Contact, *based on the outstanding novel by Carl Sagan, the protagonist is Dr. Ellie Arroway, an intrepid female astronomer wonderfully portrayed by Jodie Foster. After she is seemingly 'transported' to a beautiful planet in the Vega system for a brief few minutes, a vision of her long-dead father appears to her. At the very end of their touching, emotionally bittersweet scene, he embraces her and says to her softly, "Small steps, Ellie, small steps." The older I get, the more profound this advice becomes. I now use it as a mantra when I am forced to realize how little progress we have made since Apollo and how long the journey will likely take to truly become a celestial species.*

Is there an issue or concern of yours that I have not touched upon?

I've been in this business a long time. In my opinion, one of the biggest threats to progress on any space front today is 'Magical Thinking'. Now epidemic in space circles (including the halls of NASA), MT has become a cultural phenomenon not limited to space, science or politics. MT and 'Truthiness' (a term coined by humorist Stephen Colbert several years ago) are closely related. As originally defined by Colbert, Truthiness is something that has the quality of seeming or being felt to be true or something that, in the opinion of the believer, ought to be true even if it is not necessarily true or even true at all. It 'sounds' like truth to the believer but usually doesn't stand up to intellectual scrutiny or fact checking.

I come across it so often in space discourse I have derived a mathematical formula for Magical Thinking. Magical Thinking is equal to Ignorance to the power of Arrogance:

$$MT = Ignorance^{(Arrogance)}.$$

When you combine Magical Thinking with a media culture that prizes 'clicks' over 'credibility', the results are pretty predictable. Think MARS ONE.

Magical thinkers love Truthiness. The ascent of Magical Thinking in space circles parallels the rise of what I (and others) euphemistically refer to as the Space Cadet. The Space Cadet is an amateur, more zealot than visionary, understandably frustrated with the lack of real progress in space exploration and enraged (in many cases appropriately) at what they perceive as timidity and 'risk aversion' in people and organizations who are making the decisions and setting priorities.

In many ways, they parallel the political discourse so rampant on both sides in America these days. They love to lob metaphorical 'hand grenades' at the 'status quo' (whatever they perceive that to be). Absolutely fearless about spewing stuff that, at the end of the day, doesn't make much sense, they are into the promise *of space, but not the* problems.

My observation is the people who most often embrace, even celebrate, Magical Thinking are driven by ideology rather than fact. When you enter the realm of Magical Thinking, you leave the world of evidence and logic and enter the weird Twilight Zone of psychology. The mind of the ideologue sees space as a blank screen upon which they project their greatest hopes, deepest fears, strangest quirks, favorite philosophies and preferred political persuasions. Space Cadets have read way too much Ayn Rand, Robert Heinlein and Elon Musk (in his most exuberant states) and not enough Greg Easterbrook or Greg Klerkx. They tend live in an echo chamber of like-minded sentiments and ideas. Sound familiar?

Neither reasoned nor seasoned, what little knowledge they possess is broad but hardly ever deep. They are the epitome of "a little knowledge is a dangerous thing." It's not that they necessarily overhype potentials or the possibilities, it's that they vastly underestimate the complexities and oversimplify the solutions. Usually, but not always, they have no operational experience or any real scientific or engineering credentials and are disdainful or even outright hostile to those who do. They tend to be deniers (e.g., Mars 'radiation deniers'). And, if you dig deep enough, there's almost always at least a whiff of paranoia or conspiracy theory. Unfortunately, they have NOT yet learned that any perception or Truth that is isolated and removed from its larger context ceases to be True.

These guys (more than 99.5 percent of them are male) troll the Internet on legitimate sites (such as The Space Review – one of my favorites) attempting to dismiss, 'throw shade' or in some way countermand rational discourse by anyone more knowledgeable. In their own mind, they truly believe they can neutralize an articulate, referenced stream of logic with a generalized 'one liner' meant to evoke applause from likeminded zealots rather than provide empirical (or even verifiable) evidence to the contrary.

In many ways, they are mainstream Americans! As my German friends have quizzically observed, "Only in America do you find people who think the Moon landing was fake and wrestling is real."

A Space Cadet can be recognized by these signs:

(1) They react very negatively to any inconvenient truth that doesn't align with their belief system;

(2) They interpret any kind of legitimate dissent as betrayal deserving of punishment and retribution;

(3) They have difficulty with detachment – the ability to let the difficult facts of reality work their way into the mind; and

(4) They believe anything they state often enough with unwavering conviction, especially en mass, becomes Truth.

The mantra I live by is the one I learned in Gene Kranz's Mission Control: "In God We Trust, All Else Bring Numbers." Numbers can be debated. The assumptions, interpretations and means of observation behind numbers can be deconstructed and constructively challenged. But you can't debate ideology. Debating Space Cadets is the equivalent of debating religion. One side brings logic derived, at least in part, from empirical evidence, objective observation and verified experience to the discussion. Space Cadets spout 'truth' as they believe it to be (usually in the form of one-liners, as if they are citing verse from their own space Gospel). It's true because they say it is. Period. I have learned to exit the conversation when I sense it has degraded to the equivalent of discussing religion. Even so, every time I hear this kind of Magic Thinking rhetoric, I get a feeling of Deja Moo, which is the sinking feeling I have heard this 'Bull' before ...

Thank you.

3.3 RADIATION

"An alphabet soup of high-energy particles."

The lack of atmosphere and negligible magnetic field of the Moon raises challenges for the design of structures with shielding against various forms of radiation from the Sun and deep space. There are high-energy galactic cosmic rays (GCR) composed of heavy nuclei, protons and alpha particles, and there are the products of solar flares, or solar particle events (SPE), which are a flow of high-energy protons that result from solar eruptions. Such radiation has serious implications for human and plant survivability, as well as possible effects on materials and electronics.[11]

Various categories of radiation require shielding. Solar particle events originate in the Sun and are correlated to solar activity. These pose few hazards except during solar storms, at which time living beings on the Moon would require extra protection. Galactic cosmic radiation originates from the stars and the short-term effects are not hazardous, but long-term exposure can increase the risks of cancer. A third type of radiation are the X-rays that result from high-energy electron collisions with metal conductors or passive radiation shields. In this case, the shielding can cause more biological damage than the original particles being shielded.

Galactic cosmic radiation is very difficult to shield against.[12] A 1 GeV proton[3] has a range of about 2 m in regolith, with secondary particles – released due to the primary particle collisions – penetrating deeper. The high-energy particles, from these collisions within spacecraft materials and lunar regolith, produce secondary radiation that is more dangerous than the primary radiation. Also, galactic cosmic radiation peaks when solar activity is at a minimum.

There is a need for more precise models that characterize the solar cycles so that solar activity can be better predicted.[13] With improved characterization and understanding of radiation sources in space, missions on the lunar surface can be planned in a precise way with a high probability of safety. Space 'weather' prediction tools can reduce uncertainties and significantly increase the number of allowable days for human crews to be without higher levels of shielding. We can envision the possibility that a large network of satellites placed in orbit around the Sun will increase lead times before dangerous solar activity.

Space radiation also affects our structures and tools. Solar cells are degraded by radiation, while spacecraft operations and communications are sensitive to radiation levels.

Various units are used to measure radiation dosages. The 'gray' (Gy) equals 100 rads. Radiation absorbed dose, and effective dose in the international system of units (SI system) for radiation measurement, use gray and 'Sievert' (Sv), respectively. In the United States, radiation absorbed dose, effective dose, and exposure are sometimes measured and stated in units called 'rad', 'rem', or 'roentgen' (R). For practical purposes with gamma and X-rays, these units of measure for exposure or dose are considered equal. This exposure can be due to an external source irradiating the whole body, an extremity, or other organ or tissue, resulting in an external radiation dose. Alternately, internally deposited radioactive material may cause an internal radiation dose to the whole body, an organ, or a tissue. Smaller fractions of these measured quantities often have a prefix, such as milli (m) that means 1/1000. For example, 1 sievert = 1000 mSv. Micro (μ) means 1/1,000,000. So, 1,000,000 μSv = 1 Sv, or 10 μSv = 0.000010 Sv.

Conversions from SI units to older units are as follows:

$$1\,Gy = 100\,rad$$

$$1\,mGy = 100\,mrad$$

$$1\,Sv = 100\,rem$$

$$1\,mSv = 100\,mrem.$$

A dose of 1 Sv is considered large, with a recommended maximum of 0.05 Sv/yr. A dose of 10 Sv will result in death in a matter of days or weeks. A dose of 100 mSv leads to

---

[3] GeV, a gigaelectronvolt, is a unit of energy, equivalent to 1.60218 x 10^{-10} Joules, which is a little more than the rest mass of a proton or a neutron. The amount of energy needed to melt 1 g of ice is approximately 3.3 x 10^2 Joules.

a one-half percent increase in chance of cancer later in life. Radiation workers are allotted a 50 mSv annual dose. Human doses of 0.1 Gy (10 rad, 10 rem, 0.1 Sv, 11.4 R) are considered high doses. Of course, these are averages over large populations, and other factors such as an individual's genes or lifestyle can result in variations.

It is again emphasized that the reduction in primary radiation due to shielding might be counteracted by secondary species, such as neutrinos, formed and released when the primary radiation interacts with the shielding material, sometimes resulting in more damaging radiation.

In the Nealy study (1988), it was stated that for "relatively short-duration missions (2 to 3 months), the most important radiation hazards are the very high fluxes of energetic solar flare protons that are produced. Solar flares may also contain X-rays and gamma rays, electrons, neutrons, and some heavier charged particles as well as protons." The high-energy protons are the critical components to be shielded against, and if this is done, the effects of the remaining sources become insignificant. In this reported study, a radiation transport code was used to study both specific lunar base structural configurations and radiation fields resulting from large solar flares for which there is data, and to determine the high radiation locations within the structure. In particular, a half-buried cylinder with constant thickness regolith shielding was studied. The cylinder chosen for study was based on the ISS module (originally the *Freedom*), with dimensions 12.2 m in length and 4.6 m in diameter. Two cases of regolith thickness were reported: 0.5 m and 1.0 m. Isodose contour plots were created. Spherical structures were also studied.

This study focused on short-duration missions, and it is noted that longer duration missions (4 to 6 months or more) require consideration of GCR dosages. These are low when compared to solar flares, but are approximately constant with time and of very high energies (> 1 GeV), with a dominant dose component by the HZE contribution. It is further noted here that for thermal variation control, a regolith thickness of 2 m is found to be adequate. Figures 3.3 and 3.4 show the structural configurations and predicted doses.

There are two sources of uncertainties regarding the assessment of radiation dangers to astronauts. One is that we cannot predict linear energy transfer (LET) radiation magnitudes. LET refers to the energy deposited into the matter that has been struck by radiation, such as protons and high-mass high-energy (HZE) ions. The other is that the human risk estimates are also uncertain, even for known magnitudes.[14] An acceptable risk level for operations in low Earth orbit is that "the increase in lifetime probability, above and beyond the natural incidence ... will result in fatal cancer ... with an excess probability of three percent." The radiation exposure limits are most severe for blood-forming organs, less so for the eyes, and less so again for the skin. The respective ratios are approximately 1:4:6. There are 30-day limits, annual limits, and career limits. Sometimes the shorter duration limits are the critical ones, while the career limits depend on age and gender. Uncertainties lead to significant increases in costs, since additional shielding results in requirements for more mass, more time to construct habitats, and more launches for supplies and equipment.

A subsequent study expanded the Nealy study to include space travel as well as habitats on both the Moon and Mars, and included significant data on the levels of radiation to be expected.[15] Of course, radiation doses are considerably higher during spaceflight than when on the surface of a planetary body, where the planet provides shielding against half the environment. It was noted that blood forming organ (BFO) doses incurred from the

Flare data	Regolith thickness, cm	Predicted dose, rem	
		Cylinder (center)	Sphere (center)
1956	50	7.48	7.04
	100	2.70	2.94
1960	50	1.60	1.90
	100	.16	.23
1972	50	0.25	0.30
	100	.03	.04

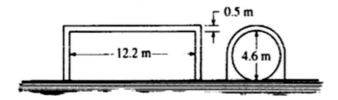

(a) Cylindrical module (side and end views).

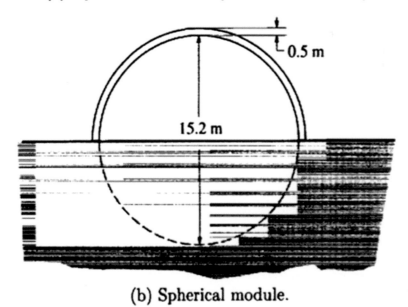

(b) Spherical module.

Figure 3.3. (Simonsen and Nealy, 1991. Courtesy: NASA)

Habitat geometry	Regolith thickness		BFO dose equivalent, rem/yr
	cm	g/cm^2 (a)	
Cylindrical	50	75	12
Spherical	50	75	12
	75	112.5	10

^{a} Assuming a regolith density of 1.5 g/cm^3.

Figure 3.4. (Simonsen and Nealy, 1991. Courtesy: NASA)

GCR at solar minimum, when added to medium doses from large solar flare events, may reach the annual dose limit and become career limiting for long-duration missions. Clearly, a Mars mission is much more dangerous from this perspective (as well as other perspectives) when compared with a Moon mission. Similar analyses to the Nealy lunar base study were performed for a Mars base. It was noted that a 0.5 m regolith layer (75 g/cm^2 assuming a regolith density of 1.5 g/cm^3) reduces the dose levels to slightly less than half of that annual dose limit. A spherical habitat of 15.2 m diameter was also modeled as a half-buried sphere, with the above-ground portion shielded with a 0.5 m, 0.75 m, and a 1 m regolith layer. It was noted that "hydrogenous materials of low atomic weight are substantially superior to heavy metals for energetic ion shielding."

Radiation risks are best modeled probabilistically due to the variability in predictions of solar cycles, SPEs, and even large changes in GCR.[16] The frequency distribution of SPEs is strongly correlated to the phase within the solar activity cycle, but their individual occurrences are random in nature. There are also temporal variations in GCR of about two-fold in intensity over the approximately 11-year solar cycle. Kim *et al.* analyzed and compared the shielding effectiveness of aluminum and polyethylene for exposure to historically large SPEs in a spacecraft on a Mars mission, as well as for a conceptual lunar habitat and rover. It was concluded that future space missions should be launched at solar maximum rather than at solar minimum in order to minimize the possibility of exceeding the acceptable three percent REID (Risk of Exposure-Induced Death) at the upper 95 percent confidence interval.

Space missions should be at solar maximum based on the following argument. SPEs are greater at solar maximum (consisting mainly of protons) but are much easier to protect against than GCR (whose constituents are high charge and high energy), which are greater at solar minimum. Thus, a space mission during solar maximum is preferable due to lesser GCR. GCR is inversely related to the solar cycle, since the Sun's magnetic field is stronger during the solar maximum (hence stronger solar particle events). Such analyses can be used to plan and optimize missions, and establish the effectiveness of shielding technologies and designs.

Given the importance of plant life to the success of a space mission or extra-planetary settlement, how plants fare in a radiative environment is of interest and concern.[17] Plants are more resistant to radiation than mammals. In particular,

"Ionizing radiation may have different effects on plants and mammals depending on the radiation quality/quantity and/or cell characteristics, as well as the developmental stage of the organism, which is generally more sensitive during the initial phase of the life cycle. Moreover, the biological effects of ionizing radiation are not only limited to the organism exposed to irradiation but may also involve next generations, although in humans, no direct evidence of hereditary effects caused by radiation has been reported so far.

"There are strong differences in the structure and metabolism between mammalian and plant cells which might be responsible for the higher resistance of plant cells. Thus, non-lethal doses for plants may be very dangerous or fatal for humans, causing death within days or weeks, mostly due to infections resulting from a depletion of white blood cells. ... The crucial radiation target in both mammalian and plant cells is the DNA."

Arena *et al.* also discussed the effects of microgravity on radiation-induced DNA damage. Experimental results in simulated microgravity do not match those from actual microgravity, where cells are able to repair radiation-induced damage almost normally. Interestingly, and counter-intuitively, "it is important to note that no published data have unambiguously shown an increased cancer risk in populations exposed to high natural background radiation. On the contrary, they seem healthier and live longer than people living in areas with normal radiation background, an observation which has been termed 'the radiation paradox'." However, no general conclusions can be drawn either for microgravity or lunar or Martian gravity. One conclusion that can be drawn, however, is that ambiguities can begin to be removed once we have long-term facilities on the Moon.

Horneck provided an extensive discussion on the lunar (and Martian) environment, the life sciences research opportunities, and the challenges to life on these extra-planetary bodies.[18] In particular,

"At the lunar surface, the galactic cosmic radiation gives rise to a dose equivalent of about 0.3 Sv/a compared to 1-2 mSv/a on the Earth. ... Solar flare events, that occur sporadically (up to about 10 events/a) and last over several days, will cause temporarily substantial increases in dose (up to 0.6 Sv/h). At a depth of 1 m in lunar soil, the annual radiation dose equivalent due to cosmic-ray particles decreases to about 2 mSv, which is comparable to terrestrial surface conditions; however, due to interactions of cosmic-ray particles with the shielding material, secondary particles, mainly neutrons, are produced, which contribute an additional dose equivalent of 0.1 Sv/a. In addition to cosmic ionizing radiation, the solar ultraviolet radiation reaches the surface of the Moon without any filtering, whereas the surface of the Earth is protected from the biologically harmful UVC and short wavelength band of UVB by the atmosphere and especially the ozone layer."

The biologically effective irradiance on the lunar surface, and on the surface of Mars, is about three orders of magnitude higher than on Earth.[19]

Further, Horneck discussed the significant opportunities for radiation biology experiments on the Moon that cannot be performed on Earth, especially considering the possibility of couplings between low gravity and biological response and adaptations, and that many radiation sources cannot be found on Earth.

Horneck emphasized the need to measure the production of secondary radiation for different shielding materials, with special emphasis on spallation events, nuclear fragments and neutrinos. Ideally, it was recommended that we "establish a long-term colony of biological systems – from tissue cultures to plants and animals, including small vertebrates. Such colonies would allow us to study long-term effects of the combined action of protons and HZE particles in conjunction with reduced gravity. The main emphasis should be placed on genetic and teratogenic alterations, as well as tumor formation and reduced life expectancy."

Solar flare alert systems on Earth and in orbit are critical to allow people in exposed situations to seek shelter quickly. Given the intense radiation at the lunar surface, and its hazardous effects on living beings and electronic equipment, shielding is a critical aspect of the design of facilities and shelters. A regolith layer of 2–3 m has been determined to be sufficient against the nominal radiation on the lunar surface. Horneck, however, warned that "a regolith layer of 2–3 m would provide a radiation shielding of about 400 g/cm^2 – in comparison, the shielding by the Earth's atmosphere is approximately 1000 g/cm^2. For a successful protection against radiation from a rare giant solar flare, shelters with more than 700 g/cm^2 are required." This suggests a structural design philosophy where a 'normally-shielded' habitat can shield the inhabitants from the usual or average radiation threats, and strategically-placed shelters can protect the inhabitants from the most severe flares. If this can be accomplished in an optimal way, there could be considerable savings in cost and time fabricating the habitats.

There have been numerous studies of lunar and Mars habitation modules devoted to shielding strategies. One such volume of studies was edited by Wilson et al.[20] Of particular note in that volume is the chapter by Simonsen.[21]

Lindsey discussed the lunar environment, in particular the thermal, radiation and meteoroid environment, and suggested that the lunar regolith can be a near-ideal material for surface structure shielding. The lunar surface is exposed to an annual radiation dose of 25 rem, as compared to the Earth annual dose of 0.36 rem. In addition, a single solar event can expose the lunar surface to a radiation level of up to 1000 rem over a short period of time.

> "Solar wind particles have such low energies that they are stopped in less than a micrometer of regolith, while solar event particles will pass through approximately 50–100 cm of regolith before being significantly mitigated. In addition, heavy nuclei GCR particles are stopped by approximately 10 cm of regolith while all other GCR (GeV) particles are stopped by 1000 g/cm^3 of material, which equates to 5 m of lunar regolith (2 g/cm^3) or the Earth's atmosphere. ... Based on the maximum protection required ... 1–2 m of regolith appears to be adequate for effective shielding of a lunar habitat to avoid radiation sickness in the crew."

Rapp stated, in an extensive study, that "for lunar sortie missions, the duration is short enough that GCR creates no serious risks.[22] For lunar outpost missions, the probability of encountering an SPE during Solar Maximum in a six-month rotation is one to ten percent, depending on the assumed energy of the SPE. Even with >30 g/cm^2 of regolith shielding, the 95th percentile confidence interval dose from a major SPE would exceed the 30-day limit. The GCR during Solar Minimum for a six-month stay on the Moon is marginal against the annual limit, but this can be mitigated somewhat by use of regolith for shielding the habitat."

Tripathi et al. further characterized the lunar surface environment and considered how it would affect different possible lunar mission scenarios.[23] Miller et al. "have measured the radiation transport and dose reduction properties of lunar soil with respect to selected heavy ion beams, with charges and energies comparable to some components of the galactic cosmic radiation, using soil samples returned by the Apollo missions and several types of synthetic soil glasses and lunar soil simulants."[24] They concluded that a modest amount of regolith would provide substantial protection against primary GCR nuclei and SPE. In particular, less than a half-meter of regolith, compacted to a density of 1.4 g/cm^3, could effectively stop primary GCR ions.

In a follow-up to the Nealy et al. study of 1988, cited previously, Lin et al. calculated the space radiation exposure of blood-forming organs everywhere inside a hemispherical dome on the lunar surface, and a spherical dome in space, and came to slightly different conclusions than the earlier paper.[25]

> "In both cases the radiation exposure is the highest at the center of the sphere and decreases as a function of the distance from the center, with faster decreases towards the inner edge of the shell. This conclusion agrees with an earlier study on a spherical shield. However, it differs from the earlier study on a hemispherical shield because that earlier study incorrectly included solid angles that are below the horizon plane containing the observation point. The reduction in the radiation exposure from the center to the inner edge of the hemispherical or spherical shell is relatively small for galactic cosmic ray environments but much larger for solar particle events. For soft solar particle events, the exposure at points on the inner surface of the hemispherical dome or the spherical shell can be smaller than that at the center by a factor of three or more."

With these results, we can position inanimate objects in locations of high radiation exposure, and humans and plants at locations of low exposure.

Pham and El-Genk studied the expected dose estimates in a lunar shelter with regolith shielding.[26] Reitz et al. studied radiation exposure in the lunar environment for an astronaut in an EVA suit.[27] Rojdev et al. considered the use of fiber-reinforced composites for potential radiation shielding in long-duration lunar missions.[28] A key result was that after accelerated radiation exposure to reflect a 30-year mission, the fiber matrix was weakened due to scission effects; that is, breakage of chemical bonds especially in long chain molecules, resulting in two smaller chains. Not considered in this study was the coupling of radiation exposure with natural material aging.

A recent review of radiation impact on humans in space was by Freese et al.[29] The health risks were discussed, as were the differences between terrestrial and space radiation, and recent increases in understanding about space radiation and potential countermeasures. The latest career exposure limits set by NASA for blood forming organs range from 1.50 Sv for a 25-year-old male to 4.00 Sv for a 55-year-old male, and from 1.00 Sv for a 25-year-old female to 3.00 Sv for a 55-year-old female. Limits for the eye and skin are higher. The European Space Agency and the Russian Federation Roscosmos State Corporation have set career limits of 1 Sv independently of age and gender. There are still significant gaps in knowledge, especially for low doses of radiation and for high-LET radiation types to humans.

Cucinotta and Cacao showed, in a study on Mars mission durations, that "the non-targeted-effects NTE model predicts a two-fold or higher cancer risk compared to a TE model for chronic GCR exposures, with predictions highly dependent on the number of bystander cells susceptible to oncogenic signals from cells directly traversed by heavy ions."[30] They pointed to several uncertainties in estimating space radiation cancer risks, but emphasized that the lack of human data, for the heavy ions and other high LET radiations that occur in space, is the largest uncertainty. It appears that it will not be possible to rectify these uncertainties except by living in space. Given how much more risky Mars missions are in comparison to lunar missions, the latter are a safer first step towards developing benchmarks for human radiation safety.

Active and Passive Shielding

While regolith shielding and the burial of habitable structures provides the foremost protection against radiation and micrometeorites, other passive and active concepts are being considered.[31] Passive concepts are those designed for a particular environment. Once placed on site, the passive shielding is not easily altered. Regolith shielding placed atop a structure is a passive design solution. Active concepts are more robust and can adapt to the changing environment, but active concept designs, relying on sensors to detect environmental changes and actuators to alter the design attributes, are more complex and expensive, and need extra maintenance. Both active and passive concepts will eventually be utilized.

Charged particles are deflected by active shielding using electric or magnetic fields. The charged particle can be deflected by a force that is proportional to the charge, its velocity, and the electric and/or magnetic field strength utilized. An electrostatic shield utilizes only a static electric field, a magnetic shield utilizes a static magnetic field, and a plasma shield utilizes both. Dynamic fields – time-dependent fields – can also be utilized.

Early shielding devices have a number of design constraints. Potential electrical breakdown of structural materials occurs if the field strength is too high. In order for the shielding devices to perform effectively, 100 MV voltages are needed, but this strength is not currently available. The electrodes need to have the mechanical strength to withstand the electrical forces between them. Electrode designs need to avoid attracting regolith particles, and need to avoid generating X-rays when impacted by high speed electrons. There are many conflicting design challenges.

The sphere tree design shown in Figure 3.5 utilizes weak, negatively charged spheres distributed along the shield's outer regions to deflect electrons with a negative potential of less than −40 MV at approximately 75 m, while strong, positively charged generators that are clustered at the center deflect high-energy protons. The spheres are placed high above the surface structure so that the electric and magnetic fields they generate do not draw electrons from the structure or surrounding area.

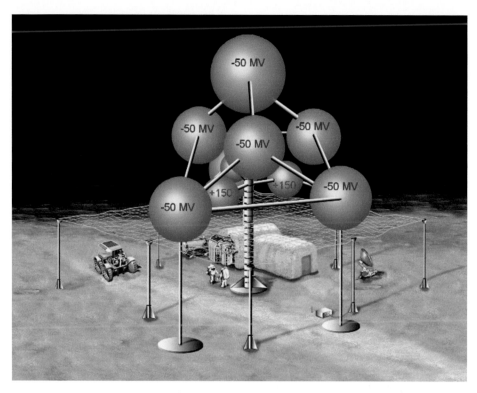

Figure 3.5. A sphere tree consisting of positively charged inner spheres and negatively charged outer spheres. The screen net is connected to ground potential. (Courtesy: NASA)

Figure 3.6 depicts a tree of positive-ion-repelling spheres. A magnetic field is created by the solenoids, deflecting the electrons. This concept also prevents the electrons from colliding with the material of the electrostatic spheres and the solenoids, thus preventing the lethal X-rays, as discussed above.

Figure 3.7 shows an early settlement, where a passive shield wall is used to deflect low-angle radiation. In each of these design case studies, the deflection properties are different. As we review these three images, it is worth noting that in each design, structures are needed to support the charged spheres. While the fabrication and placement of these structural elements will not be an issue eventually, they do pose challenges early in the life of the facility.

The Apollo astronauts were fortunate that they were not subjected to major radiation events during their spaceflights and lunar sojourns. Had there been a major solar flare, the astronauts would have had no safe haven from the spike in radiation. Being on a planetary body like the Moon, however, offers some natural protection against cosmic radiation, since the planetary body provides shielding from half the sky. But the Apollo astronauts all visited the lunar surface during daylight, and were therefore subject to the Sun's full radiation load. When astronauts are on a trajectory between planets, the radiation is striking from all directions and the radiation is twice as potent when compared to being on a planetary body. For this reason, the three-day trip to the Moon is much safer than the nine-month trip to Mars.

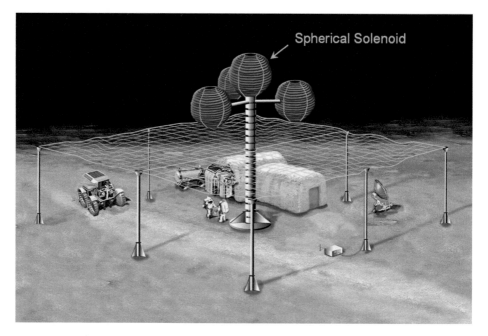

Figure 3.6. Electrostatic shield concept using only positively charged spheres, wrapped with a current-carrying wire. Electrons are repelled by the magnetic field generated in the spherical wire loops. (Courtesy: NASA)

Figure 3.7. The protected habitat volume is grounded, the low-angle radiation from the horizon is not sufficiently impeded by the shield's electrostatic field. Thus, a passive shield wall is needed in this concept to mitigate that low-angle radiation (Courtesy: NASA)

3.4   REGOLITH

"Dangerous, but vital."

The lunar surface has a layer of fine particles, referred to as regolith, that has formed over billions of years, primarily due to continued impact by meteoroids and larger rocks. This regular action against the surface of the Moon has homogenized, pulverized and compacted the local lunar rocks. As a practical matter, the lunar regolith is a serious environmental hazard to human and machine. A photoelectric charge in the conductivity of the dust particles causes them to levitate and adhere to surfaces. The dust is also abrasive, toxic if breathed, and poses serious challenges for the utility of construction equipment, air locks and all exposed surfaces.

The thickness of the regolith varies from about 5 m on the mare surfaces to about 10 m on highland surfaces.[32] The majority of the regolith is a fine gray soil with an average density of 1.5 g/cm^3. But there are also breccia (rock consisting of angular fragments cemented together) and rock from the bedrock. The continuous bombardment of micrometeoroids breaks up all the particles on the surface and melts portions of the soil. This melt mixes with sand-sized particles to form irregular clusters called agglutinates. The solar wind implants large quantities of hydrogen and helium, and trace amounts of other elements into this mixture. Evidence of large motion projectiles are seen in the lunar crater rays. Some prefer to limit the application of the term 'soil' to where contents include organic matter. While that does not apply to the lunar regolith or dust, we do not make such a distinction.

The lunar soil, or dust, is the sub-centimeter fraction of the lunar regolith.[33] "Five basic particle types make up the lunar soils: mineral fragments, pristine crystalline rock fragments, breccia fragments, glasses of various kinds, and the unique lunar structured particles called agglutinates." Robens *et al.* investigated the parameters relevant to the capability of the lunar surface to store water, and the amount of kinetics adsorption/desorption in relation to the environmental conditions. The interest in the presence of water on the Moon is fundamental to mission and settlement strategies. That the south polar region has permanently shadowed areas leads to the possibilities of cold traps for volatiles such as water ice.[34] "Permanently shadowed floors of craters Faustini and Shoemaker ... and the lower portion of Shakleton crater" suggest possible sites for water mining facilities. However, the ruggedness of the terrain, the difficult lighting conditions where the Sun is perpetually at or near the horizon, and possibly other serious challenges, makes these sites less likely for a first human presence.

The lunar dust is ubiquitous and dangerous to human and machine. Lunar dust was identified as a problem when the Apollo astronauts found the grey powder clinging to everything during their sojourns on the Moon. It choked even the vacuum cleaners that were designed to clean the spacesuits and craft of the dust.

From a physiological point of view, the concern about the dust is that it can enter the lungs, interact with the blood and be carried throughout the body. The reduced gravity keeps the fine particles suspended in the airways longer and allows them to get deeper into the lungs and reach the bloodstream. Particles that are smaller than 2.5 microns (0.0025 mm) can cause the most damage and are also most affected by the reduced gravity on the lunar surface. Lunar dust is the portion of the regolith that is less than 20 μm in size, and it makes up about 20 percent of the volume by weight. A special simulant of the lunar dust – known as JSC-1Avf – was created to study its potential effects.[35]

NASA formed the Lunar Airborne Dust Toxicity Advisory Group (LADTAG) of experts in September 2005 to address the problem of setting health standards for astronaut exposure to lunar dust. Expertise came from astronauts, flight surgeons, inhalation toxicologists, particle physicists, lunar geologists, and pathologists. The group was evenly balanced between NASA experts and others from academia or industry. The goal of LADTAG was to identify research questions, suggest a means of answering those questions, guide research to answer the questions, and then set short-term and long-term human exposure levels that are safe. These environmental standards would guide vehicle engineering decisions for the Crew Exploration Vehicle and the Lunar Surface Access Module, form the basis of flight rules, and dictate the need for environmental control and monitoring.

The group had "concerns about the toxicity of the chemically reactive lunar dust grains, which also contain nanoparticles of natural metals and glass shards formed from a combination of chemical reactions, meteorite impacts and solar wind bombardment."[36] The goals, quoted below, were to assess human exposure and potential health effects:

- To assess human exposure, it is necessary to create activated simulant and/or lunar dust and characterize their passivation in a life support habitat. Data on lunar dust with size <10 μm needs to be reviewed for surface area, mineralogy, size distribution, surface morphology, chemistry, and electrostatic properties, with special attention to differences from Earth analogs. *In-situ* assessments of dust at the proposed landing site(s) focus on size distribution, chemical composition, chemical reactivity, and passivation in a habitat atmosphere. A question requiring attention is how different is the dust at the South Pole as compared to the lunar dust and simulant samples that we have from the Apollo missions.

- An assessment of potential health effects requires a review of databases on human and animal exposures to materials similar to lunar dust, such as volcanic ash, mineral dusts, and occupational dust exposures. We need a review of the consequences of the Apollo astronauts' exposure to lunar dust. Are the filters from the Apollo capsules informative? [We also need to] perform a critical review of existing lunar dust studies; conduct in-vitro studies of cellular response to simulants and lunar dusts; conduct intratracheal instillation studies of simulants and lunar dusts in one rodent species; conduct a six-hour inhalation study of a simulant and one lunar dust in one rodent species; conduct a subchronic (28-day) inhalation study using a simulant or lunar dust in one rodent species; and conduct a study of brief human exposure to activated simulant or lunar dust to assess the acute human response, contingent on the above results.

The detrimental health effects and toxicity of the lunar regolith dust is well known, and not to be minimized by mission planners.[37] Linnarsson *et al.* provided an extensive discussion of this toxicity, including effects not as widely reported on the skin and eyes.[38] The Moon's hard vacuum leaves the dust grains covered with chemically reactive bonds that would be neutralized on Earth by interaction with atmospheric oxygen. On the Moon, the dust can react in toxic ways when breathed into the lungs.

The regolith dust has a low electrical conductivity, allowing individual grains to remain electrostatically active and to attach themselves to everything.[39] Any piece of equipment will be covered in a matter of days. Due to the charge, the dust can migrate and crawl up surfaces, which it can do without being driven by any local physical activity, and has highly adhesive characteristics. This charging can be attributed to the solar wind, UV radiation, X-rays, and cosmic rays.[40]

One of the Surveyor sightings of such activity was a glow above the western horizon about one hour after sunset.[41] It was observed that positively charged, 5–10-micron particles levitate due to local electric fields and can move up to 1 m above the lunar surface. After sunset, light scatters from these particles, resulting in the glow. These are called lunar horizon glow particles. Particles of sub-micron size can move to much higher altitudes on ballistic trajectories at speeds of about 1 m/s. Such mobile particles can impact lunar surface activities, which also adds to the quantity of charged suspended particles. That the regolith particles are jagged and cling add to the design challenges for spacesuits, pressure doors and seals, machine joints, and just about anything else that the particles can come into contact with.

One strategy is to place structures and delicate equipment at least a meter above the surface, and to microwave the surface around any facilities to create a melt. Habitats need to be over-pressured around the openings to prevent ingress of dust.

3.5 SOIL MECHANICS

Soil mechanics is the engineering and scientific study of the behavior of soils under all environmental forces. Soil and regolith are very complex media. They are uneven in geometry and strength, with properties changing when compressed or, in the case of Earth soil, when pores become filled with water. Some do not categorize regolith as soil, considering that term to imply a process of biological decomposition.

The foundation engineer needs a precise mathematical model of the soil at the site where the structure is to be erected. Soil properties vary over small distances, which is why two adjacent structures respond very differently to the same earthquake. Slight differences in soil properties lead to earthquake energy being transmitted with very different magnitudes and directions to the structures above. One structure will survive and the other nearby will collapse.

Assuming that we have access to the soil where we will build a structure, we can test for its properties – density, stiffness, damping, void ratios – and determine its bearing capacity, thus allowing us to design a foundation that supports the structure placed atop it for all environmental conditions. As we gear up to return to the Moon, precursor missions are needed for direct access to the site where we will place the first lunar bases.

An interesting example of the difficulty of site characterization on the Moon takes us back to the Apollo era. There was little knowledge of the regolith properties where Apollo 11 was to land, and therefore much uncertainty about how deep the lander would sink into the regolith. Figure 3.8 shows one of the lander's legs, with the large saucer at the base designed to distribute the lander weight.

Figure 3.8. Astronaut Edwin E. Aldrin, Jr. walks on the surface of the Moon near a leg of the Lunar Module during the Apollo 11 extravehicular activity. Astronaut Neil A. Armstrong, Apollo 11 commander, took this photograph with a 70-mm lunar surface camera. The astronauts' footprints are clearly visible in the foreground. (AS11–40-5902, July 20, 1969. Courtesy: NASA)

Clearly and fortunately, the lander did not sink very much. To gather additional information about the regolith, a picture was taken of an astronaut footprint – see Figure 3.9 – so that the depth of penetration could be used to determine its bearing strength. Other such tests were performed during Apollo.

Figure 3.10 shows the effects of the bounced landing of Surveyor 3, and also that the regolith is supportive of significant mass.

Early studies have compared Earth soil mechanics to lunar regolith mechanics and discussed how the analysis and design procedures would differ.[42] There is a large literature on simulated lunar regolith. In the following paragraphs, we summarize some of the properties of lunar regolith of interest to engineering applications, based on Colwell *et al.*[43] The lunar surface is regularly impacted by micrometeoroids of typical mass 10^{-10} kg to 10^{-8} kg, with speeds up to approximately 72 km/s. Such impacts result in the delivery of significant kinetic energy, creating impact craters and ejected mass. The yield:

$$Y = \frac{M_{ej}}{M_{imp}},$$

Figure 3.9. A close-up view of an astronaut's boot print in the lunar soil, photographed with a 70-mm lunar surface camera during the Apollo 11 extravehicular activity (EVA) on the Moon. The first footprints on the Moon will be there for a million years. There is no wind to blow them away. Only direct micrometeoroid impacts and ejecta can damage the print. (AS11–40-5878, July 20, 1969. Courtesy: NASA)

is a measure of the ratio of ejected mass to impact mass. A hypervelocity impact can result in a yield anywhere from 10^3 to 10^6, depending on the properties of the target region and the micrometeoroid. Impacts on solid rock have lower yields but higher ejected velocities than impacts on unconsolidated particles or powder.

Lunar regolith particles are of jagged shape and much sharper than terrestrial counterparts, due to the bombardment of the surface by micrometeoroids over millions of years. Sharper means more abrasive, a property that affects all aspects of engineering designs. In particular, regolith is commonly agreed to be the material of choice to cover lunar surface structures for shielding purposes. Whether the regolith is placed in bags or placed directly on top of the structure, relative motion and abrasion are design considerations. The sharp particle shape affects the bulk modulus of the material due to the increase in interlocking. Some regolith properties are listed in Table 3.2.

The high vacuum and absence of oxygen and water affect the surface properties of the regolith grains, and thereby also affect the bulk modulus. Due to the lower gravity, there is a resulting reduction in confining stresses. The high vacuum can also affect the surface cleanliness of the particles, resulting in an altered shear strength due to surface friction changes.

Figure 3.10. A close-up view of a footpad and surface sampler with scoop (arm, out of frame) on the Surveyor 3 spacecraft that was photographed by the Apollo 12 astronauts during their second extravehicular activity (EVA) on the Moon. The Apollo 12 Lunar Module (LM), with astronauts Charles Conrad Jr. and Alan L. Bean aboard, touched down in the Ocean of Storms only 600 feet from Surveyor 3. The television camera and several other pieces were taken from Surveyor 3 and brought back to Earth for scientific examination. The unmanned spacecraft soft-landed on the Moon on April 19, 1967. Note the extra footpad imprint in the lunar soil that was caused when the Surveyor 3 bounced upon landing. (AS12–48-7110, November 20, 1969. Courtesy: NASA)

Table 3.2. Engineering properties of lunar regolith [Colwell *et al.*]

Depth cm	0–15	0–30	30–60	0–60
Avg bulk density (±0.05) g/cm³	1.50	1.58	1.74	1.66
Void ratio (±0.07)	1.07	0.96	0.78	0.87
Relative density (±3) %	65	74	92	83
Avg cohesion kPa	0.52	0.90	3.0	1.6
Avg friction angle deg	42	46	54	49

While the relative density in the top 10–30 cm of regolith is low, it is very compacted below that surface layer. Due to the micrometeoroid impacts and the regular low-level seismic activity, there has been a continuous densification of the regolith at all but the shallowest layer. A useful result is that the regolith is relatively stable against lunar quakes and nearby manned activities. On the other hand, it is therefore exceedingly difficult to excavate regolith below the upper 10–30 cm. Heavy equipment that is used to excavate foundations on Earth would be of less use on the lunar surface and its 1/6 *g*. Percussive and other impulsive, vibratory digging devices may prove more useful.

Once excavated, lunar regolith will go from 90 percent relative density to a less compact 30–40 percent relative density. This results in an increase in volume of regolith. For geotechnical structures, for example habitats made from blocks of excavated regolith, there could be significant settlement and more susceptibility to even low magnitude seismic activity.

It has been noted that there is increased dust activity at the terminator where the finest particles, of approximate size 1 *μ*m, may be launched into ballistic trajectories, reaching heights of a few to several hundred meters. As per Colwell *et al.*, future manned and unmanned lunar exploration activities will face a complex and time-variable plasma environment, where micron-sized dust can lift off the surface to great altitudes. Engineers will need to develop new technologies that are nearly immune to the dense, angular lunar regolith, and that can mitigate dust contamination of machines and people. Spacesuits will need to be able to repel the clingy regolith, unlike the Apollo astronaut spacesuits that became coated with it. Optical components will also need to be protected against the fine dust.

"The landing of the Apollo 12 lunar module 183 m from the Surveyor 3 lander resulted in a 'sandblasting' of the Surveyor spacecraft. New in-situ experiments are needed to explore the mechanical properties of the regolith at depths below the first few decimeters and to understand the relationship between charged dust mobility and the surface plasma environment. Langmuir probes can measure the electron and ion densities and temperatures, and deployable booms can measure the near-surface electric field over the course of several lunar days. In-situ dust detectors, including piezoelectric sensors, charged dust detectors and optical experiments, can be deployed on the lunar surface to measure dust transport simultaneously with the plasma measurements. A better understanding of the mechanical properties of lunar dust, as well as its dynamical response to the changing lunar plasma environment, will help in devising dust mitigation strategies for future lunar exploration."

Similar, but perhaps more severe consequences are likely as rockets land on the lunar surface.[44] Without proper controls, the high temperature, supersonic jet of gas that is exhausted from a rocket engine is capable of damaging both the rocket itself and any hardware in the surrounding environment. Rocks, gravel and dust can be propelled, not only causing damage to the landing craft and other hardware, but potentially misleading the landing craft sensors, resulting in an unsuccessful landing.

Since before Apollo data became available, engineers have been developing theories on regolith properties, as well as how to move it around and how to use it as a building material. Data is available from the Surveyor and Lunar Orbiter spacecraft. There are multiple challenges, not the least of which is that the lunar sites are not accessible for testing, and it has not been possible to recreate the lunar environment on Earth. Therefore, our understanding is segmented. While that understanding is significant, it is necessary to get back to the Moon in order to calibrate our theories and gain practical experience.

In Earth soil mechanics, there are key parameters that define the soil. These are used in the design of foundations of all types, and in the design of machinery of all sizes for excavation. Key mechanical properties are the medium's density, porosity, cohesion, and angle of internal friction. Based on these are material strength, compressibility, and stress-strain characteristics. Generally, these properties are all highly variable with location. In some ways, the lunar soil or regolith is similar to Earth soil, and in other ways there are significant differences due to the history and environment of the Moon.

The soil shear strength s is given by:

$$s = c + \sigma \tan \phi,$$

where c equals the soil cohesion, σ is the stress normal to the failure plane, and ϕ is the internal friction.[45] Thus, any understanding of the soil must include the coupling of all these parameters. Mitchell et al. summarized available data about the lunar soil, including the following plots:[46]

- the grain size distribution ranges from different Apollo sites
- friction angle as a function of porosity for a ground basalt lunar soil simulant
- cohesion as a function of porosity for a ground basalt lunar soil simulant
- Apollo 15 penetration records
- Lunokhod 1 penetration records
- Lunar soil density estimates from Surveyor, Luna, Apollo and Lunokhod data (1964–1972), and
- bulk density of lunar soil from Apollo 15 core tube sites.

Estimates were given of lunar soil cohesion (kN/m²) and friction angle (degrees) based on pre-Apollo data:

$$0.1 < c < 1.7$$
$$10 < \phi < 60,$$

and based on data from Apollo 11, 12 and 14:

$$0.03 < c < 2.1$$
$$35 < \phi < 47.$$

The relation between cohesion c and angle of internal friction ϕ is highly nonlinear.

Best estimates and range values from Station 8 of the Apollo 15 mission were given for the following soil properties, assuming an average specific gravity of 3.1:

$$\text{Porosity \%} = 36.5 \ \left(35-38\right)$$
$$\text{Void ratio} = 0.58 \ \left(0.54-0.61\right)$$
$$\text{Density g/cm}^3 = 1.97 \ \left(1.92-2.01\right)$$
$$\text{Friction angle deg} = 49.5 \ \left(47.5-51.5\right).$$

Mitchell et al. summarized their findings. The fine-grained soil on the lunar surface has a grain size distribution that corresponds to the silty fine sands of Earth. Lunar soil mechanical properties are similar to those of Earth for a comparable gradation. "This is a direct reflection of the fact that for soils in this particle size range, density and particle size and shape distribution exert a larger influence on mechanical properties than does composition." Lunar soils are generally more cohesive. For a given grain size, porosity is the parameter that most influences cohesion and friction angle.

Earlier studies of lunar bases assumed that the construction of lunar surface structures would follow the general procedures of Earth-based construction. That is, heavy machinery would be used to dig and move the lunar regolith as the site is prepared for the laying of a foundation, or the digging of piles, for the future structure and eventual city. Studies representative of this work were by Bernold, Matsumoto *et al.*, Florek, and Wilkinson and DeGennaro.[47-50] Willman and Boles summarized the work of various researchers who derived certain excavation parameters.[51]

Ettouney and Benaroya studied how gravity differences between Earth and the Moon can affect the mechanical and dynamic characteristics of the regolith. In order to remove the effects of gravity in a number of parameters, the use of mass properties was recommended. The regolith bulk modulus, and the moduli of rigidity and elasticity, have lunar values of about one-sixth their Earth values for the same soil sample. The bulk modulus is the slope of the curve that relates the hydrostatic soil pressure to the volumetric strain. That curve identifies the pressures at which the regolith grains begin interlocking, and when they begin to undergo severe crushing. A consideration of regolith wave dynamics concluded that the longitudinal and transverse wave speeds are $\sqrt{1/6} = 0.408$ times their value on the Earth.

Damping in the form of energy dissipation is $1/6 = 0.167$ times the equivalent damping on Earth, meaning that elastic wavelengths are much longer and their attenuation rates are slower in the lunar environment that on Earth. This has seismic implications, but there are additional implications to the behavior and design of longer structures. Bearing capacity of footings and their vibration were discussed.

The properties of regolith indicated that an accurate evaluation of regolith stresses and strains would require a greater use of nonlinear stress-analysis methods than such calculations for the Earth environment.[52] This is because regolith interlocking stress levels are very small, about 0.73 psi (5 kPa), and crushing stress levels may be as small as 4.35 to 10.1 psi (30 to 70 kPa). These numbers are considerably smaller than their equivalents on Earth, even after accounting for the differences in gravitational accelerations. The nonlinearity of regolith stress-strain behavior will need to be taken into account in lunar engineering projects. The dominance of nonlinear effects in regolith, as compared with Earth soil materials, is not necessarily an unwanted engineering effect. Since interlocking occurs for very small stress levels, the regolith will become stiffer and relatively smaller strains will develop, resulting in better system behavior at these stress levels. On the other hand, lower crushing stress levels can cause problems in larger projects.

3.6 ABRASION

Of the challenges that the lunar dust offers settlers on the Moon, abrasion is one that has been observed since the days of Apollo.[53] Lunar soil formation is primarily due to innumerable micrometeorite impacts, transforming smooth particles into angular and irregular shaped silicate glass particles. Pulverization of the lunar materials creates small independent dust particles and can also lead to rapid melting and solidification, resulting in agglutinate formation, or completely melt the materials to form glass. Due to the absence of an atmosphere or fluid erosion, the particles maintain their sharp edges and points, resulting in their abrasive properties. Kobrick *et al.* summarized the physical properties of the regolith dust, and how different groups segregate sizes differently:

> "Various definitions are used by different groups to describe what size particles constitute 'dust'. Lunar regolith occupies the upper several meters (in some cases up to 15–20 m) of the Moon and consists of unconsolidated rocks, pebbles, and dust over lunar bedrock … [they can be] categorized [as follows:] regolith smaller than 1 cm as 'soil', less than 1 mm as 'fines', and smaller than 100 μm (effective particle radius of 50 μm) as 'dust'. Over 95 percent of the regolith particles are under 1.37 mm, 50 percent are under 60 μm (the thickness of a human hair strand), 10–20 percent are finer than 20 μm, and five percent are less than 3.3 μm. Dust accounts for 10–20 percent of the regolith's bulk mass. NASA's Constellation program uses a definition of less than 10 μm for dust, while the Dust Management Project in NASA's Exploration Technology Development Program has heretofore been using 20 μm. The health exposure programs refer to dust as less than 5 μm, which is the respiratory cutoff. The properties and composition of dust particles of less than 20 μm are not well known, as this portion of lunar samples was not well preserved, partially because the dust grains in that range adhered to the sample bags and were not removed or analyzed."

It is interesting to note that the severity of the dust problems was consistently underestimated. Lunar Extravehicular Activities (EVA) led to issues such as vision obscuration, false instrument readings, dust coating and contamination, loss of traction, clogging of mechanisms, abrasion, thermal control problems, seal failures, inhalation and irritation, excessive crew time being used to clean EVA suits and equipment, and electrical conductivity. These problems began before touchdown on the lunar surface, when jet-blasted dust obscured vision, leading to issues such as a landing that straddled a crater, and eye irritation during the trip back to Earth.

Dust interactions can occur due to direct contact (two-body problem), or when dust is trapped between two surfaces in relative motion (three-body problem). In both problem classes, we can see compression, rolling and sliding. In the two-body problem there is also the bending of flexible materials.

Examples of two-body interaction include compression under boots and landers, and sliding of boots and wheels. For three-body interactions, dust can be compressed within seals such as airlocks, suits and mechanical joints. It can be rolled within bearing and joints.

By defining zones around a habitat – outside, transitional area, and inside – it is possible to better characterize the environment in terms of design parameters. For example, each zone may suggest the use of a different material.

Tests are suggested to estimate probabilities of dust grains entering a critical location, whether an astronaut's suit or a mechanical component. Such quantitative assessments can be used in trade studies of all components, such as mechanical interfaces and airlocks, as well as material characteristics.

Kobrick *et al.* identified numerous approaches for dust mitigation, but noted that none of these have been tested under actual lunar conditions. They suggested a non-dimensional abrasion index of the functional form:

$$\text{Abrasion Index} = f(R, H, S, F),$$

where R denotes the risk to mission success (of a material, part or system), H is a measure of the local mineralogy hardness, S is the severity of abrasion, and F is the frequency of dust particle interactions. A higher abrasion index implies greater potential impact to the system, and can provide designers a systematic methodology for assessing various lunar sites for the risks they pose regarding abrasion, so that these can be factored into calculations of expected life and performance.

Quotes

- "There are people who do things, and there are people who write about people who do things. Both are important, but not equally." Haym Benaroya.
- "It's a very sobering feeling to be up in space and realize that one's safety factor was determined by the lowest bidder on a government contract." Alan Shepard.
- "I believe that this nation should commit itself to achieving the goal, before this decade is out, of landing a man on the Moon and returning him safely to the Earth." President John F. Kennedy announces his intention to put a man on the Moon before a joint session of Congress on May 25 1961.
- "10, 9, ignition sequence start, 6, 5, 4, 3, 2, 1, zero. All engines running. Liftoff! We have a liftoff! Thirty-two minutes past the hour. Liftoff on Apollo 11!" Jack King, NASA Chief of Public Information, on the launch of the Apollo 11 over a live television broadcast on July 16, 1969.
- "Houston, Tranquility Base here. The Eagle has landed." Neil Armstrong tells NASA's Mission Control base in Texas that the Eagle landing module has reached the Moon's surface on July 20, 1969.
- "This is the LM pilot. I'd like to take this opportunity to ask every person listening in, whoever and wherever they may be, to pause for a moment and contemplate the events of the past few hours and to give thanks in his or her own way." Buzz Aldrin's broadcast, shortly after landing on the Moon and before he took a private communion on board the Eagle landing module.
- "The surface is fine and powdery. I can kick it up loosely with my toe. It does adhere in fine layers, like powdered charcoal, to the sole and sides of my boots." Neil Armstrong describes the surface of the Moon.

References

1. G. Heiken, D. Vaniman, and B.M. French: *Lunar Sourcebook, A User's Guide to the Moon*, Cambridge University Press, Cambridge 1991.
2. B. Sherwood and L. Toups: *Technical Issues for Lunar Base Structures*, <u>Journal of Aerospace Engineering</u>, Vol.5, No.2, pp.175–186, April 1992.

3. A.M. Jablonski and K.A. Ogden: *Technical Requirements for Lunar Structures*, Journal of Aerospace Engineering, Vol.21, No.2, pp.72–90, April 2008.

4. C.E.H. Scott-Conner, D.R. Masys, C.T. Liverman, and M.A. McCoy, Editors: *Review of NASA's Evidence Reports on Human Health Risks*, 2014 Letter Report, Institute of Medicine of the National Academies.

5. N.J. Lindsey: *Lunar Station Protection: Lunar Regolith Shielding*, International Lunar Conference 2003.

6. R.D. Irons, R. Eberhardt, and J. Schultz: *Risk Characterization and the Extended Spaceflight Environment*, Acta Astronautica, Vol.27, pp.243–250, 1992.

7. G. Horneck: *Life Sciences on the Moon*, Advances in Space Research, Vol.18, No.11, 1996, pp.95–101.

8. V.M. Aponte, D.S. Finch and D.M. Klaus: *Considerations for Non-invasive In-flight Monitoring of Astronaut Immune Status with Potential Use of MEMS and NEMS Devices*, Life Sciences, Vol.79, 2006, pp.1317–1333.

9. *Pharmacokinetics and Pharmacodynamics in Space*, Report of a Workshop Sponsored by the Biomedical Operations and Research Branch, Medical Sciences Division, August 1988, NASA, LBJ Space Center, Houston, NASA Conference Publication 10048.

10. D.R. Adamo, and J.S. Logan: *Aquarius, A Reusable Water-Based Interplanetary Human Spaceflight Transport*, Acta Astronautica 128 (2016) pp.160–179.

11. J.E. Nealy, J.W. Wilson, and L.W. Townsend: *Solar Flare Shielding with Regolith at a Lunar Base Site*, NASA Technical Paper 2869, 1988.

12. T. Straume: *Ionizing Radiation Hazards on the Moon*, 2167.pdf, NLSI Lunar Science Conference, 2008.

13. *Space Radiation Hazards and the Vision for Space Exploration: Report of a Workshop*, Ad Hoc Committee on the Solar System Radiation Environment and NASA's Vision for Space Exploration, National Research Council, 2006.

14. J.W. Wilson and J.E. Nealy: *Effects of Radiobiological Uncertainty on Shield Design for a 60-Day Lunar Mission*, NASA Technical Memorandum 4422, 1993.

15. L.C. Simonsen, J.E. Nealy: *Radiation Protection for Human Missions to the Moon and Mars*, NASA Technical Paper 3079, February 1991.

16. M-H.Y. Kim, G. de Angelis, and F.A. Cucinotta: *Probabilistic Assessment of Radiation Risk for Astronauts in Space Missions*, Acta Astronautica 68 (2011) pp.747–759.

17. C. Arena, V. de Micco, E. Macaeva, R. Quintens: *Space Radiation Effects on Plant and Mammalian Cells*, Acta Astronautica 104 (2014) pp.419–431.

18. G. Horneck: *Life Sciences on the Moon*, Advances in Space Research, Vol.18, No.11, pp.(11)95-(11)101, 1996.

19. G. Horneck, R. Facius, G. Reitz, P. Rettberg, C. Baumstark-Khan, and R. Gerzer: *Critical Issues in Connection with Human Planetary Missions: Protection of and From the Environment*, Acta Astronautica Vol.49, No.3–10, pp.279–288, 2001.

20. J.W. Wilson, J. Miller, A. Konradi, and F.A. Cucinotta: *Shielding Strategies for Human Space Exploration*, NASA Conference Publication 3360, December 1997.

21. L.C. Simonsen: *Analysis of Lunar and Mars Habitation Modules for the Space Exploration Initiative*, in Wilson *et al.*, pp.43–77.

22. D. Rapp: *Radiation Effects and Shielding Requirements in Human Missions to Moon and Mars*, MARS: The International Journal of Mars Science and Exploration, MARS 1, 1–13, 2005, doi:https://doi.org/10.1555/mars.2005.1.0.

23. R.K. Tripathi, J.W. Wilson, F.F. Badavi, and G. De Angelis: *A characterization of the Moon radiation environment for a radiation analysis*, Advances in Space Research 37 (2006) 1749–1758.

24. J. Miller, L. Taylor, C. Zeitlin, L. Heilbronn, S. Guetersloh, M. DiGiuseppe, Y. Iwata, and T. Murakami: *Lunar soil as shielding against space radiation*, Radiation Measurements 44 (2009) pp.163–167.

25. Z.W. Lin, Y. Baalla, and L.W. Townsend: *Variation of space radiation exposure inside spherical and hemispherical geometries*, Radiation Measurements 44 (2009) pp.369–373.

26. T.T. Pham, M.S. El-Genk: *Dose estimates in a lunar shelter with regolith shielding*, Acta Astronautica 64 (2009) pp.697–713.

27. G. Reitz, T. Berger, and D. Matthiae: *Radiation exposure in the Moon environment*, Planetary and Space Science 74 (2012) pp.78–83.

28. K. Rojdev, M.J.E. O'Rourke, C. Hill, S. Nutt, and W. Atwell: *Radiation effects on composites for long-duration lunar habitats*, Journal of Composite Materials, 2014, Vol. 48(7) pp.861–878.

29. S. Freese, A.P. Reddy, and K. Lehnhardt: *Radiation impacts on human health during spaceflight beyond Low Earth Orbit*, REACH - Reviews in Human Space Exploration 2 (2016) pp.1–7.

30. F.A. Cucinotta, and E. Cacao: *Non-Targeted Effects models predict significantly higher Mars mission cancer risk than Targeted Effects models*, Scientific Reports, May 12, 2017, DOI:https://doi.org/10.1038/s41598-017-02087-3.

31. C.R. Buhler and L. Wichmann: *Analysis of a Lunar Base Electrostatic Radiation Shield Concept*, NIAC CP 04-01, 2005.

32. NASA Lunar Petrogtaphic Educational Thin Section Set, C. Meyer 2003.

33. E. Robens, A. Bischoff, A. Schreiber, A. Dabrowski, and K.K. Unger: *Investigation of surface properties of lunar regolith*, Applied Surface Science, Vol.253, Issue 13, April 30, 2007, pp.5709–7714.

34. B.A. Campbell and D.B. Campbell: *Regolith properties in the south polar region of the Moon from 70-cm radar polarimetry*, Icarus 180 (2006) pp.1–7.

35. J. Park, Y. Liu, K.D. Kihm, and L.A. Taylor: *Characterization of Lunar Dust for Toxicological Studies. Part I: Particle Size Distribution*, Journal of Aerospace Engineering, Vol.2, No.4, pp.266–271, October 2008; Y. Liu, J. Park, D. Schnare, E. Hill, and L.A. Taylor: *Characterization of Lunar Dust for Toxicological Studies. Part II: Texture and Shape Characteristics*, Journal of Aerospace Engineering, Vol.2, No.4, pp.272–279, October 2008.

36. J. Hsu: *Scientists to Set Lunar Health Standards*, space.com, June 10, 2008.

37. N. Khan-Mayberry: *The lunar environment: Determining the health effects of exposure to Moon dusts*, Acta Astronautica 63 (2008) pp.1006–1014.

38. D. Linnarsson, J. Carpenter, B. Fubini, P. Gerde, L.L. Karlsson, D.J. Loftus, G.K. Prisk, U. Staufer, E.M. Tranfield, and W. van Westrenen: *Toxicity of lunar dust*, Planetary and Space Science 74 (2012) 57–71.

39. T.J. Stubbs, R.R. Vondrak, and W.M. Farrell: *Impact of Dust on Lunar Exploration*, Proceedings 'Dust in Planetary Systems', Kauai, Hawaii, September 26–30, 2005 (ESA SP-643, January 2007).

40. M.M. Abbas, D. Tankosic, P.D. Craven, J.F. Spann, A. LeClair, and E.A. West: *Lunar dust charging by photoelectric emissions*, <u>Planetary and Space Science</u>, Vol.55, Issues 7–8, May 2007, pp.953–965.

41. J.E. Colwell, C.J. Grund, and D.T. Britt: *Mechanical and Electrostatic Behavior of Lunar Dust*, 2143.pdf, NLSI Lunar Science Conference, 2008.

42. M. Ettouney, H. Benaroya: *Regolith Mechanics, Dynamics, and Foundations*, <u>Journal of Aerospace Engineering</u>, Vol.5, No.2, April 1992.

43. J.E. Colwell, S. Batiste, M. Horanyi, S. Robertson, and S. Sture: *Lunar Surface: Dust Dynamics and Regolith Mechanics*, <u>Reviews of Geophysics</u>, 45, RG2006/2007, Paper number 2005RG000184.

44. P.T. Metzger, J.E. Lane, C.D. Immer, and S. Clements: *Cratering and Blowing Soil by Rocket Engines During Lunar Landings*, 6th International Conference on Case Histories in Geotechnical Engineering, Arlington, VA, August 11-16, 2008.

45. There are numerous excellent texts on soil mechanics. An example is by R.F. Craig: **Craig's Soil Mechanics**, Seventh Edition, 2004, Spon Press, an imprint of Taylor & Francis.

46. J.K. Mitchell, W.N. Houston, R.F. Scott, N.C. Costes, W.D. Carrier, III, and L.G. Bromwell: *Mechanical properties of lunar soil: Density, porosity, cohesion, and angle of internal friction*, Proceedings of the Third Lunar Science Conference, Vol.3, pp.3235–3253, The MIT Press, 1972.

47. L.E. Bernold: *Experimental Studies on Mechanics of Lunar Excavation*, <u>Journal of Aerospace Engineering</u>, Vol.4, No.1, January 1991, pp.9–22.

48. S. Matsumoto, T. Yoshida, H. Kanamori, and K. Takagi: *Construction Engineering Approach for Lunar Base Development*, <u>Journal of Aerospace Engineering</u>, Vol.11, No.4, October 1998, pp.129–137.

49. J.R. Florek: *Excavating and Using Regolith for Lunar Base Applications - Review*, Rutgers University Technical Report, April 2007.

50. A. Wilkinson and A. DeGennaro: *Digging and pushing lunar regolith: Classical soil mechanics and the forces needed for excavation and traction*, <u>Journal of Terramechanics</u> 44 (2007) pp.133.152.

51. B.M. Willman and W.W. Boles: *Soil-Tool Interaction Theories as They Apply to Lunar Soil Simulant*, <u>Journal of Aerospace Engineering</u>, Vol.8, No.2, April 1995, pp.88–99.

52. S. Johnson: *Extraterrestrial Facilities Engineering*, <u>Encyclopedia of Physical Science and Technology; 1989 Year Book</u>, Academic Press, 1989.

53. R.L. Kobrick, D.M. Klaus, and K.W. Street, Jr.: *Defining an abrasion index for lunar surface systems as a function of dust interaction modes and variable concentration zones*, <u>Planetary and Space Science</u> 59 (2011) pp.1749–1757.

4

Structures

"Structures to protect us,
and where we can thrive."

4.1 CLASSES

These lists organize structures, their infrastructure, and their applications. They show how the interrelated systems are integrated and considered in the design of habitats and other structures. The lists were developed and refined over a series of technical meetings at space conferences during the 1990s.

Structures are grouped into habitats, storage facilities or shelters, and supporting infrastructure. These are then further subdivided according to types of structures, types of applications and application requirements, material considerations, structures and technology drivers, and requirement definitions and evaluations.

We note that there is a considerable overlap in structures and their designs, and this suggests that designers maximize the commonality in the designs. Perhaps there can be two classes of structures; those for habitation, and all the rest. Commonality helps both with the design efforts and the fabrication efforts. It also positively affects maintainability by minimizing the number of different parts and components. This is true for structures and machines. The downside of commonality is that a single part defect can then become a systemic problem. At some point, when 3D printing becomes a viable technology, the whole process of structural and parts fabrication will be rewritten.

Types of structures

Habitats
- landed self-contained structures
- rigid modules (prefabricated/*in-situ*)
- tensile-cable

© Springer International Publishing AG 2018
H. Benaroya, *Building Habitats on the Moon*, Springer Praxis Books,
https://doi.org/10.1007/978-3-319-68244-0_4

- inflatable modules/membranes (prefabricated/*in-situ*)
- tunneling/coring
- layered fabrication (*in-situ*)
- exploited caverns/lava tubes

Storage facilities/shelters/science
- open tensile (tents/awning)
- "tinker toy"
- modules (rigid/inflatable)
- trenches/underground
- ceramic/masonry (arches/tubes)
- mobile
- shells

Supporting infrastructure
- all of the above
- slabs (melts/compaction/additives)
- trusses/frames

Types of applications

Habitats
- people (living/working)
- agriculture
- airlocks (ingress/egress)
- temporary storm shelters for emergencies and radiation
- open volumes

Storage facilities/shelters/science
- cryogenic (fuels/science)
- hazardous materials
- general supplies
- surface equipment storage
- servicing and maintenance
- temporary protective structures

Supporting infrastructure
- foundations/roadbeds/launchpads
- communication towers and antennas
- waste management/life support
- power generation, conditioning and distribution
- mobile systems
- industrial processing facilities
- conduits/pipes

Application requirements

Habitats
- pressure containment
- atmosphere composition/control
- thermal control (active/passive)
- acoustic control
- radiation protection
- meteoroid protection
- integrated/natural lighting
- local waste management/recycling
- dust control
- airlocks with scrub areas
- emergency systems
- psychological/social factors

Storage facilities/shelters/science
- refrigeration/insulation/cryogenic systems
- pressurization/atmospheric control
- thermal control (active/passive)
- radiation protection
- meteoroid protection
- dust control
- hazardous material containment
- maintenance equipment/tools

Supporting infrastructure
- all of the above
- regenerative life support (physical/chemical/biological)
- industrial waste management

Material considerations

Habitats
- shelf life/life cycle
- resistance to space environment (UV/thermal/radiation/abrasion/vacuum)
- resistance to fatigue (acoustic and machine vibration/pressurization/thermal)
- resistance to acute stresses (launch loads/pressurization/impact)
- resistance to penetration (meteoroids/mechanical impacts)
- biological/chemical inertness
- reparability (process/materials)

Operational suitability/economy
- availability (lunar/planetary sources)
- ease of production/use (labor/equipment/power/automation/robotics)
- versatility (materials and related processes/equipment)

- radiation/thermal shielding characteristics
- meteoroid/debris shielding characteristics
- acoustic properties
- launch weight/compressibility (Earth sources)
- transmission of visible light
- pressurization leak resistance (permeability/bonding)
- thermal and electrical properties (conductivity/specific heat)

Safety
- process operations (chemical/heat)
- flammability/smoke/explosive potential
- outgassing
- toxicity

Structures and technology drivers

Mission/application influences
- mission objectives and size
- specific site-related conditions (resources/terrain features)
- site preparation requirements (excavation/infrastructure)
- available equipment/tools (construction/maintenance)
- surface transportation/infrastructure
- crew size/specialization
- available power
- priority given to use of lunar material/material processing
- evolutionary growth/reconfiguration requirements
- resupply versus reuse strategies

General planning/design considerations
- automation/robotics
- EVA time for assembly
- ease and safety of assembly (handling/connections)
- optimization of teleoperated/automated systems
- influences of reduced gravity (anchorage/excavation/traction)
- quality control and validation
- reliability/risk analysis
- optimization of *in-situ* materials utilization
- maintenance procedures/requirements
- cost/availability of materials
- flexibility for reconfiguration/expansion
- utility interfaces (lines/structures)
- emergency procedures/equipment
- logistics (delivery of equipment/materials)
- evolutionary system upgrades/changeouts
- tribology

Requirement definitions and evaluations

Requirement/option studies
- identify site implications (lunar soil/geologic models)
- identify mission-driven requirements (function/purpose/staging of structures)
- identify conceptual options (site preparation/construction)
- identify evaluation criteria (costs/equipment/labor)
- identify architectural program (human environmental needs)

Evaluation studies
- technology development requirements
- cost/benefit models (early/long-term)
- system design optimization/analysis

4.2   CONCEPTS AND DESIGNS

"It's more than a structure,
it's a place to survive and thrive."

As humanity prepares for the settlement of the Moon, we recognize that everything that our civilization has learned over thousands of years needs to be brought to bear on this greatest of challenges. It is not enough for scientists and engineers to build rockets for travel and structures for habitation. We need to be planning for a manned and permanent presence on the Moon, and then Mars. An urban environment is what we are looking to create, in a place that is hostile to human life.

Humans are complex. Given life support, we can survive a round trip to the Moon, as we did with Apollo. Close quarters and the lack of recreation or distraction were not a problem; the color of the interior of the space capsule was not a concern. But then those were trips of eight to twelve days. Table 4.1 shows the different time durations for three of the Apollo trips to the Moon.

Table 4.1. Sample Apollo mission durations

h-min-s	Trip duration	Time on Moon	EVA duration
Apollo 11	195-18-35	21-38-21	2-31-0
Apollo 12	244-36-24	31-31-0	7-45-0
Apollo 17	301-51-59	74-59-40	22-4-0

As groups on Earth move forward with plans to settle the Moon, considerations of the "other issues" that need attention grow in importance. Humans in close quarters tend to be stressed and require some way to relax. The interior environment needs to be a psychologically positive experience.

Evaluating structural concepts for a lunar base is a very difficult task. There are so many things regarding construction on the Moon about which we lack knowledge. Most of the criteria given below are also highly qualitative, so applying them to a structural concept has to be done very carefully. Objectively weighting the different criteria is difficult, because they are all important and they inform the development of suitable concepts for lunar habitats that consider the complete lifecycle of the structures.

Among the criteria for evaluating lunar base concepts are:

Transportation:
The mass and volume that has to be transported greatly affects the total cost and time. Second- and third-generation structures will be increasingly based on 3D printing technologies utilizing *in-situ* resources.

Ease of construction:
Does the erection process of the structure need a shirt-sleeve environment? Simple modular connections result in less astronaut construction time and astronauts should not be construction workers. The ISS is a lesson. What kind of heavy construction equipment is needed? Heavy construction equipment is unlikely for the Moon, and the use of deployable/self-erecting structures can help to minimize EVA activities. Given that 3D printing technologies were not even on the radar screen several decades ago, it is not beyond reason to expect that in one or two decades – the current timeline for a manned return to the Moon – such technologies will offer completely autonomous, *in-situ*-created, structural fabrication in conjunction with advances in automation and robotics.

Experience with structure and materials:
Lunar construction is accompanied by numerous unknowns and variables. Some of the data needed to map lunar construction will only be available after a certain presence on the Moon is established. Therefore, it is not desirable for early concepts to include structural systems that themselves inherit too many uncertainties. Also, it will be easier to react to problems arising during construction on the Moon if there is experience with that type of structure from former projects. Extensive prototype testing can help create that experience. Implicit is that structural concepts with which we are experienced can be designed in a robust way, leading to long lifetimes.

Expandability:
Designing a viable and relatively easy expansion of existing facilities is valuable. With 3D printed facilities on track to become viable for lunar surface fabrication, expandability may become less of an issue. We may expect that autonomous, *in-situ*-created, structural fabrication will be an ongoing activity on the lunar surface.

Excavation:
It is unlikely that structures designed for the lunar surface will require heavy construction equipment. Such equipment is not viable on the Moon. Lava tubes may provide us with subsurface volumes within which to place habitats. Some appear to have caverns extending from the base of the tubes.

Foundations:
The concrete foundations that are widely used for almost every construction project on Earth will not be available for early lunar bases, so new foundation types will have to be developed. Reducing or completely avoiding horizontal and tensile support reactions is essential. Self-balancing structures that do not rely on foundation reaction forces for equilibrium are ideal. Maximum use of the lunar soil strength is mandatory for other types of structures.

Human factors:
The structure must accommodate lighting, ventilation, insulation and acoustics. Crew psychology, and therefore effectiveness, will greatly depend on the quality of life in the lunar habitat. One important factor is the perceivable volume. A series of structural cylinders attached to each other may provide sufficient volume, but they can never provide the inhabitants with a perception of large spaces. Inflatable structures offer this possibility.

Recycling:
New generations of structures lead to the deconstruction and recycling of old structures. It is a worthwhile design goal to have structures that are easily assembled and also easily disassembled. Metals are easier to recycle than composites, for example. But we may come to a time when 3D printed structures, autonomously fabricated, will dot the lunar landscape, and, when no longer useful or beyond their design lives, will become the feedstock for new structures.

Maintenance:
The longer the desired lifespan of a lunar structure, the more important maintenance requirements become. Maintenance costs can greatly affect overall costs and will therefore have to be minimized. At the same time, maintenance is always necessary for a safe structure, so access to, and easy replacement of, structural components is crucial. Self-repairing capabilities are likely to advance, especially in light of advances in 3D technologies, robotics and sensors. Clearly, the more complex the systems, the more difficult it will be to estimate and assure reliability.

When we think of cities on the Moon and Mars, outposts on asteroids and the moons of the planets, what we see in our mind's eye are structures. These structures are unique to their environments, just as are structures on Earth, and when people start to imagine the possibility of settling these final frontiers, they envision structures and cities.

Just as cities on Earth are centers of commerce and art, and structures house all of our creative endeavors, cities and structures are places where we live and work but are also creative ends of their own merit. Structures are a reflection of our civilization. They are engineered as a culmination and merging of science and aesthetics, serving many functions for people. On Earth, they house us and are a place for our belongings. On the Moon, they will also keep us alive.

Structures are designed, at a minimum, to function satisfactorily in the environment where they will be built. There are environmental forces to contend with. On Earth, those environmental forces include wind and rain, sometimes earthquakes and other times ocean waves. In addition, there are the 'loads' inside; those due to people and equipment that are sometimes stationary and other times in motion.

On the Moon, there are environmental characteristics that are very different to those found on Earth. There is no wind since there is no atmosphere, and the lack of atmosphere also leads to a number of serious human survival issues that the structure needs to counter by way of its design. The first is the need for an atmosphere within the structure, pressurized to an acceptable level for humans. While the atmosphere on Earth protects us from galactic and solar radiation as well as meteoroids, the lack of one around the Moon means that lunar structures – at least those on the surface – will need to be designed with shielding against these dangers.

4.3   LUNAR ARCHITECTURE AND ENGINEERING

A 2008 study to consider how space architecture can address the broader habitation issues, made the following definition: "Space architecture is the theory and practice of designing and building the human environment in outer space."[1] Clearly an all-encompassing definition.

How does space architecture – a discipline born in the space age – look at the human environment? "Many considerations familiar to terrestrial architects – productivity, privacy, assembly, aesthetics, place identity, sensations, view, mood, safety, utilities, and adaptive use, to name just a few – are increasingly relevant to the design of habitable environments for outer space. In addition, living in space brings into sharp focus some new considerations increasingly important to architecture on Earth; sustainability, material recycling, regenerable life support."

This study also explored the roles that artists can play as members of human space project teams. The artist's aesthetics are a valuable resource in the design of facilities for long-duration habitation. It is the rare person that can envision what it means to live in very close quarters for long periods of time. A prisoner lives that life – even though the prisoner

Figure 4.1.   The deployment of the United States flag on the surface of the Moon is captured on film during the first lunar landing mission, Apollo 11. Here, astronaut Neil A. Armstrong, commander, stands on the left at the flag's staff. Astronaut Edwin E. Aldrin, Jr., lunar module pilot, is also pictured. The picture was taken from film exposed by the 16 mm Data Acquisition Camera that was mounted in the Lunar Module. While astronauts Armstrong and Aldrin descended in the Lunar Module "Eagle" to explore the Sea of Tranquility region of the Moon, astronaut Michael Collins, command module pilot, remained with the Command and Service Module "Columbia" in lunar orbit. (S69-40308, July 20, 1969. Courtesy NASA)

was recruited to that life involuntarily. For the lunar facility, colors can greatly enhance the psychological state of the occupants. The windowless lunar structure can initially have 'virtual' windows composed of flat-screen images, giving the impression that the outside is not so far away. Artists can create powerful sceneries – not necessarily lunar landscapes – that provide a supportive ambiance to the inhabitants.

Space architecture is not only for architects. Engineers also practice space architecture when they integrate human factors into structural designs. The needs of human occupants of structural systems are incorporated into the technical analysis. While there are many goals for a human lunar settlement, a primary goal must be the well-being of the people. All else flows from a satisfied population. At this time, space architecture is focused on the support of human life in small pressure vessels in LEO.[2] As we move into space and settle on the Moon, we will go beyond survival and toward living a full life.

As humanity evolves into a spacefaring and space inhabiting civilization, the development of a space urbanism becomes a critical component of living beyond Earth. Architecture can be viewed as a coordinating framework for the development of facilities in space, and on the Moon and Mars, where humans can build comfortable and enriching lives.[3] The physiological and psychological challenges that humans will have to meet can be partially addressed by a well-designed urban environment. Given the lunar extra-planetary environment, the wide, open spaces will exist, but they will be experienced through spacesuits. The shirt-sleeved environment will be comparatively small, but made to feel large and welcoming, and as Sherwood summarized:

> "By starting from a few accurate principles – that lunar urbanism will be densely populated, hermetic, interior, kinesthetically expansive, visually lightweight, and based on indigenous materials; that it will be non-sterile; and that lunar wilderness will become irreplaceably precious – those who do plan can contribute meaningfully to the responsible realization of one of the grandest projects ever imagined in human history."

As we introduce and discuss lunar habitat concepts, we first note that many *are* concepts and not designs. There are artists' renderings that suggest possible concepts without addressing the quantitative process known as design engineering, where a structure is conceived (material, dimensions, mechanical and electrical components, reliability estimates, construction sequencing) with specific consideration of the constraints of the environment, and the survival characteristics of the structure over significant periods of time (design life). The artists' concepts developed in conjunction with engineers are important, however, since they help generate and develop nascent ideas, and set up a conceptual framework that can be used by designers to engineer a particular structure to make it work in the lunar environment. Often, concept and design work together and iteratively.

4.4   INTERVIEW WITH MARC COHEN

This interview with Marc Cohen was conducted via email beginning on June 23, 2017 and concluding on July 16, 2017. We begin with a brief professional biography and then continue with the interview.

Brief biography

Marc M. Cohen is a licensed architect who was the second Space Architect in NASA, after Maynard Dalton. Marc is a founder of Space Architecture as a field of research and development as an organizer of the American Institute of Aeronautics and Astronautics Space Architecture Technical Committee (SATC). Marc began his aerospace career at NASA Ames Research Center, where he served in the Facilities Engineering Branch, the Space Human Factors Office, the Advanced Space Technology Office, and the Advanced Space Projects Branch. Following early retirement from NASA, Marc worked for Northrop Grumman Integrated Systems, primarily on the Altair Lunar Lander. He started his own business, Astrotecture®, and won two NASA Innovative and Advanced Concepts Grants as PI for the Robotic Asteroid Prospector and as 80 percent CoI for the Water Walls Life Support Architecture. Currently, he is a member of both the SATC and the AIAA Life Sciences and Systems Technical Committee.

Marc earned a D.Arch from the University of Michigan, on the NASA Full Time Graduate Fellowship and the Saarinen-Swanson Essay Prize. At UM, his dissertation was *Problem Definition in a Participatory Design Process* about the creation of the Human Exploration Demonstration Project at Ames. Marc earned a M.Arch from Columbia University in the City of New York, with a Kinne Summer Travelling Fellowship, where his thesis was *Greenpoint Waterfront*. Marc earned an AB cum laude in Architecture and Urban Planning from Princeton University, where his thesis was *Village Square Community Center*.

Our interview

> *Before we start, I would like to follow up on a prediction I made in our last interview that appears in your book,* **Turning Dust to Gold.**
> *I said, "In human spaceflight, I expect that we will see a blossoming of private launch and flight vehicles. … It is therefore with deep regret that I submit that one or more of these small space startups will kill a crew."*
>
> *Alas, I was right. We are seeing an expansion of NewSpace private and commercial launch and spacecraft companies, some of whom are close to successfully flying a crewed orbital mission. Unfortunately, I was also right about loss of crew. Virgin Galactic's VSS Enterprise broke up during reentry/descent from its first suborbital flight, killing the pilot, Michael Alsbury and severely injuring the copilot, Peter Siebold. I deeply regret that my prediction may have seemed glib when I made it; I regret even more that it was spot-on. I hope that the record of commercial crew safety improves and that they do not suffer another loss.*

Lunar structures have been grouped into three categories: I – habitats brought from Earth, II – habitats with components from Earth, but also components made from lunar resources, and III – habitats primarily made from lunar resources, ISRU, perhaps 3D printed structures. The longer it takes for us to get back to the Moon to stay, the more certain technologies advance, for example 3D printing. Would you keep the above categories, or modify them in some way?

> *It is awkward and unfortunate to begin your first question with a passive verb construction, Lunar structures have been grouped …, but who grouped them?*

You must recall that in the 1997 NASA Habitats and Surface Construction Roadmap – 20 years ago – Kriss Kennedy and I identified three classes of habitat:
I. Pre-Integrated,
II. Deployable, and
III. Using ISRU components.
Subsequently, we republished, refined, and expanded upon this concept several times, one of which you coauthored with me.[4] [5]
Your troika blurs these distinctions by incorporating ISRU into Class II. However, you make an important point; eventually we will need to consider the proportion of ISRU materials that comprise the total habitat. Perhaps we should start characterizing Class III habitats in terms of percentage of ISRU components or materials, e.g., a habitat covered by regolith shielding might constitute ten percent ISRU by component count or complexity, but, say, 50 percent by mass.

Given that we have been waiting almost 50 years to return to the Moon, do you have a feeling as to when we might return, with people, for an extended stay?

That would depend on whom you mean by "we." If by "we" you mean the United States, let's not hold our collective breath. The last President who fought to fund the human spaceflight program was Bill Clinton. He committed to building the ISS as a cornerstone of his foreign policy. After the Space Shuttle Columbia disaster, President George W. Bush announced the Constellation Program with the Orion to replace the Shuttle and start back to the Moon. However, W. never asked Congress for any appropriations to make Constellation possible, leaving his NASA Administrator Sean O'Keefe no option but to try to cannibalize the budget of all the rest of NASA to pay for it. Except Congress would not allow him to cannibalize science. So he left, and Mike Griffin came in as Administrator with the solution of eliminating nearly all NASA-funded space technology development to pay for Constellation. That move eliminated most of the advances that could make future exploration more cost-effective, efficient, reliable, safe, and timely.

When President Obama came into office, he saw the true ESAS cost estimates that had stayed hidden under Bush and Griffin: $10 billion more per year in the NASA budget to pay for the Constellation lunar project. President Obama said, "no way," cancelling Ares I and Ares V, and the Altair Lunar Lander. The heavy lift requirement crept back into the budget as the Space Launch System (SLS). Meanwhile, the Orion and SLS limp along on an insufficient budget that appears to consist mostly of overhead, taking at least a decade longer than originally planned. Compare that to the Apollo Program that NASA accomplished in eight years, with peak funding at 4.0 percent of the federal budget. The NASA budget today teeters below 0.02 percent of the federal budget. Now, the internal NASA cost estimates to launch an SLS with any payload begins in the range of $2 billion.[6] *This price tag is so expensive that NASA could probably afford to launch the SLS only once every two years (not counting the overhead costs of keeping MSFC and KSC at the ready). The Trump Administration keeps proposing cuts to the NASA budget, which fortunately Congress keeps resisting.*

The NewSpace companies such as Bigelow, Blue Origin, Orbital ATK, and SpaceX, pose a different discussion. Thus far, they have enjoyed success in pressurized cargo flights to LEO for the paying customer: NASA. The Bigelow BEAM experiment attached to ISS also marks a positive sign. The challenge of going to the Moon on one's own $10s of billions is that the paying customer may not exist.

Ironically, perhaps the most hopeful news is that China has been preparing development of a human lunar landing. In June 2017, the Chinese Space Program announced they were close to approving a human lunar program.[7]

Not coincidentally, Japan announced its own plans to land humans on the Moon by 2030.[8] Japan has more of a head start than most people might realize with its HTV pressurized cargo vehicle that they can upgrade to a human spacecraft.

Alas, the time is long past when NASA could rely upon its rivals to save its freeze-dried bacon by instigating a new space race. In your mind, what is a likely scenario and timeline for manned space?

I anticipated this question in my previous response. You use a highly ambiguous phrase in "likely scenario." I stand firmly in the camp that believes that humans need to establish a solid presence on the Moon first, and use it to test exhaustively all out technology before we venture beyond cislunar space. I said to you in our previous interview that I heard Krafft Ehrike say: "if God wanted man to go to Mars, He would have given him a Moon."[9]

I accept the reality that I will not see a human landing on Mars in my lifetime. I just hope I will see a human return to the Moon, by whatever country or company can do it. Regrettably, the Space Community has divided itself into two camps, the 'Mars or Bust' camp and the 'Return to the Moon' camp. Each sets up an unreasonable objection to the other's goal: "If we stop at the Moon before we go to Mars, we will never go there in my lifetime," versus: "If we try to go to Mars, we will never return to the Moon in my lifetime." Fact: Mars appears out of the question in terms of a sustainable base or settlement until we learn how to create and operate one on the Moon. If the 'Mars or Bust' advocates mean a sprint 'flag and footsteps' scenario like Apollo, they must accept for a hiatus of more than half a century like we still see after Apollo. Also, they must accept losing more crews in space – at a higher rate than we have yet seen – without the cislunar testing ground.

Do you agree with those who have made a case for settling the Moon first and then using it as a base for travel to Mars?

Didn't I just address that very thing? However, I do not believe that it is correct to think of the Moon, per se, "as a base for travel to Mars." Instead, I think of the development of cislunar space as the springboard. That includes a platform to act as a transit node. We hear a lively debate as to which Lagrange point or lunar retrograde orbit (LRO, not to be confused with Lunar Reconnaissance Orbiter) or lunar distant retrograde orbit (LDRO) serve best for the job. NASA road mapping efforts call for a Deep Space Gateway at one of these space-time-trajectory coordinates. This campaign would include one or more bases on the surface with extraction activities to produce water, H_2 and O_2 to use for propellant and life support. Transported to the Deep Space Gateway, these resources would supply outbound to Mars or the asteroids.

Is ISRU oversold for near-term lunar development?

Now you are using two ambiguous terms in your question! What do you mean by "oversold" and what do you mean by "near-term?" How can a commodity be "oversold" when nobody has ever sold it and nobody has ever bought it? By "near-term," I can only imagine you refer to a lander that lands with a fully fueled ascent module, which need not rely upon ISRU propellant refueling. In that case, yes, the earliest human landings in this century will land with fully fueled ascent modules until we can provide ISRU fuel production for Ascent vehicles that land empty.

Key challenges to engineers and scientists who will be tasked to develop the lunar mission and structural design are the knowledge gaps that cannot be filled before being on site. For example, the lunar 1/6 *g* environmental effects on biology, or systems reliability in the lunar dust environment. Also of great concern is our understanding of the very complex systems that will not be integrated before they are placed on the Moon. Even if the behavior of all the components are fully understood, the complete integrated system will generally have unanticipated characteristics. Is there a way of dealing with these knowledge gaps before being there?

First, stop trying to speak bureaucratese, as in engineers and scientists who "will be tasked …" It is unbecoming as a scholar and a gentleman. Not only is that another passive verb construction in your opening line, but also it illuminates the pitfall of corporate and government hierarchies. If everyone does things only because someone up the food chain 'tasks' them to do it, they feed the stagnation of the organization. To succeed in such an endeavor, the people must find their motivation and much of their direction within themselves. They should not rely on being tasked. Also, the notion that virtue abides in telling people what to do becomes counter-productive, creating an overpopulation of management wannabes.

Seriously though, I can tell you more things not to do than what we should do. That imbalance arises because "glass box methods" such as System Engineering as they are known in Design Methodology are predominately preventative to avoid making errors; they are prophylactic in their essence. I have seen people screw up projects so many different ways that greatly outnumber the successful ways.

On the positive side, I can state two precepts that constitute necessities for the successful design of deep space habitats which, except for cislunar space, are by definition long-duration habitats. 'Get the Physics Right' and 'ECLSS Comes First'.

First, get the physics right and extract the implications for habitat and spacecraft design. Understand the physics of the space environment, orbital mechanics, spacecraft propulsion and stabilization, and what they mean for the safe functioning of the space habitat on whatever platform it resides. The atmosphere, dust, gravity, radiation, and thermal environments stand out among these regimes.

Second, design the Environmental Control and Life Support System (ECLSS), including thermal control FIRST, before designing or selecting a module. By "selecting," I mean picking an existing pressure vessel as 'primary structure', whether it is an existing ISS module or an upper stage tank (à la Skylab). ECLSS always becomes primary to the design of healthy and safe habitation. However, throughout the history of human spaceflight ECLSS engineers and scientists have been compelled to shoehorn ECLSS and thermal into predetermined boxes, bays, or racks.

These forced physical installations impair ECLSS' accessibility, adjustment, efficiency, functioning, maintenance, reliability, repair, replacement, and upgrade.

It may seem paradoxical that as a Space Architect I would place an engineering system ahead of architectural design.[1] However, the way I understand how Maslow's hierarchy applies to a space habitat, it requires the addition of a new survival *bottom layer, below* Safety *because it's more fundamental than* Safety. *I call this new foundation 'Oxygen', short for "Oxygen, CO_2 removal, pressure, temperature control (the lunar surface varies from −200°C to 140°C), buffer gas, and clean water."*

Dust mitigation is a critical issue generally on the Moon. How far are we from getting a handle on that problem, both the biological risks and the engineered systems risk?

The answer to this question requires an explication of how NASA's Human Research Program (HRP) works. The way HRP conducts calls for proposals at this time covers five topic areas for competition:

1. *Exploration Medical Capability (ExMC)*
2. *Human Factors and Behavioral Performance (HFBP)*
3. *Human Health Countermeasures (HHC)*
4. *International Space Station Medical Projects (ISSMP)*
5. *Space Radiation (SR).*

Lunar or Martian dust would fall under Human Health Countermeasures; dust proposals must compete against research proposals in all the countermeasure topics, including radiation, microgravity, and partial gravity. In my view, NASA would not become serious about research and countermeasures for dust until it can establish a directed research topic, focused solely on dust effects on health and countermeasures to prevent it. Also, NASA would provide sufficient lunar dust from its 382 kg in lunar samples from Apollo to these researchers, so that they can work with the real article, including destructive testing, not just with very imperfect simulants.

The isolated environment that astronauts will face on the Moon, and even more isolated on Mars, makes critical the need for high levels of reliability. Part of that implies a self-healing capability. Such technologies are being studied and developed but are far from being usable. Do you see self-healing as a critical technology?

I would need to ask you: "self-healing" for what? Certain technologies, perhaps habitats, might benefit from a self-repair or 'self-healing' capability. However, what is more important is developing truly excellent reliability data, so that failures of various components, subsystems, or systems become probabilistic and predictable. Thus, it might be a better investment to design systems to degrade gradually and gracefully, rather than just fail suddenly, regardless of whether there is a self-healing capability. On the Water Walls Life Support Architecture *project, our team tried to envision such a system that would predictably and gracefully degrade so that the crew or robots could repair, replace, or resupply it before there was a threat of a life-critical failure.*

---

[1] Just as I am a member of the AIAA Space Architecture Technical Committee, I am equally a member of the AIAA Life Sciences and Systems Technical Committee.

Human space settlement has been promoted as a justified expenditure, financed by taxpayers, for many reasons: economic development, scientific advances, SETI, manifest destiny, building bridges between nations, limitless energy production, for example. What is your case for manned space settlement? Do you have an overarching reason for supporting such an endeavor?

> To me, it is purely an 'article of faith'; it comes down to a single existential and spiritual precept. We live on a planet in space. It is our destiny to explore and live on many more planets. Tomorrow the Stars.
>
> Some people in the Space Community still struggle to find THE magic formula. This special sauce mixes commerce, economics, science, partisanship, nonpartisanship, science, nationalism, internationalism, militarism, anti-militarism, science, lunar/planetary exploitation, lunar/planetary protection, science, cooperation, and competition. And did I mention science? Then, react them together in a galactic crucible to make the public and the politicians all magically support human exploration.
>
> Double, double toil and trouble. I am tired of arguing justifications; nothing could be more unprofitable. The great advantage of private and commercial space projects is that they need not worry about most of that controversy.

It appears that missions to settle the Moon or Mars have two underlying potential show stoppers: (1) assuring human survival and good health, and (2) assuring very high reliability for all engineered components. While we have shown with the ISS and Apollo that we can manage short forays to the Moon, or longer ones to LEO, satisfying (1) and (2) in limited ways, the bar is an order of magnitude higher for long-duration lunar stays, and an order of magnitude higher than that for Mars human missions. Are guaranteeing (with high probability) human and machine survival and high operability on the horizon, or beyond it?

> First of all, I apologize for my (still unpublished) drafts entitled "Five Showstoppers for Mars." I was using 'showstopper' incorrectly. The dictionary meaning of showstopper is that the audience applauds so enthusiastically and loudly that it stops the performance. Engineers have misused showstopper to mean something that is infeasible or unviable. I changed that title to "Five Environmental Barriers to Mars."
>
> Second, while testing engineering components in the lab is essential, ultimately it is necessary to test these components and the subsystems in the hostile space environment. In NASAspeak, that is called Technology Readiness Level 7 (TRL-7) – Testing in a Space Environment – before it can be considered approved for production as spaceflight-rated. Only through using, testing, and thoroughly wringing out all the hardware and software systems in the space environment, can we develop truly safe and reliable mission capability, which applies equally to all crew systems and habitats.
>
> Indeed, the bar for reliability on a Human Mars mission is easily an order of magnitude higher than for a short-term lunar mission. However, the only way we will accomplish that order of magnitude improvement would be through exhaustive testing on the Moon and in cislunar space.

Here are some numbers. On the (unpublished) Lunar Lander Development Study (LLDS) on which I worked at Northrop Grumman for NASA JSC, NASA gave us these target Figures of Merit (FOM): Probability of (not) Loss of Mission = 0.995 and Probability of (not) Loss of Crew = 0.999. These FOMs apply to a one to two-week lunar surface stay, plus up to about 20 days in flight under the "Go anywhere, go anytime, return anytime" regime.

For Mars, there is no 'anytime' option (which would substantially reduce risk for lunar missions). The mission must hit precise launch and return windows or the crew may find themselves stuck for 26 months until the next launch window. The Mars mission is about 1000 days instead of a lunar maximum ~ 34 days.[10] *A Mars mission involves aerothermal entry, descent, and landing at Mars. The challenge for reliability on Mars is to extend that "three nines" or better for a mission about 30 times longer than the Moon and much greater complexity.*

Who would you consider as your key influence in your pursuit of space studies? Who inspired you?

I suppose I can name a few influences: Buckminster Fuller, Frank Lloyd Wright, Erich Mendelsohn, and other early Modern architects. I wrote about Bucky in my 2014 paper The Continuum of Space Architecture, *describing how his reorganization of the Platonic Solids by vertices spurred my analysis for the Triangular-Tetrahedral Space Station configuration.*[11] *Of course, the Mercury, Gemini, and Apollo programs played a formative role in my early thinking that the space program was the story of my life. Buzz Aldrin stands out for me among that generation.*

The first time I met Buzz was at the 1996 ASCE Space Construction Conference in Albuquerque, NM. I was presenting my paper on the difference between an interplanetary vehicle habitat and a Mars surface habitat. Buzz came prancing into the room with the pre-published conference proceeding in hand. He had read my paper, and proceeded to quiz me about it on the spot. I was thrilled!!!

We talked at length, and developed a friendship. A year or so later, he called me to say he was coming to NASA Ames and wanted to tell me about his latest advances in his Mars Cycler concept. After meeting at Ames, he invited me to come to his sister's country club, the Circle Club in Atherton. He introduced me to her as if I was his oldest friend, and then he commenced drawing and describing his system on the tablecloth. I had a chance to talk to his sister, who regarded him with a wry and knowing grin.

I have seen Buzz on numerous occasions since, and it has always been a delight, although often it may feel like I smoked I-don't-know-what.

All of us working on space recognize that much of what we do and are interested in promoting will not happen in our lifetimes. Of course, we have no choice but to accept this reality, but how do you view this?

Didn't I say already that I accept that I will not see humans on Mars in my lifetime? I still hope for an American return to the Moon. The Trump/Pence administration made some positive tweets, but the real test will be whether they ask for the budget increase necessary for NASA to do it. If they repeat what W. did – direct NASA to return to the Moon without funding – it will fail too.

Figure 4.2.   Photo of (l to r) Buzz Aldrin, Marc M. Cohen and Carl Meade (Shuttle astronaut and Director of Exploration at Northrop Grumman) at an AIAA dinner in Los Angeles. (Courtesy M.M. Cohen, with permission of all)

You have developed and designed many habitats for extra-planetary sites, accounting for the severe environment, especially the radiation threat. If we want to send people to the Moon for semi-permanent stays as a prelude to permanent settlements, what would your first issue to solve be?

The first step would be to revive the Altair Lunar Lander studies that we were con-ducting in 2009 at Northrop Grumman and also at Boeing and Lockheed-Martin. No return to the Moon scenario is possible without a lander that can bring a crew to the surface and stay for at least one week.[12] *A key feature was the* Crew Productivity Figure of Merit *that the Northrop Grumman team of the astronaut Ken Cameron, Paul Houk, and I developed and applied to the Altair.*[13]

Despite that cancellation, I believe that there have been some promising lander development studies in the past decade, including the Morpheus project at NASA JSC, the Robotic Lunar Lander Development Project (RLLDP) at NASA MSFC, and the Lunar-X Prize teams' efforts, notably Moon Express' collaboration with RLLDP.

Do you think that a fleet (say 4-6) of pressurized rovers could be our first habitats, assum-ing they could be shielded properly? Something Habot-like?

You refer to the Habitat Robot (Habot) project that I did for John Mankins at NASA HQ in 2003–2004. The context for that project was the absence of agreement among the lunar scientists about where it would be best or most productive scientifically to place the first lunar base. For that reason, John gave me 'direct funding' to sub-

stantiate the Habot concept for the capability of moving multiple pressurized modules as a mobile base that moves from site to site and then self-assembles to form the base.

You developed a concept using water as a radiation shield for space travel. Has that concept evolved?

I taught you the last time to exercise caution about using the biological concept of 'evolution' to describe technological development. Evolution takes place over hundreds, if not thousands of years, whereas technological change can be abrupt in response to the economy. Yet there has been nothing sudden about adopting water shielding for human spacecraft.

You refer, I believe, to my 1996-1997 Interplanetary Vehicle (IPV) Habitat project, in which I designed a water shield 30 cm thick, with a mass of about 42 tonnes of water.[14] *That concept emerged as part of the debate at that time stemming from the first NASA Mars Design Reference Mission (of which I am a coauthor) about whether the IPV Habitat should serve also as a crew lander.*[15] *To me, it was blindingly obvious that if we put sufficient shielding on the IPV, it made no sense to drag it all down to the surface with the habitat. Therefore, the crew lander and surface habitat should constitute different elements from the IPV habitat. The TransHab concept a year later incorporated a 10 cm water shield around the crew sleep quarters. I recount the debate in my 2015 "First Mars Habitat Architecture."*[16]

In terms of my 1997 IPV Habitat design study, I have not seen another since then that went into a comparable degree of detail in the design of the water tankage. Water is a contender but the challenge arises in launching tens of tonnes to fill the shielding tanks on a habitat. Perhaps the most promising prospect would be to extract the water from the lunar surface and launch it to a "Deep Space Gateway" platform where it would fill the shielding tanks. However, as you noted earlier, the whole concept of ISRU water could be "oversold."

Consider that regolith may contain a very small proportion of water as a fraction of mass, on the order of two to six percent. Please correct me if I am wrong, Dr. Concrete, but isn't the residual water of hydration in terrestrial concrete on the order of two to three percent? In other words, just think about going to the Moon to mine concrete to extract the water.

More recently, I worked with Michael Flynn and his life support engineering team at NASA-Ames on a NASA Innovative and Advanced Concepts (NIAC)-funded project to conceptualize the Water Walls Life Support Architecture *that would afford a measure of water shielding. However, the shielding aspect is secondary in terms of the biochemical research and development challenges in Water Walls. Whereas the water shield in the 1997 IPV Habitat comprised a 30 cm thickness of water, in this embodiment of Water Walls, the thickness of the algae-grown units is only 5 cm. However, it is possible to design a double layer around a common light source, increasing the thickness to 10 cm of water.*

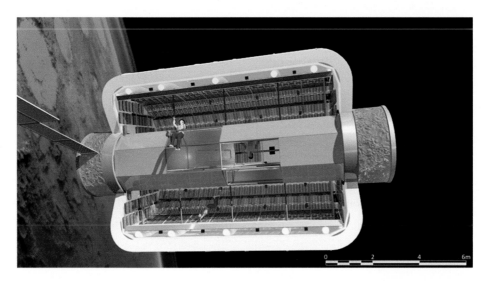

Figure 4.3.   Longitudinal Section: Water Walls Life Support Architecture installed around the inner wall of a space habitat module, based upon the Bigelow Aerospace BA330 inflatable module concept. Drawing by François Levy. (Courtesy M.M. Cohen)

Do inflatable structures have a large role in lunar and Martian development?

Inflatables have a potential role, although what that would be precisely remains to be determined. As a Class II deployable habitat, inflatables offer the potential to provide a pressurized volume that is significantly larger than a Class I pre-integrated habitat that must fit within the fairing of a launch vehicle. The big challenge for an inflatable remains how to outfit it with all the hardware necessary to make it useful. It is not feasible to attach load-bearing hard-points to the inflatable fabric without installing a complete interior secondary structure. That is why the main prototype for inflatables, the TransHab that Kriss Kennedy and Constance Adams developed at JSC, uses a rigid hub to support all the internal structures as cantilevered decks. Constance and Georgi Petrov in Synthesis-International have proposed an adaptation of the TransHab to lunar and planetary surfaces with their Surface Endoskeletal Inflatable Module (SEIM).[17]

Meanwhile, Bigelow Aerospace licensed the NASA TransHab patent, as the basis for their proposed BA330 module that we adopted for the Water Walls module implementation concept.

As a space architect, do you take insights from the biological world as you consider extra-planetary structures?

Your question is much broader than just Space Architecture. Biological analogy has been one of the organizing principles for Architecture and Urban Planning for millennia. The most common example of biological analogy is bilateral symmetry. All vertebrate bodies are bilaterally symmetrical. Just think about how many buildings you know that are bilaterally symmetrical. Typical examples include ancient

Egyptian cities and temple sites such as the Great Temple of Ptah in Luxor, (18th Dynasty, circa 1550–1290 BCE); the Beit HaMikdash, Solomon's Temple on the Temple Mount in Jerusalem (circa 1000–997 BCE); Angkor Wat's Baphuon Temple, Siem Reap, Cambodia (circa 1100–1200 CE). The addition of a dome at the center of symmetry suggests a head on the bilateral body. St. Peter's Basilica in the Vatican City is one of the best-known examples (1506–1626 CE), where Bernini's colonnade forms the arms. These buildings exist to serve one function: religion. Expand the function and many capitols, county seats, and city halls, schools, office buildings and even some housing join the mix.

There are many other aspects of biological analogy and biomimetic concepts relating to function in Architecture. I could prepare a year-long course in this subject, but I think you get the idea.

You are one of the founders of the discipline known as Space Architecture. It is really a multi-discipline. How is Space Architecture defined and how is it well suited to the new challenges of creating extra-planetary habitats that are not only livable but a place where people can thrive?

Guilty! But I worked and continue to work with a brilliant group of architects, engineers, and industrial designers to develop this discipline as its own field of design, development, research and study. Unfortunately, 'multidisciplinary' is one of those poorly understood and widely misused terms, like 'evolution.' At most universities, nearly all the people give lip service to being 'multidisciplinary'. However, when I try to talk to professors and deans at many of these institutions, they cannot grok *how one person can work across a range of specializations such as Architecture, ECLSS, Human Factors, and Structures. It seems that to most of them, 'multidisciplinary' means once a year an Engineering professor has lunch with an English professor, and they talk about their kids. However, to me, I am doing only one thing: the practice of Architecture as an integrative discipline.*

As for the definition of Space Architecture, we have founding documents, that are available online at http://spacearchitect.org/resources/. The Millennium Charter (12 Oct 2002, 43 signatories from 16 countries), states:

"Space Architecture is the theory and practice of designing and building inhabited environments in outer space."

Still, I put my own caveat on this 'mission statement'. For anything to constitute 'architecture', as I explain in The Continuum of Space Architecture, *it must involve* physical problem solving. *Apologies to my friends in our allied fields, it may be art, or economics, or environmental psychology, or sociology, or system engineering, but if it does not involve physical problem solving it is not Architecture.*

The AIAA Space Architecture Technical Committee's founding charter, created by Ted W. Hall, A. Scott Howe, and Brent Sherwood, includes a diagram, which I modified slightly to convey my view of space architecture. This diagram represents the 'multidisciplinary' nexus of Space Architecture at the center of designing for humans in space. The three circles define the root disciplines as understood in academia or industry: Life Sciences and Systems, Physical Sciences, and Space Systems and System Engineering. The intersections between these circles form vesica that denotes the domain of Space Architecture.

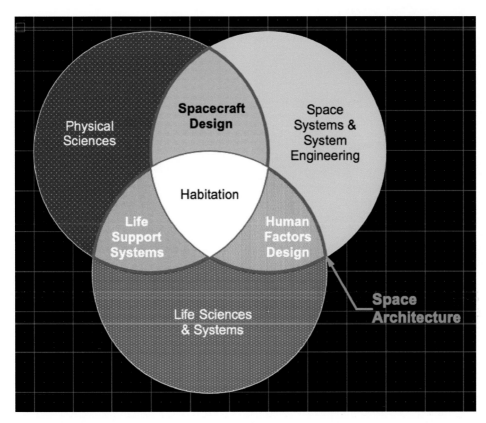

Figure 4.4.   Modified vesica diagram of the AIAA Space Architecture Technical Committee's concept of Space Architecture as a multidisciplinary nexus. (Courtesy M.M. Cohen)

Architecture is the most integrative of disciplines, coordinating art, building design, engineering, landscape, and urban scale. Thus, it is logical and fitting that Space Architecture should extend this integrative heritage into space.

Thank you.

4.5   RIGID STRUCTURES

"Our likely first structure."

During the late 20th century, there was still hope that a manned American return to the Moon could proceed rapidly if given the go-ahead of the President of the United States. A quick return could have utilized proven technologies and the National Space Transportation System (NSTS) for the early development of a lunar outpost.[18] Transfer vehicles and surface systems would have been developed so that the payload bay of the Shuttle could have been used for transport. The lunar outpost structure would have been easily broken into parts, separating radiation protection from module support, allowing easy access, installation, and removal of the elements attached to the Shuttle trusses.

Figure 4.5. A 1963 Boeing Concept. The LESA (Lunar Exploration System for Apollo) initial concept was designed to accommodate six people for six months. It would hold 46,000 pounds of payload, a 10 kW nuclear reactor, a 3765 lb. rover, and equipment to move regolith for shielding use. (Courtesy Lunar and Planetary Institute)

Preliminary designs of permanently manned lunar surface research facilities were also a focus, with criteria for the base design to include scientific objectives as well as the transportation requirements to establish and support its continued operations.[19]

Figure 4.5 shows us how relatively simple the first lunar settlements may be, although there are many challenges to place even such simple structures on the Moon. We see regolith shielding placed atop the cylinder on the right side of the figure. In this concept, regolith also flows down the inside of the structure, in between the inner and outer walls, providing lateral shielding.

Figures 4.6 and 4.7 depict a pressurized module, and access, that forms the backbone of the more ambitious settlement seen in two perspectives in Figures 4.8 and 4.9, where multiple cylindrical structures are placed on their sides and fixed to a truss framework. These figures depict an intermediate stage of construction where regolith shielding is being placed on the framework, which is covered by a corrugated material. Eventually, the complete network of cylinders is covered. Figure 4.10 is a detailed sketch of the hatch used for access to, and egress from, the cylinders.

The following three studies discussed concepts for lunar bases by the Spacecraft Analysis Branch of the NASA Langley Research Center, examining the potential application of the Space Transportation System (STS) external oxygen tank as a lunar habitat. Concepts for the basic lunar habitat used the liquid oxygen tank portions of the Space

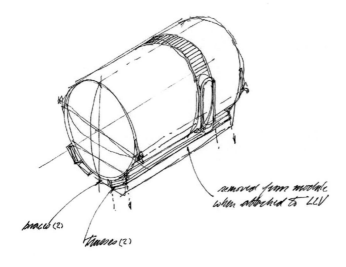

Pressurized Module delivered to lunar surface with building materials attached

braces (2)

trusses (2)

removed from module when attached to LLV

BG 12-87 ©

Figure 4.6. Pressurized module delivered to the lunar surface with building materials attached. (Courtesy Brand Griffin)

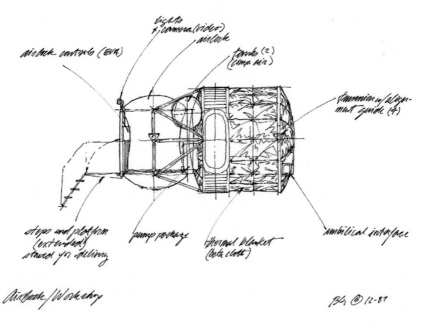

airlock controls (EVA)

lights + camera (video)

airlock

tanks (2) (comp air)

extension w/ alignment guide (4)

steps and platform (extended) stowed for delivery

pump package

thermal blanket (beta cloth)

umbilical interface

Airlock / Workshop

BG © 12-87

Figure 4.7. Airlock and attached workshop. Steps are shown in an extended position on the left. (Courtesy Brand Griffin)

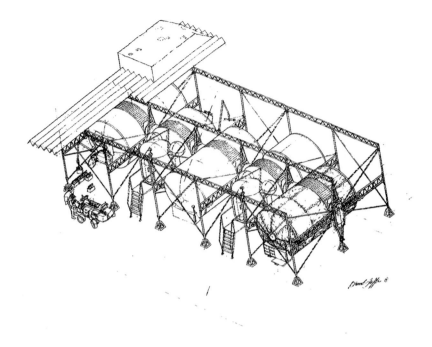

Figure 4.8. This sketch shows an isometric view from above of how a number of structural cylinders, on their sides, are connected by way of a framework and then covered with a corrugated material. Regolith shielding is then placed on the cover. (Courtesy Brand Griffin)

Shuttle external tank assembly left in orbit, thus saving launch costs.[20] [21] The modifications of the tank, to take place in low Earth orbit, included the installation of living quarters, instrumentation, air locks, life support systems, and environmental control systems. The habitat would then be transported to the Moon for a soft landing. Unfortunately, this never happened, as the Shuttle was retired.

"Based upon these previous efforts, the JSC Planet Surface Systems Office requested the Langley Research Center to study alternate concepts for surface habitats and to recommend one or more habitats as candidates for more detailed study. ... The design philosophy that will guide the design of early lunar habitats will be based on a compromise between the desired capabilities of the base and the economics of its development and implantation. [The] preferred design will be simple, make use of existing technologies, require the least amount of lunar surface preparation, and minimize crew activity. Three concepts for an initial habitat supporting a crew of four for 28 to 30 days are proposed. Two of these are based on using Space Station Freedom structural elements modified for use in a lunar gravity environment. A third concept is proposed that is based on an earlier technology, based on expandable modules. The expandable offers significant advantages in launch mass and packaged volume reductions.

Figure 4.9. Here we see a ground-level view of cylindrical habitats being placed in a structural framework, as well as the initial placement of the corrugated material upon which regolith shielding would be subsequently placed. (Courtesy Brand Griffin)

"It appears feasible to design a transport spacecraft-lander that, once landed, can serve as a habitat and a stand-off for supporting a regolith environmental shield.

"A permanent lunar base habitat supporting a crew of twelve for an indefinite period can be evolved by using multiple initial habitats. There appears to be no compelling need for an entirely different structure of larger volume and increased complexity of implantation."[22]

The design philosophy was to be as simple as possible, utilizing as much existing technology as possible. No grandiose design concepts that would require extensive new technologies, completely new launch systems and massive logistics operations. "With the human crew in a hostile environment, far from help, and encumbered by the limitations of long periods of time in pressure suits, the implantation of the habitat must be as simple as possible. The implantation must involve the minimum amount of surface preparation, consume the least amount of time until becoming operational, require the fewest number of specialized equipment units, and minimize strenuous efforts by crewmen. This approach may not be quite as valid for the implantation of the permanent habitat since the crew will have access to a safe, operational initial habitat."

This concept is shown in Figure 4.11, where the structure is braced by a framework and covered by regolith, with the interior work space pressurized to 10 psia (68.95 kPa).

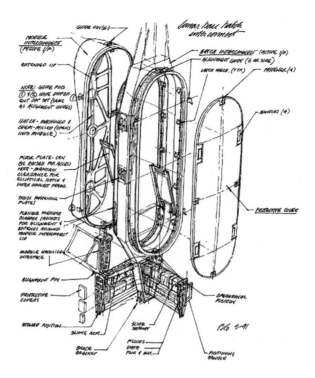

Figure 4.10.   Lunar base hatch interconnect, along with the structural and other attachments. (Courtesy Brand Griffin)

The roof structure is supported on columns at the four corners. This structure can be partially buried or placed within a crater.

We can understand the desire and need to utilize existing technologies, as much as possible, in the creation of the next structure. This pattern enhances reliability and assures us that by using small increments in technology and processes, the possibility of large deviations in behavior from earlier systems is less likely. As the space and lunar infrastructure is developed over a series of decades, all the elements and components will also be used, reused, recycled, and used again.

A sub-class of rigid lunar structural concepts is the mobile planetary base:[23]

"The idea of the mobile base addresses several key challenges for extraterrestrial surface bases. These challenges include moving the landed assets a safe distance away from the landing zone; deploying and assembling the base remotely by automation and robotics; moving the base from one location of scientific or technical interest to another; and providing sufficient redundancy, reliability and safety for crew roving expeditions. The objective of the mobile base is to make the best use of the landed resources by moving them to where they will be most useful to support the crew, carry out exploration and conduct research."

Of particular note is the Mankins Habot (Habitat Robot) concept, depicted in Figure 4.12. The Habot would land autonomously on the Moon, and then move away from

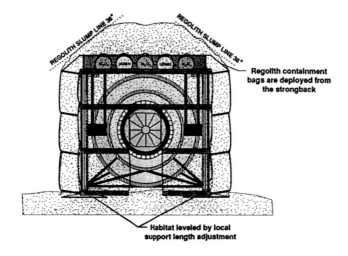

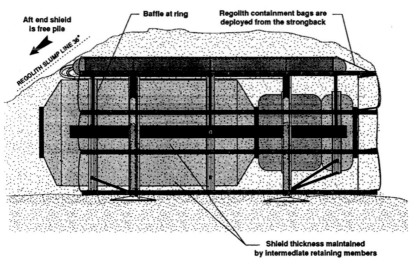

Figure 4.11. Habitat with regolith shield from Hypes *et al.*, designed with a 1.5 factor of safety. (Courtesy: NASA)

the landing site so that more Habots can land. Once a critical number have landed, they would then couple to form a base that pressurizes, awaiting the arrival of the first lunar expedition. In this concept, once the crew completed its tasks at this initial location, they could depart for Earth in another dedicated vehicle, or stay with the Habots as they decoupled and moved to another site of interest, repeating the initial deployment. If the original crew departed for Earth, then a second crew would be sent to the new Habot location. Individual Habot units could be utilized as pressurized rovers for local exploration.

Figure 4.12. Pat Rawlings rendering of the Habot mobile lunar base concept, proposed by J. Mankins and substantiated by M. Cohen. (Courtesy NASA)

The suggested crew size was in the 4–8 range, occupying several of the Habots, noting that three crew members would be needed to operate and maintain the units, and the remaining crew members would focus on exploration, science, and engineering. The Habots would not all be identical, and one of the design challenges would be to distribute capabilities between the units in order to add depth, breadth, and robustness to the colony's cluster.

In a related study, Cohen proposed and used a modeling method, the Habitat Multivariate Design Model, to estimate the volume, size, shape, and configuration required for the design of a habitat.[24] In an example, a single cylindrical module was used to demonstrate the relations between the independent variable, the diameter, and the dependent variables, including the number of pressure ports, the floor area, the height of the end dome, the height of the cylindrical portion of the module, the number of floor decks, the floor-to-floor height, and the volume of vertical circulation. The goal of this study was to develop abstract programming expressions from concrete architectural properties.

A preliminary design study by Grandl for a modular lunar base built of at least six cylindrical modules, as shown in Figure 4.13, was proposed for launch by an ARIANE Rocket with a payload of 12 tons.[25] To land the modules on the Moon, a design was proposed for a Teleoperated Rocket Crane that would be assembled in lunar orbit. The modules would be made of aluminum sheets, using a double-shell structure to protect a crew of eight astronauts from radiation, micrometeorites, heat, and low temperatures during the lunar night. Lunar regolith would be used for shielding. Each cylinder would be 17 m in

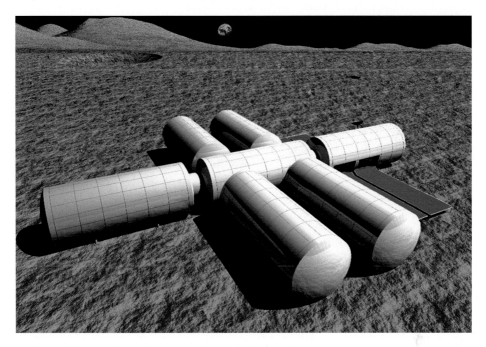

Figure 4.13.   Double-shell concept. (Courtesy W. Grandl)

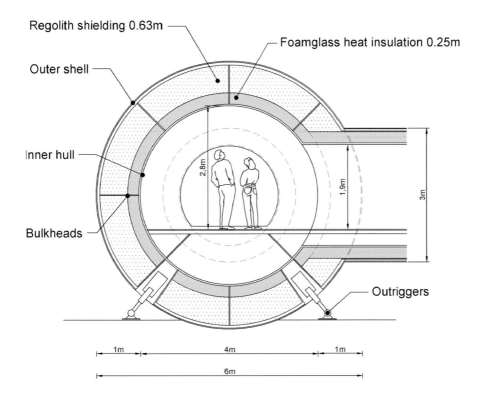

Figure 4.14.   Cross-section of double-shell habitat. (Courtesy W. Grandl)

length and 6 m in diameter, with an inner diameter of 4 m, as depicted in Figure 4.14. The outer wall of each cylinder would be a double-shell system stiffened by radial bulkheads. Within the double shell would be 0.25 m foamglass heat insulation and 0.65 m of regolith shielding, as shown in Figure 4.15. Some of these concepts have been advanced recently for use in the far reaches of the Solar System.[26]

While rigid structures are still likely to be the first and the early lunar surface habitats, we expect that rigidized inflatables and ISRU-derived, layered, manufactured structures are increasingly likely early-on.

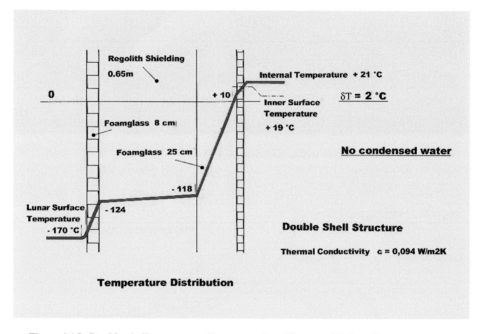

Figure 4.15   Double-shell structure wall cross-section. (Courtesy W. Grandl)

4.6   INFLATABLES

"Less volume and mass to carry,
and much more volume at the end."

There have been a number of technology reviews of inflatable and deployable structures over the past two decades, as that technology has advanced to the point where there is confidence for its use in extreme environments, including the lunar surface. While the terms 'inflatable' and 'deployable' do not have a standard use in the literature, it appears preferable to use 'inflatable' for soft and foldable materials that are balloon-like, and 'deployable' for rigid but storable components that are mostly mechanisms. Such structures have had limited use on Earth and in LEO. Our goal is to extend and utilize that expertise for similar structures on the Moon. It is a reasonable expectation that many of the technologies, procedures and processes will translate from use on Earth and in space to use on the Moon, with the development of suitable transportation and deployment technologies.

The design of inflatable structures can be more of a challenge than the design of rigid structures for two reasons. The first is that inflatables are made of more exotic materials that are relatively new to us when compared to the metals used in rigid structures. We understand them less well since our experience base is much shorter. The second reason is that the behavior of inflatable structures is nonlinear. They deploy from a folded and small volume into a large volume, in a nonlinear geometrical motion undergoing large deflections. Such motions are difficult to model mathematically. After deployment, these structures are rigidized, in different possible ways, where the inflated structure is subjected to various forms of energy or where the structure is pushed through its yield point. These are all complex processes that are understood through a mix of experiments and fundamental materials theories. There is no doubt that much has been learned about inflatables technology and science in the last several decades, but much remains to be understood before the design process is at the level of that for rigid structures.

Various works have identified the classes of lunar structural developments. A possible classification, where there is an early deployment of inflatables, can be along the lines: rigid, then inflatable, and finally in-situ resourced structures.

Experience with inflatable space structures includes space-based reflectors and antennas.[27] Low weight and ease of packaging are identified as their primary attractions. The experience in space is likely to be different to what we will see on an extra-planetary body. In space, the following advantages of inflatables over rigid structures are identified:

- they have a 50 percent weight advantage since very thin materials are used (0.25 to 10 mil)
- they have a 25 percent packaging advantage
- the inflatable is inherently strong since loads are resisted over large surface areas
- they offer reduced costs up to a factor of ten for structures such as space antennas
- they can be inflated repeatedly
- they may be up to 50 percent less expensive than other deployables
- they can be inflated to a variety of shapes
- they have favorable dynamics regarding resonances, since motion is not harmonic under the constant internal pressures, and
- they offer favorable thermal behavior given the ability of the large surfaces to reject heat.

Deployable space structures for astrophysics applications also have a significant base of experience.[28] They have sometimes been challenging to deploy since the kinematics are generally complex. There are hopes to deploy space habitats for habitation.[29]

Haeuplik-Meusburger and Ozdemir have provided an overview of deployable lunar habitat structures and space suits.[30] The concepts discussed included those by NASA and the Jet Propulsion Laboratory, the Russian Space Agency, Goodyear, ILC Dover, L'Garde, ESA, and others. It was noted that multi-layer deployable structures can offer enhanced radiation protection, when compared to traditional rigid structures that have secondary radiation effects from their metallic components.

Recent research on flexible composites opens the door to expandable space and lunar structures:[31]

"One approach that has been under development for many years is to utilize soft-goods – woven fabric for straps, cloth, and with appropriate coatings, bladders – to provide this expandable pressure vessel capability. The mechanics of woven structures is complicated by a response that is nonlinear and often nonrepeatable due to the discrete nature of the woven fiber architecture. This complexity reduces engineering confidence to reliably design and certify these structures, which increases costs due to increased requirements for system testing. [This] study explores flexible composite materials systems as an alternative to the heritage softgoods approach. Materials were obtained from vendors who utilize flexible composites for non-aerospace products to determine some initial physical and mechanical properties of the materials. Uniaxial mechanical testing was performed to obtain the stress-strain response of the flexible composites and the failure behavior. A failure criterion was developed from the data, and a space habitat application was used to provide an estimate of the relative performance of flexible composites compared to the heritage softgoods approach. Initial results are promising, with a 25 percent mass saving estimated for the flexible composite solution."

Deployable textile hybrid structures may also see lunar applications.[32] It is feasible to envision using deployable flexible composites to create habitats in regions on the lunar surface where the topology is non-planar, thus saving the effort of clearing the site.

As part of a lunar deployable greenhouse thesis design study, Vrakking provided an extensive overview of the lunar environment, a systems analysis for the lunar greenhouse module, potential structural concepts, and a preliminary design.[33] A state-of-the-art overview was given on inflatable structures technology: packaging, deployment, rigidization, materials, integrated electronics and modeling.

Packaging is accomplished by a series of rolling and folding operations, with the goal of minimizing volume and to ensure optimal deployment. Structural symmetry needs to be maintained in the packed state. Accordion folds facilitate a fast inflation and spherical surfaces should be folded along lines of longitude, avoiding orthogonal folds that result in high stress concentrations.

Deployment can be either free (structure is free to move in space once deployed even if not yet inflated), or controlled (only fully inflated segments are allowed to move freely in space).

Rigidization (following Cadogan's classification, discussed subsequently) depends on material properties that can be classified as thermosetting composite materials, thermoplastic (and lightly cross-linked thermoset) composite materials, or aluminum/polymer laminates. Cross-linking is a technical term for hardening the matrix resin. Thermoplastic composites can be reset into a new shape by reheating them above their forming temperatures. There are thermal rigidization, chemical rigidization, solvent boil-off rigidization, foam rigidization, and work-hardening rigidization.

No single material fulfills all the requirements of an inflatable structure or its environment. As in TransHab, discussed below, multi-layers of materials are designed so that, as a complete structure, they resist the harsh space and lunar environment. Materials design also integrates self-healing processes, of which there are a number of possibilities. Self-healing implies that specialized pockets or layers of materials are embedded in the structural layers, or shells, with properties that are enacted if a particular damage state occurs or is sensed.

An ability to embed self-healing components within the structural layers implies an ability to embed electronics as well. Integrated electronics offer opportunities to generate power and develop active control forces. The complexity of embedding self-healing components and electronics challenges system reliability. Complexity and reliability can be a zero-sum game, but this does not mean that we avoid complexity. Rather, designers need to balance that complexity carefully in a suboptimal way, recognizing that there is likely a point of diminishing returns.

Mathematical and computational modeling is required of the stowed, deploying, and deployed structure. Engineering (based on first principles) and phenomenological (based on an experimental dataset) models are needed to represent the mechanisms, dynamics, and material behavior, as well as the inflation mechanism; for example, the inflation gas flow into the packaged structure.

Early Concepts

An inflatable lunar habitat proposed by Roberts, with a discussion on the many architectural and structural design issues, is depicted in Figures 4.16 and 4.17, in artist renderings with a cutaway that shows the interior structure and purposing, as well as the habitat entrance.[34]

This structure has the advantages of inflatable structures in general. Transportation costs are reduced, and the size of the volume created on the Moon would be significantly larger than a rigid structure transported at a comparable cost. Operations and growth are more easily accomplished within the large volume of the inflatable. Roberts summarized the advantages of the inflatable concept:

"The greatest advantage of large inflatable habitats may be the most difficult to quantify: habitability. Habitability is the sum total of those qualities that make an environment a pleasant place to live and a productive place to work. Studies have shown that personal space, for work and for leisure, is an important factor in the psychological wellbeing of isolated groups. The inflatable would not only provide a large volume, it would provide *perceptible* volume. A base built up out of rigid modules might be expanded indefinitely by adding more modules, but no single element could ever be larger than the payload bay of the launch vehicle that lifted it from Earth. At no point in the lunar complex could a person perceive (or utilize) a volume larger than that of a single module. The space inside the inflatable, on the other hand, may be divided up in any way desired or left open to create large chambers."

The open concept provides a large amount of volume without a large impact on the transportation system. Inflatables can be designed to deploy in almost any shape, but the spherical shape is the most volumetrically efficient.

The chief structural considerations are the membrane stresses, leakage (which may be unavoidable over time), and puncture resistance. Roberts suggested cables and hoops in order to reduce the membrane stresses and reduce local radius of curvature. While the inflatable material may be impermeable to air, connections to hatches and doors are

Figure 4.16. A 16 m diameter inflatable habitat is depicted and could accommodate the needs of a dozen astronauts living and working on the surface of the Moon. Depicted are astronauts exercising, a base operations center, a pressurized lunar rover, a small clean room, a fully equipped life sciences lab, a lunar lander, selenological work, hydroponic gardens, a wardroom, private crew quarters, dust-removing devices for lunar surface work, and an airlock. The top level shows joggers required to run with their bodies almost parallel to the floor as a result of the low gravity. This artist's concept reflects the evaluation and study at Johnson Space Center by the Man Systems Division and Johnson Engineering personnel. S89-20084, July 1989. (Courtesy G. Kitmacher, NASA)

susceptible to leakage. If the inflatable is designed to deploy and remain flexible, then an internal structure is needed in case of a catastrophic collapse.

Figure 4.18 is a schematic of this structure that labels the interior levels, of which there are five. Some of the level descriptions are different to what was shown in the cutaway illustration of Figure 4.16. For example, rather than locating the crew quarters at the top level as in the schematic, this illustration shows the quarters at a level located below ground, where the shielding is more substantial. The schematic shows the habitat wall, shielding, and interior structure. Level one and zero are below ground level.

This inflated sphere design is 16 m in diameter, containing 2145 m^3 of open volume. The internal pressure is assumed to be 101.4 kPa (standard sea-level atmospheric pressure on Earth). The inflatable envelope is made of Kevlar-29, a high-strength aramid fiber with a

Figure 4.17. As much dust as possible must be removed before re-entering the habitat. The astronauts might pass through wickets (far left) that remove much of the dust. A perforated metal porch would allow dust to fall through. Once inside the dust lock (center) the astronauts would remove their white coveralls. This outer garment would provide an extra layer of dust control and protection for the precision moving joints of the space suit from gritty dust. An air shower could remove remaining dust with strong jets of air. An astronaut at right, after having removed as much dust as possible, would be able then to move into the airlock to don his suit. The airlock could accommodate up to four astronauts at one time. Suits could be stored there when not in use. This artist's concept reflects the evaluation and study at the Johnson Space Center by the Man Systems Division and Johnson Engineering personnel. S89-20088, July 1989. (Courtesy G. Kitmacher, NASA)

thickness of 0.114 mm and a breaking strength of 525 N/cm. The envelope mass is approximately 2200 kg, with a structural safety factor of 5. A structural rib placed under the envelope is estimated at 5 mm, giving a material volume of 4 m^3 Assuming a 10:1 ratio of packaged volume to material volume, the packaged volume of the inflatable will be 40 m^3.

The habitat is assumed to be covered with 3 m of lunar regolith contained in sandbags. The regolith imposes a maximum load of 7.8 kPa on the surface of the habitat, easily supported by the internal pressure of 101.4 kPa. However, placing the regolith is very labor intensive, taking up much of the crew's time during the early missions. An alternative might be to provide a few inches of shielding to protect the habitat from galactic cosmic

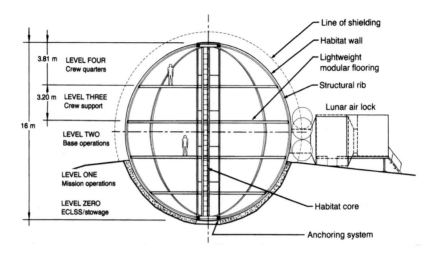

Figure 4.18. A schematic of the interior. The inflatable habitat, as currently envisioned, consists of a spherical pneumatic envelope with an interior structural cage to support the floors, walls and equipment, and to hold up the envelope if pressure is lost. (Courtesy NASA)

radiation, with a separate, smaller, storm shelter to protect the crew in the event of a serious solar flare.

Issues of concern include:

- the sensitivity of the inflatable material to radiation exposure
- abrasion between the habitat material and the shielding regolith caused by the relative motion between the two, due to activity within the structure and outside, and
- exposure of the inflatable material to extreme temperature and vacuum conditions during transport and deployment.

A follow-on preliminary design of the interior structure of this inflatable concept utilized high-strength structural aluminum 2219 for the interior framing and exterior support structure.[35] The aluminum frame was sized and designed based on the AISC design code.[36] The foundation mat was designed with a factor of safety of 4 against the yield strength of the aluminum and the ultimate bearing capacity of the lunar soil. In this design, as mentioned above, the crew quarters are placed beneath ground level. The interior aluminum framing includes cross-bracing between floors.

Another inflatable concept, the pillow-shaped inflatable structure shown in Figure 4.19, is an early concept for a permanent lunar base.[37–39] The proposed base consists of quilted inflatable pressurized tensile structures using fiber composites. Shielding is provided by an overburden of regolith, with accommodation for sunlight ingress. This structure is composed of several basic modules, each consisting of a roof and floor membranes, four columns and footings, four arched ribs, and external wall membranes. Each module is 6.1 x 6.1 x 3.0 m (20.0 x 20.0 x 10.0 ft. length by width by height), with the radius of curvature of the roof membrane at 6.1 m (20.0 ft.). This modular approach includes a minimal number of structural components to facilitate manufacturing, expandability through any of the exterior wall membranes, modularity, and a low ratio of volume to usable floor space. An internal pressure of 69 kPa (10 psi) is selected. This corresponds to an equivalent Earth elevation of approximately 3050 m (10,000 ft.). Kevlar 49, with an allowable tensile stress of 690 MPa (100 ksi), is the material chosen.

Figure 4.19. A pillow-shaped inflatable structure. This large grid of inflatable structures is partially covered by regolith shielding, with light being directed inside via mirrors. (Courtesy P.S. Novak)

Inflatable structural concepts for a lunar base have been considered as a means to simplify and speed up the process of creating a habitable lunar facility while lessening transport costs.[40] The inflatable structure is suggested as a generic test bed structure for a variety of application needs for the Moon.[41] Design criteria have been developed, since few structural engineers have experience with inflatable structures.[42] Because of the possibility of deflation during construction, due to accident or meteorite strike, temporary columns can be used to support the inflated structure. In some concepts, the supporting columns can be removed if the structure is made rigid subsequent to inflation.

Inflatable concepts are also attractive because they can be rolled up or folded into a small volume and easily transported to the surface of the Moon. There, the 'balloon' structure is inflated at its final location – the surface has been smoothed out and otherwise made ready for the structure – and once fully inflated, various ways are possible to make the structure rigid. One way is by injecting foams that harden with time within the interstices and within the space between the layers of the inflatable structure. Once the inflated structure is fully pressurized and rigidized, a structural framework can be erected inside for use by inhabitants.

The deployment of the inflatable structure requires the design of a foundation, and reliability concerns are significant for such structures.[43] While the concept of folding a deflated membrane is simple, the practicalities are major. If there is too much pressure on any fold for too long a period of time, that region may be weakened. If the folded structure shifts, abrasion may occur. Such concerns lead to studies to minimize the risks of damage during the packaging of the structure for shipment and during its inflation.

An example of a more elaborate pressurized membrane structure for a permanent lunar base is shown in Figure 4.20.[44] (The concept has been patented. See Figure 4.21.) It is

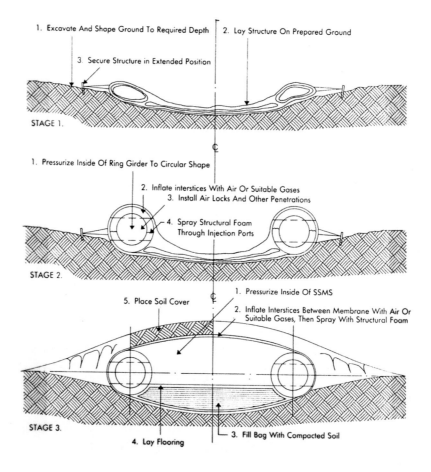

Figure 4.20. This inflatable membrane structure concept is shown from the time it is placed on a prepared site on the lunar surface, inflated, and then rigidized in its final form. This concept was also patented by Phil Chow and T.Y. Lin International; U.S. Patent 5058330. (Courtesy T.Y. Lin International)

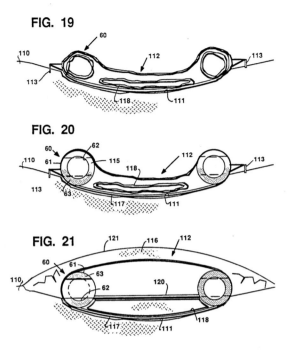

Figure 4.21. Chow U.S. Patent 5058330.

constructed of a double-skin membrane filled with structural foam. A pressurized torus-shaped substructure provides edge support and shielding is provided by an overburden of regolith. The deployment requires shaping the ground and spreading the uninflated structure upon it, after which the torus-shaped substructure is pressurized. Structural foam that hardens quickly is then injected into the inflatable components, and the internal compartment is pressurized. The bottom of the inflated structure is filled with compacted soil to provide stability and a flat interior floor surface.

Figure 4.22 shows the torus-shaped compression ring, having an outside diameter of 60 ft. (18.3 m). The compression ring contains the structure, and it is the fundamental structural component, establishing structural height among other characteristics. The Pressurized, Self-Supporting Membrane Structure (PSSMS), as shown, will enclose a net usable volume of about 20,000 ft.^{3} (566.3 m^3). The 12 in. (30.5 cm) thick double-skin membrane walls forming the roof and the floor are 12 ft. (3.66 m) apart at the edges and 20 ft. (6.1 m) at the center. The membranes are rigidized by the injection of structural foam after the PSSMS is inflated.

Figure 4.23 depicts the PSSMS partitioned to provide four-apartment residence units. The apartments open to the circular corridor within the ring beam that leads to the entrance. The entrance consists of a separate structure that accommodates an air lock and a storage area for the space suits.

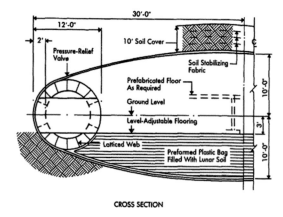

CROSS SECTION

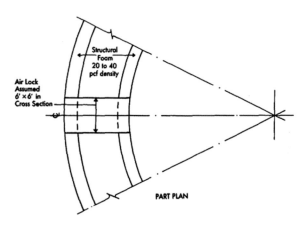

PART PLAN

Figure 4.22. Typical 60 ft. (18.5 m) diameter PSSMS. (Courtesy T.Y. Lin International)

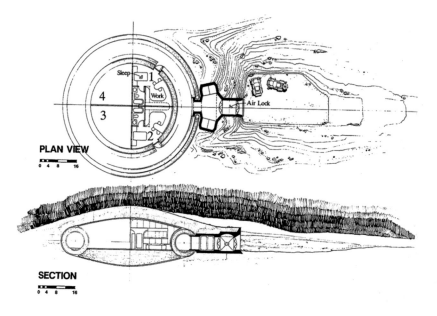

PLAN VIEW

SECTION

Figure 4.23. Layout of Four-Apartment PSSMS. (Courtesy T.Y. Lin International)

Moore *et al.* proposed two inflatable concepts, one based on Chow and Lin, and the other based on the work of Buckminster Fuller.[45] This conceptual study suggested possible interior configurations for the Chow and Lin structure.

TransHab

TransHab is a unique hybrid structure that has an inflatable shell surrounding a central hard structure core.[46] [47] It has the packaging and mass efficiencies of an inflatable structure, and the advantages of a load-carrying hard structure. It is viewed as a first step to creating habitable structures in space and on the Moon. Hybrid structures such as TransHab combine light weight with high strength, are reliable and repairable, perhaps self-healing, and provide protection from the radiation and micrometeoroids of space and potentially on extra-planetary surfaces.

The central core hard structure observed in Figure 4.24 is comprised of longerons, launch shelves, bulkheads, radiation shield water tanks, utility chases, and integrated duct-work. The inflation system spherical tanks can be seen below Level 1.

The multi-layer shell is TransHab's primary structure that is folded and compressed around the core at launch. It is deployed once the structure is in orbit. The inflatable volume contains the crew quarters, and provides orbital debris protection and thermal insulation. Figure 4.25 identifies the multi-layer system. The Micrometeoroid Orbital Debris

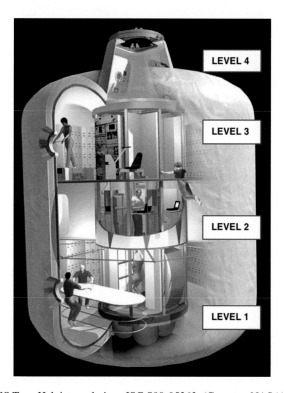

Figure 4.24. ISS TransHab internal view. JSC S99-05363. (Courtesy NASA)

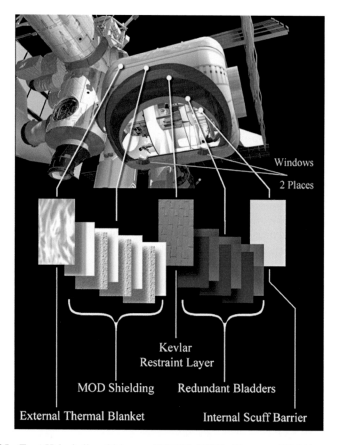

Figure 4.25. TransHab shell multi-layers. JSC S99-05362. (Courtesy NASA)

(MMOD) protection shielding is composed of multiple layers that can absorb particles traveling at hypervelocities. The Kevlar restraint layers are designed to contain up to four atmospheres of air pressure. The bladders and inner barrier provide fire retardance and abrasion protection, respectively.

Research on TransHab was abandoned by NASA, but licensing was picked up by Bigelow Aerospace, which placed a redesigned TransHab into orbit and attached it to the ISS on April 10, 2016. It was inflated and pressurized on May 28, and will spend two years in orbit being tested.

Three important goals for the TransHab were set by NASA, and then Bigelow Aerospace:[48]

- Determine how to protect an inflatable structure from being ruptured by micro-meteoroid or orbital debris.
- Show that TransHab can be folded, packaged and then deployed in space.

• Demonstrate that a large fabric inflatable structure can be pressurized to one atmosphere in the space vacuum.

One can envision a structure similar to TransHab deployed to the Moon, landing on and attaching itself to an anchoring base that houses lunar surface access ports via the central hard core. The central hard core would support the structure, though it may have to be strengthened to support the structure in the 1/6 g lunar gravity field. The inflatable structure may also require design enhancements due to the added gravity beyond that experienced when the TransHab was docked at the ISS, but otherwise, the general design is very adaptable to the lunar surface, insofar as shielding capabilities and the requisite volume needed for comfortable habitation are concerned. TransHab was designed to house a crew of six and, if rotations are prescribed, it can house a crew of 12 to accommodate the overlap during rotations.

Inflatable and Rigidizable

A NASA structures and mechanisms study has been carried out to explore the development of lightweight structures and low temperature mechanisms for lunar and Mars missions.[49] The goals of the study were to significantly improve structural systems options for man-rated pressurized structures by:

• lowering mass and/or improving efficient volume for reduced launch costs
• improving performance to reduce risk and extend life, and
• improving manufacturing and processing to reduce costs.

The focus of the study was the primary structure of the pressurized elements of the lunar lander for both sortie and outpost missions, and surface habitats for the outpost missions. Concepts for habitats that support a minimum six-month lunar outpost mission were considered and both rigid and flexible habitat wall systems were discussed. There were a number of important observations. Technical challenges currently exist for materials, bearings, lubricants, sensors, actuators and motors, and thermal control systems that can operate on the Moon, where temperatures are over 100°C colder than current lower design limits. These components are the backbone for lunar surface rovers, robotics, and other mechanized operations, such as those envisioned for ISRU machinery, cranes, deployment systems and airlocks.

The study offered a number of conclusions regarding structural options:

"[The] most significant concern for rigid habitat composite systems is the hermeticity of the skins with loading and non-visible damage. For the metals, it is the formability, weldability and maintenance of properties in the desired product forms. ... [Furthermore,] inflatable habitat materials systems have not shown appreciable mass savings to date due to the large number of layers needed for hermeticity, durability, thermal management, GCR and SPE absorption and MMOD protection. While flexibility of the materials allows the structure to be packaged into a small volume for launch, advances in materials and load path concepts are needed to reduce the weight of these flexible material systems. Advanced flexible polymers and high strength fiber reinforcement should enable lower mass structures."

Materials research progresses at a very rapid rate, however, and many options can be foreseen for the use of inflatable structures as habitats.

Inflatable structures have a long history of use on Earth.[50] The challenges they pose to engineering analysts and designers originate from their complex topology (shapes that are mathematically complicated to model), and that their behavior is nonlinear (the equations that define structural shape and deformation are nonlinear, requiring either simplified modeling or more realistic computational modeling). Yet the advantages of such structures, as outlined above, stipulate their use for challenging environments such as space or an extra-planetary surface. Gruber et al. [51] investigated the possibilities of human-inspired concepts for deployable structures.

Uncertainty quantification underlies all engineering activities, and none more so than aerospace and space engineering. We devote more discussion to such issues in a later chapter, but mention that inflatable and rigidizable structures are prime candidates for such studies.[52] Sources of uncertainties include changes in temperature and loading conditions and magnitudes, and the effects of random measurement noise in electronics and control systems for deployment, inflation, and rigidization.

Cadogan et al. provided an overview of how recent composite materials technology advances have promoted the use of inflatable, composite structures for a variety of space applications.[53] [54] Several concepts were reviewed: the Livermore Habitat Module; the NASA lunar inflatable, deployable, cylindrical structure with rigid end caps; and the NASA JSC TransHab concept that was originally viewed as a potential transfer vehicle to Mars, or as a habitat module for the Space Station (a concept brought to fruition by Bigelow Aerospace).

A comparison was made between all-flexible structural modules and flexible modules with fixed end plates. The benefits of the hybrid inflatable with rigid end plates include: "lower mass, simplified attachment of equipment to floor, flexible assemblies are simplified for cylindrical modules, pressure compartmentalization is simplified with rigid end-points, lower permeation rates with less flexible material area." For the all-flexible module, benefits include: "greater usable volume, lower leakage (less hardware/soft-goods transition length), lower cost, better impact tolerance, simplified packing and deployment."

Inflatable, rigidizable orbital structures designed to gather and transfer space solar power have also been conceptually studied.[55] A rigidizable material is one that is initially flexible for purposes of transport and then can be rigidized by applying a thermal or radiation load, for example. Case studies for space solar power have explored various design options; they are modeled essentially as large beams with design constraints such as buckling loads, stiffness bounds, and natural frequency and resonance conditions. The beam cross-sections are varied (bundled tubes, trusses, tapered tubes), and interestingly, most designs are driven by structural natural frequencies and long column compression, rather than bending. The structures studied are deployed as inflatables, and then rigidized using any of a variety of technologies:

- using a packaged thin aluminum laminate that is deployed by inflation, over-pressurized to remove any wrinkles, then evacuated to leave a smooth, stiff shell structure

- sub-Tg composite rigidization, referring to cooling the material below the glass transition temperature (Tg) of the material so that the matrix becomes hard, thus rigidizing the composite, where Tg can be varied depending on the application
- thermoset rigidization of composites that occurs by the application of heat, generally requiring much power
- UV-cured composites that can be sourced from the Sun, or lamps, and require an order of magnitude less power than thermosets for rigidization, and
- a water-based matrix composite, known as hydrogel, that can be rigidized after deployment by allowing the water to evaporate into the vacuum of space, leaving behind a rigid composite.

While these are applied to orbital space structures, many of the same ideas and technologies can be applied to a surface lunar structure. A key technology requirement to enable the use of inflatable structures in space, and for the Moon, is that the constituent materials must be both foldable and rigidizable. This technology has been developing rapidly during the past couple of decades. An extensive set of materials and techniques are becoming available for such inflate-rigidize procedures.[56] [57] Numerous concurrent structural, environmental, manufacturing and testing requirements must be met for the design of such materials. For example, a rapid and predictable rigidization process reduces the probability of micrometeoroid penetration while the structure is still soft. Also of importance is the minimization of performance degradation with thermal aging and cycling, and with radiation exposure over time.

Non-traditional structural materials, combined with the need for safety in a harsh environment, led to the consideration of an integrated structural health management system for future habitats, to ensure their integrity.[58] Prototype sensing technologies and self-healing materials address the unique requirements of habitats comprised mainly of softgoods. "Addressable flexible capacitive sensing elements and thin film electronics in a matrixed array detect impact damage that is repaired by micro-encapsulated self-healing elastomers. Passive wireless sensor tags are used for distributed sensing, eliminating the need for on-board power through batteries or hard-wired interconnects." Such structures could be deployed autonomously, prior to the arrival of astronauts. A question to address is whether such systems need to be shielded against lunar surface radiation levels.

Embedding adaptive capabilities into a structure, to respond autonomously to information provided by integrated sensing technologies, offers possibilities for self-repair. Ideally, these can increase crew safety and reduce maintenance time. "Some of the capabilities under consideration include coatings which can vary their emissivity signatures to control temperature, and actuators based on piezoelectric materials, shape memory alloys or magnetostrictive materials for modifying and maintaining the proper shape of the habitat, following deployment." Impact damage due to micrometeoroids are prime candidates for self-repair. Also, once the structure is covered by regolith, access to damaged locations may be nearly impossible, and self-healing may be the only option short of excavation.

While it is reasonable to expect that such capabilities will be developed for lunar structural systems, it is important to note that their complexity can lower system reliability unless great care is taken by designers. Similarly, designing redundancies into the system can also reduce reliability. We see this issue in aircraft structures that have numerous active redundant elements. There is an inherent trade-off between automated safeguards and reliability that must be understood. Furthermore, the inclusion of sensors within structures alters their mechanical behavior, and introduces stress concentrations at those locations.

4.7 INTERVIEW WITH DAVID CADOGAN

This interview with David Cadogan was conducted via email beginning on June 23, 2017 and concluded on July 7, 2017. We begin with a brief professional biography, and then continue with the interview.

Brief biography

Dave Cadogan is Director of Engineering & Product Development at ILC Dover, leading one of the most dedicated and capable softgoods development teams in the world. (Softgoods are also known as textile and film-based structures, and flexible composites.) He earned a BSAE from Western Michigan University in 1986. He has been with ILC Dover for over 30 years (since 1986) and has worked on numerous inflatable space structure programs, including space suits (Shuttle, ISS, CSSS, Z series, etc.), space habitats (TransHab, InFlex, Toroidal, Extendable, etc.), Mars Pathfinder & MER landing airbags, Boeing CST-100 capsule landing airbags, reentry aeroshells (IRVE & SIAD), and numerous deployable antennas & solar arrays.

He has actively contributed technical solutions in the form of technology, product and systems architecture development, and has collaborated with many people in government, industry and academia in the development of inflatable and membrane products for aerospace, military and industrial applications.

He has served as a Principle Investigator or Technical Leader for numerous NASA, DoD, DARPA and other government programs, and has authored or contributed to over 80 technical papers / book chapters, and 15 patents.

He enjoys being creative and supporting the creativity of others in solving complex problems through the application of textile and film-based structures. He draws on three decades of experience, bringing innovation to structural design, softgoods technology, and the configuration of systems architecture.

Our interview

Lunar structures have been grouped into three categories: I – habitats brought from Earth, II – habitats with components from Earth, but also components made from lunar resources, and III – habitats primarily made from lunar resources, ISRU, perhaps 3D printed structures. The longer it takes for us to get back to the Moon to stay, the more certain technologies advance, for example 3D printing. Would you keep the above categories, or modify them in some way?

The grouping makes sense to me. This is a good systematic way of linking the cost of development and transportation of equipment to the surface vs habitat volume. This metric, and others like it, can be used to show which approach works best along the timeline of colonization. Metal structures of fixed volume first, then inflatable habitats to house more people, then ISRU equipment to make air and water, then ISRU equipment to make habitats, etc.

You have been at the center of the development of inflatable structures for use on Earth, in extreme environments for example, and for space, in particular inflatable space structures. You have also led studies on the use of inflatable and hybrid inflatable structures as lunar habitats. Where are we with inflatable structures technologies? Is it likely that they will be our first habitats on the Moon?

As much as I think that deployable technology makes sense for exploration, I don't think it will be the first technology used for the next lunar habitats. If anything, it could be a hybrid of rigid and inflatable volumes for various purposes. My experience of people's fear of inflatable structures leads me to believe decision-makers will prevent it from being the primary technology. Inflatable structures are highly engineered, just like metal structures, but the engineering community in general has limited experience with the technology, so it typically isn't their first choice. When I enter these fear-based discussions with people, I often remind them that they entrust their lives to inflatable structures every day in the form of tires, automotive airbags, aircraft escape slides, or even angioplasty devices to name a few instances. It gets them thinking. So, inflatable habitats probably not the first technology used, but definitely soon after because the benefits are considerable (stowed to deployed volume in particular). I could be wrong about being first if Bigelow Aerospace keeps having success. They are doing great work, with Mr. Bigelow's considerable fortune, that has really advanced people's acceptance of the technology through demonstration. Von Braun wrote about inflatable space habitats early in his life, and he seemed to be a pretty sharp guy. Maybe the community will come to accept the technology more rapidly than I think.

If we want to send an inflatable-rigidizable structure to the Moon autonomously and have it self-deploy, inflate and rigidize, can we do that given the current technology level? If not, what critical elements are missing from our engineering knowledge base?

The short answer is yes. Inflatable structures are ready now. Rigidization may require a little more development work depending on the application, but I don't think it is necessary to use rigidization for a habitat. If you need to depressurize the entire internal volume (unlikely), you can hold-up the material with a deployable external structure of poles. Rigidization is a feasible possible approach, but it brings some complexity that might not outweigh its benefits. We aren't quite to the A&B allowables point with rigidization yet as far as predicting performance, but we have a good amount of test data to back-up design work if that design path is chosen.

With inflatable structures, one of the challenges is how to attach the membrane to the various access ports. Is this a problem? Should inflatable habitats in general be coupled to rigid components that would have doors for access outside?

Integrating ports, windows, or hatches to inflatable habitats can be done, and has been done. Don't forget the first spacewalk by Alexi Leonov was conducted from an inflatable airlock where the inflatable materials were joined to metal hatches! However, in some cases it might make sense to integrate the hatch into a solid core to facilitate mass reduction and deployment simplification for risk reduction. Think of it like the old pop-up campers. You have the central body with all the rigid equipment, and soft parts that deploy off that rigid structure. Again, the answer hinges on the concept of operations, or CONOPS, of how the habitat is delivered, deployed, and used. If you are going to depressurize the entire volume with people in it and use it like an airlock, then you would require some means of keeping the fabric in place so as to not lay it on the astronauts. The attachment between rigid and soft materials isn't the problem. Controlling deployment (and possibly dealing with depressurization) is the bigger issue. If you have a core on the lander that has a few inflatable modules with windows or even an airlock in them, extend off that core, then you need to make sure the rigid parts don't drag or damage the softgoods during deployment. It just takes some clever design, like that pop-up camper.[2]

Inflatables can be designed for almost any shape. If you were called on to design an inflatable habitat for the lunar surface, what would its shape be?

This is driven by CONOPS (how the system is delivered, deployed, used, etc.). Are the habitats stationary, mobile, interconnected, etc.? There are a lot of trades to be conducted to determine the most efficient design. Some people favor a torus that inflates from a central hub. This packs around the central hub and lands nicely, but the walls are highly stressed since the flexible part is the major diameter. The configuration for the habitat system is usually given to us (the manufacturer) by NASA or a prime contractor because they are the system architects, but I personally like the idea of cylinders emanating from a central hub. This allows us to separate work areas (dirty, sleep, food, etc.), and include partitions that could allow you to close-off compartments in case they become severely damaged for some reason. Cylinders would also be lower stress and have less risk, but on a weight per volume could be higher than a torus. Spheres are the most efficient mass to volume structure (because the skin stress is pr/2 vs pr in one direction of a cylinder), however they are more difficult to design with (like geodesic domes). This ultimately requires a lot of study regarding mission parameters before the optimal approach is determined. Part of the benefit of inflatable structures that makes them interesting is that they are easily configured to meet the need (you can make them whatever shape you want without too much engineering). They are also impact tolerant and dampen vibrations well.

Are there any bio-inspired concepts that you think can teach us about inflatable structures?

I have examples for inflatable structures in general, but I can't think of anything for habitats except bubbles. Frei Otto made a career out of this (very interesting!).

---

[2]A concept of operations, abbreviated CONOPS, CONOPs, or ConOps, is a document describing the characteristics of a proposed system from the viewpoint of an individual who will use that system. It is used to communicate the quantitative and qualitative system characteristics to all stakeholders.

He studied the intersection of bubbles, which led to an understanding of how some inflatables should be joined. On a separate note, we have used nastic cells (how flowers track the Sun) to alter the shape of inflatable Unmanned Aerial Vehicle wings and space suit torsos via the pressurization of embedded cells in the structure. There are other examples of biomimetics of inflatable structures that make for interesting study, but perhaps not specifically for simple habitat structures.

Do you see a way to create an inflatable structure via in-situ resource utilization?

Interesting question that I have never considered. I struggle with this one making sense. I always thought of ISRU as the way to make concrete structures (and oxygen & water), but not a way to create the building blocks of inflatable structures. I don't believe the Moon would support the manufacture of polymers, but Mars might. The manufacture of fiberglass and ceramic fibers is probably possible in both locations, but without a source of hydrocarbons I don't think the manufacture of the polymers we use in the gas retention layers is feasible.

I know people are doing a lot of work on 3D printing large concrete structures, which might make the most sense in this case. However, I could also see a simpler (perhaps earlier) approach in the build-in-place technology timeline on surface, where inflatables could be used as forms to build with ISRU materials (like reusable formers for cement). This could be an evolutionary path that might make sense.

Given that we have been waiting almost 50 years to return to the Moon, do you have a feeling as to when we might return, with people, for an extended stay?

Our space exploration program needs to be de-politicized. The changes that are imposed on the agency every so often are disruptive. Beyond that, NASA seems to have several camps of thought regarding what to do and are not pushing clear missions, and in some cases groups are even competing with one another internally for the limited funding they have. In my opinion Mars should be our next target because of the inspiration it would provide, but I would be happy to see some activity on the Moon as well. However, to answer your question – who knows! I believe other nations will work hard to get to the Moon for prestige and inspiration reasons sooner than the U.S. will become fully dedicated to either the Moon or Mars.

In your mind, what is a likely scenario and timeline for manned space?

I believe we will be on ISS through 2028. Hopefully to 2032. ISS will ramp-up support of space commercialization activities over the next several years. SpaceX will make a Mars push by 2025. Sub-orbital tourism flights will begin by 2020. I don't know if there is enough political support (money) for NASA to really work a lunar or Mars program with the funding they need. I am hopeful that will change.

Do you agree with those who have made a case for settling the Moon first and then using it as a base for travel to Mars?

I don't personally know enough to understand the system benefits and trades of each approach. What I do know is that the big decisions often don't get made on technical merit alone. Since there isn't an obvious economic draw to the Moon (or Mars for that matter), the decision will be made by other factors. A direct-to-Mars program will signal focus to the masses and stir excitement, inspire people to action more quickly in support of the mission, and yield a more rapid ROI (and it seems that our society demands a nearer-term ROI more rapidly every day now, which has a negative effect on science and longer-term exploration programs).

Is ISRU oversold for near-term lunar development?

In my opinion, yes for near-term. This relates to the first question. ISRU will require a lot of untested equipment to realize goals. This gear will have to be developed, transported, and maintained to do the work. I learned how to think about system level studies like this by looking at how the NSF runs the Antarctic exploration (science) program during some work we did for them. They don't bring in heavy resources until the numbers of people increase beyond a short-stay camp. It isn't exactly an ISRU situation except for power and water, but it is a similar situation. It seems to make more sense to build commitment to the effort with increasing successes over time, working from a few people who bring their own resources, to a colony who must use ISRU. This is just a feeling on my part though, as I have not spent much time studying the technology available or the system trades.

Key challenges to engineers and scientists who will be tasked to develop the lunar mission and structural design are the knowledge gaps that cannot be filled before being on site. For example, the lunar 1/6 g environmental effects on biology, and systems reliability in the lunar dust environment. Also of great concern is our understanding of the very complex systems that will not be integrated before they are placed on the Moon. Even if the behavior of all the components are fully understood, the complete integrated system will generally have unanticipated characteristics. Is there a way of dealing with these knowledge gaps before being there?

Great point! Clever test facilities, design, and analog environment testing seem to be the only hope. However, designing a system that is robust, easily repairable or adaptable, or able to be modified in-situ, could be a more practical approach than trying to ensure perfection on Earth prior to performing the mission (good design philosophy discussion). Combined effects testing is very difficult, but very important. You can be fooled if you don't test under all threats at once as they did with the insulation blankets on the original Hubble mission. Looking at Earth analogs, such as military activities in the desert or polar exploration, is critical to inform the system design process in my opinion. However, people don't generally want to accept this for some reason. We have a unique perspective at ILC Dover because we work in all these worlds, so we try to take the best ideas from one area into another. Sometimes the space folks don't have that experience and can't see the benefits of benchmarking and leveraging what exists (either in technology or in design/operations philosophy).

I have a quick space suit story to demonstrate the point. Lunar dust was a disaster for the Apollo suit as we all know. We looked at Earth analogs of operating in dust and chem-bio warfare environments to inform solutions (around 2005).

The thing that made sense was to put something like a Tyvek® suit, like you wear while laying fiberglass insulation, over the space suit when performing surface EVA to keep the dust out of the suit, and the habitat itself, by removing the dirty cover before entering the habitat. This is how the military operates in a chem-bio warfare environment. We did some testing to show it protected the suit and didn't degrade operational performance. A pretty simple idea that makes a lot of sense in my opinion (a technical paper exists on this). However, the approach was pushed aside and researchers put effort into making more durable space suit materials, bearings with numerous dust seals, dust repulsion technologies, etc., which degrade suit performance and increase complexity of the space suit (and don't work, especially at the system level). The decision making was being done at the technology researcher level, who would want to devote more effort in technology exploration, and not from a systems perspective (cost of EVA). It isn't easy to make these decisions at a systems level, and changes need to be made to streamline the process toward more rapid, robust solution finding in my opinion.

Dust mitigation is a critical issue generally on the Moon. How far are we from getting a handle on that problem, both the biological risks and the engineered systems risk?

The answer above tells you a little about my experience and how I would answer. There is a lot of military experience to draw from to inform design for dust, but also that a certain design philosophy needs to be adopted by NASA and contractors for success. I believe the technology required is available and what is required is proper implementation. If you look at robots in chemical and dust environments on Earth (automotive manufacturing for instance), they often have fabric dust covers. Again, it seems that the technology exists but the challenge is connecting mission planners and technology developers to existing solutions.

The isolated environment that astronauts will face on the Moon, and even more isolated on Mars, makes critical the need for high levels of reliability. Part of that implies a self-healing capability. Such technologies are being studied and developed but are far from being usable. Do you see self-healing as a critical technology?

We had a program called InFlex (Intelligent Flexible Materials for Deployable Space Structures) in the early 2000s. The focus was increasing inflatable structures performance metrics from habitats, to space suits, to aeroshells. We categorized each individual threat (Micrometeoroid and Orbital Debris [MMOD], impact from external equipment, impact from inside, material degradation, etc.) in every possible environment (LEO, Moon, Mars, etc.) and looked at what technologies made sense to deal with the threats (self-healing, structural health monitoring, layered materials, etc.). We demonstrated several functional self-healing technologies on the program around the anticipated damage (from MMOD holes up to 0.06 mm in diameter, to internal cuts from dropped equipment), and I think we could apply them now if we had to.

However, the real question is if self-healing is a critical technology, and that should be driven by the threat risk specifically. It is critical to not let gas escape the habitat unless you have ISRU. The real threat to the bladder layer (gas containment layer) is MM from the outside on the Moon (low), or accidental impact from the inside. If the habitat is going to be abandoned for long periods of time then passive systems make sense for MM impacts. However, if the crew are in the habitat it might make more sense to use structural health monitoring to identify leaks that the crew can repair. This path would also bring protection in off-nominal cases (larger than planned hole, unusual impact, etc.). Structural health monitoring isn't easy either, but has more broad application in addressing several needs for the mass (and cost) incurred in my opinion. The best line of defense might be, or include, added layers of protection (thicker Whipple bumpers, cut resistant layers, etc.), to meet the robust nature of the habitat you describe.

Manned space settlement has been promoted as a justified expenditure, financed by taxpayers, for many reasons: economic development, scientific advances, SETI, manifest destiny, building bridges between nations, limitless energy production, for example. What is your case for manned space settlement? Do you have an overarching reason for supporting such an endeavor?

I believe the justification begins with inspiring our youth (as many including me were inspired during project Apollo), spin-off technologies (bigger impact to society than people understand), and international cooperation (to the point of keeping scientists and engineers engaged in peaceful activities). It is critical to be able to link the money spent to economic benefits. Manned space exploration isn't an easy venture to justify. It would be easier if we were going there to mine platinum or something, because of the tangible outcome. However, space exploration is more like a wellness program where you put in effort and you are better for it in the long run. It is just difficult to see the results rapidly, or quantify them up front.

It appears that missions to settle the Moon or Mars have two underlying potential show stoppers: (1) assuring human survival and good health, and (2) assuring very high reliability for all engineered components. While we have shown with the ISS and Apollo that we can manage short forays to the Moon, or longer ones to LEO, satisfying (1) and (2) in limited ways, the bar is an order of magnitude higher for long-duration lunar stays, and an order of magnitude higher than that for Mars human missions. Are guaranteeing (with high probability) human and machine survival and high operability on the horizon, or beyond it?

I believe it is absolutely on the horizon. Considering analogs like submarines or aircraft shows that we can do it. It is possible to design robust, redundant systems for use in extreme environments. We will need to take some risks though, like adopting nuclear power, but I think all the parts are there for us to do it. I also find it interesting from a human nature perspective that people are more than willing to take risks in exploration, or even go on a one-way mission in the case of some of the Mars programs being discussed. It is the political consequences that seem to limit our moving faster in some cases. The other part of your question is the transition between exploring and settling. To me settling implies ISRU, especially for oxygen and water. I don't have up to date knowledge on what is possible to support colonization, so the horizon on ISRU might be a little further out than I am aware of.

Who would you consider as your key influence in your pursuit of space studies? Who inspired you?

JFK, Apollo astronauts, and Shuttle astronauts were the strongest influencers. Hollywood influenced me as well (Star Trek and Star Wars). I also had indirect influence to pursue activities in space exploration from several professors. When I started work at ILC Dover, the guys who were my mentors were the guys who designed and built the Apollo suits and Space Shuttle Extravehicular Mobility Unit. The stories they told! I was definitely hooked then.

All of us working on space recognize that much of what we do and are interested in promoting will not happen in our lifetimes. Of course, we have no choice but to accept this reality, but how do you view this?

You are right, it is a reality we just have to live with. It has been worth it so far for me (especially because some of my work has seen practical application, so that helped). It is difficult to see the political climate producing another cold war JFK moment, but you never know. Personally, I wish we would shift some of our defense budget and therefore people, to space activities ... but I am not holding my breath. I am hopeful that eventually humanity will understand the futility of fighting and greed, and find a way to build a world society more like what Gene Roddenberry envisioned where we work toward positive goals (but without the Earth being threatened every 15 minutes!).

Is there an issue or concern of yours that I have not touched upon?

Your questions are well reasoned and designed to provide a larger view of exploration, so no. They invite a holistic look at space exploration that you don't regularly see in the community. There is so much more to the story of space exploration beyond the physical hardware!

Thank you.

References

1. Report of the IAA Commission VI Study Group 6.9 – *Space Architecture: Tools for the 21st Century*, December 2008.
2. K.J. Kennedy: *The Vernacular of Space Architecture*, AIAA Space Architecture Symposium, October 10-11, 2002, Houston, AIAA 2002-6102.
3. B. Sherwood, Lunar Architecture and Urbanism, 2nd Ed., ICES 2005, 05ICES-79.
4. M.M. Cohen, K.J. Kennedy, (1997 November): *Habitats and Surface Construction Technology and Development Roadmap*, in A. Noor, J. Malone (Editors), Government Sponsored Programs on Structures Technology (NASA CP-97-206241, p. 75-96), Washington, DC, USA: National Aeronautics and Space Administration; M.M. Cohen, (2002 October): *Selected Precepts in Lunar Architecture*, IAC-02-Q.4.3.08, 53rd International Astronautical Congress (IAC), World Space Congress, Houston, Texas, USA, October 10-19, 2002, Paris, France: International Astronautical Federation; K.J. Kennedy, (2002 October): *The Vernacular of Space Architecture*, AIAA 2002-6102, 1st Space Architecture Symposium (SAS 2002), Houston, Texas, USA, October

10-11, 2002, Reston, Virginia, USA: American Institute of Aeronautics and Astronautics; K.J. Kennedy, (2009): *Vernacular of Space Architecture*, in A.S. Howe, B. Sherwood (Editors), **Out of This World: The New Field of Space Architecture** (Chapter 2, p. 7-21), Reston, Virginia, USA: American Institute of Aeronautics and Astronautics.

5. M.M. Cohen, H. Benaroya, (2009): *Lunar-Base Structures*, in A.S. Howe, B. Sherwood (Editors), **Out of This World: The New Field of Space Architecture** (Chapter 15, p. 179-204), Reston, Virginia, USA: American Institute of Aeronautics and Astronautics.

6. Personal conversation with Dr. Steven Zornetzer, Ames Research Center Associate Director for Research, 2015.

7. N. Connor (2017, June 7): *China Prepares for Manned Moon Landing*, London UK: The Telegraph, http://www.telegraph.co.uk/news/2017/06/07/china-prepares-moon-landing/, retrieved July 2, 2017.

8. H. Petit, (2017, June 29): *Japan announces plans to put a man on the Moon by 2030 as 'Asian space race' intensifies*, London, UK: Daily Mail/Mailonline. http://www.dailymail.co.uk/sciencetech/article-4650524/Japan-announces-plans-man-moon-2030.html, Retrieved July 2, 2017.

9. **Turning Dust to Gold: Building a Future on the Moon and Mars**, H. Benaroya, Springer-Praxis, 2010.

10. M.M. Cohen, (2009 September): *From Apollo LM to Altair: Design, Environments, Infrastructure, Missions, and Operations*, (AIAA 2009-6404), AIAA Space 2009 Conference and Exposition, Pasadena CA.

11. M.M. Cohen, (1988 March 1): *Space Station Architecture, Module, Berthing Hub, Shell Assembly, Berthing Mechanism and Utility Connection Channel* (U.S. patent no. 4,728,060); M.M. Cohen, (2012 July): *The Continuum of Space Architecture: From Earth to Orbit*, (AIAA 2012-3575), 42nd International Conference on Environmental Systems (ICES), San Diego, California, USA, July 15-19, 2012, Reston, Virginia, USA: American Institute of Aeronautics and Astronautics.

12. M.M. Cohen, (2011 June): *Comparative Configurations for Lunar Lander Habitation Volumes: 2005-2008*, (SAE 2009-01-2366), 39th International Conference on Environmental Systems (ICES), Savannah, Georgia, USA, July 12-16, 2009, in SAE International Journal of Aerospace, Vol.4, No.1, pp.75-97, Warrendale, Pennsylvania, USA: Society of Automotive Engineers.

13. M.M. Cohen, P.C. Houk, (2010 September): *Framework for a Crew Productivity Figure of Merit for Human Exploration*, (AIAA 2010-8846), AIAA Space 2010 Conference & Exposition, Anaheim, California, USA, August 30 – September 2, 2010, Reston, Virginia, USA: American Institute of Aeronautics and Astronautics; M.M. Cohen, (2010 July): *Trade and Analysis Study for a Lunar Lander Habitable Module Configuration*, (AIAA 2010-6134), 40th International Conference on Environmental Systems (ICES), Barcelona, Spain, July 11-15, 2010, Reston, Virginia, USA: American Institute of Aeronautics and Astronautics.

14. M.M. Cohen, (1996 July): *Design of a Planetary Habitat Versus an Interplanetary Habitat*, (SAE 961466). in SAE Transactions, Journal of Aerospace, Vol.105, Sec.1, pp.574-599, Warrendale, Pennsylvania, USA: Society of Automotive Engineers;

M.M. Cohen, (1996 September): *Habitat Distinctions: Planetary versus Interplanetary Architecture*, (AIAA 96-4467), Space Programs and Technologies Conference, Huntsville, Alabama, USA, September 24-26, 1996, Reston, Virginia, USA: American Institute of Aeronautics and Astronautics; M.M. Cohen, (1997): *Design Research Issues for an Interplanetary Habitat*, (SAE 972485). In SAE <u>Transactions, Journal of Aerospace</u>, Vol.106, Sec.1, pp.967-994, Warrendale, Pennsylvania, USA: Society of Automotive Engineers.

15. S.J. Hoffman, D.I. Kaplan, Editors (1997): *Human Exploration of Mars: The Reference Mission of the NASA Mars Exploration Study Team*, Houston, TX: Lyndon B. Johnson Space Center. NASA Special Publication 610.7.

16. M.M. Cohen, (2015 August): *First Mars Habitat Architecture*, (AIAA 2015-4517), AIAA Space 2015 Conference & Exposition, Pasadena, California, USA, August 31 – September 2, 2015, Reston, Virginia, USA: American Institute of Aeronautics and Astronautics.

17. C.M. Adams, and G. Petrov, (2005 July): *Variants on the TransHab Paradigm (2): The Surface Endoskeletal Inflatable Module (SEIM)*, (SAE 2005-01-2847), 35th International Conference on Environmental Systems (ICES), Rome, Italy, July 11-14, 2005, Warrendale, Pennsylvania, USA: Society of Automotive Engineers.

18. B.N. Griffin: *An Infrastructure for Early Lunar Development*, SPACE 90 Engineering, Construction, and Operations in Space, pp.389–398, 1990.

19. S.J. Hoffman, and J.C. Niehoff: *Preliminary Design of a Permanently Manned Lunar Surface Research Base*, Lunar Bases and Space Activities of the 21st Century, Proceedings of the Lunar and Planetary Institute, Houston, pp.69–76, 1985.

20. C.B. King, A.J. Butterfield, W.D. Hypes, and J.E. Nealy: *A Concept for Using the External Tank from a NSTS for a Lunar Habitat*, Proceedings, 9th Biennial SSI/ Princeton Conference on Space Manufacturing, Princeton, May 1989, pp.47–56.

21. C.B. King, A.J. Butterfield, W.D. Hypes, J.E. Nealy, and L.C. Simonsen: *Single Launch Lunar Habitat Derived From NSTS External Tank*, NASA Technical Memorandum 4212, 1990.

22. W.D. Hypes, A.J. Butterfield, C.B. King, G.D. Qualls, W.T. Davis, M.J. Gould, J.E. Nealy, and L.C. Simonsen: *Concepts for Manned Lunar Habitats*, NASA Technical Memorandum 104114, August 1991.

23. M.M. Cohen: *Mobile Lunar and Planetary Bases*, Space 2003, September 23-25, 2003, Long Beach, CA, AIAA 2003-6280.

24. M.M. Cohen: *Habitat Multivariate Design Model Pilot Study*, 34th International Conference on Environmental Systems, Colorado Springs, CO July 19-22, 2004, SAE 2004-01-2366.

25. W. Grandl: *Lunar Base 2015 Stage 1 Preliminary Design Study*, <u>Acta Astronautica</u> 60 (2007) pp.554-560.

26. W. Grandl: *Human life in the Solar System*, <u>REACH - Reviews in Human Space Exploration</u> 5 (2017) pp.9-21.

27. C. Cassapakis, and M. Thomas: *Inflatable Structures Technology Development Overview*, AIAA 95-3738, 1995.

28. L. Puig, A. Barton, and N. Rando: *A review on large deployable structures for astrophysics missions*, <u>Acta Astronautica</u> 67 (2010) pp.12-26.

29. Vogler: *Modular Inflatable Space Habitats*, First European Workshop on Inflatable Space Structures, May 21-22, 2001, ESA/ESTEC, Noordwijk, The Netherlands.
30. S. Haeuplik-Meusburger, and K. Ozdemir: *Deployable Lunar Habitation Design*, Chapter 20 in **MOON: Propsective Energy and Material Resources**, V. Badescu, Editor, Springer 2012.
31. S.J. Scotti: *A Study of Flexible Composites for Expandable Space Structures*, NASA/TM-2016-219171, March 2016.
32. S. Brancart, L. De Laet, and N. De Temmerman: *Deployable textile hybrid structures: design and modelling of kinetic membrane-restrained bending-active structures*, Procedia Engineering 155 (2016) pp.195-204.
33. V. Vrakking: *Design of a Deployable Structure for a Lunar Greenhouse Module*, MS Thesis, Faculty of Aerospace Engineering, Delft University of Technology, in conjunction with DLR, March 12, 2014.
34. M. Roberts: *Inflatable Habitation for the Lunar Base*, 2nd Conference on Lunar Bases and Space Activities of the 21st Century, W.W. Mendell, Editor, April 5-7, 1988, Houston, published 1992, pp.249-253.
35. P.K. Yin: *A Preliminary Design of Interior Structure and Foundation of an Inflatable Lunar Habitat*, NASA, Johnson Space Center, August 18, 1989.
36. **Specification for the Design, Fabrication and Erection of Structural Steel for Buildings**, published by the American Institute of Steel Construction (AISC), 8th Edition, 1980.
37. M.D. Vanderbilt, M.E. Criswell, and W.Z. Sadeh: *Structures for a Lunar Base*, Proceedings of Space 88, S.W. Johnson, and J.P. Wetzel, Editors, ASCE, New York, pp.352-361.
38. P.S. Nowak, and W.Z. Sadeh: *Geometric Modeling of Inflatable Structures for Lunar Base*, Journal of Aerospace Engineering, Vol.5, No.3, July, 1992, pp.311-322.
39. W.Z. Sadeh, and M.E. Criswell: *Inflatable Structures – A Concept for Lunar and Martian Structures*, AIAA 93-0995, AIAA/AHS/ASEE Aerospace Design Conference, Irvine, 1993.
40. W.J. Broad: *Lab Offers to Develop an Inflatable Space Base*, The New York Times, November 14, 1989.
41. W.Z. Sadeh, and M.E. Criswell: *A Generic Inflatable Structure for a Lunar/Martian Base*, SPACE 94, Engineering, Construction, and Operations in Space, 1994, pp.1146–1156.
42. M.E. Criswell, W.Z. Sadeh, and J.E. Abarbanel: *Design and Performance Criteria for Inflatable Structures in Space*, SPACE 96, Engineering, Construction, and Operations in Space, 1996, pp.1045–1051.
43. P.S. Nowak, M.E. Criswell, and W.Z. Sadeh: *Inflatable Structures for a Lunar Base*, SPACE 90 Engineering, Construction, and Operations in Space, 1990, pp.510–519; P.S. Nowak, W.Z. Sadeh, and M.E. Criswell: *An Analysis of an Inflatable Module for Planetary Surfaces*, SPACE 92 Engineering, Construction, and Operations in Space, 1992, pp.78–87.
44. P.Y. Chow, and T.Y. Lin: *Structures for the Moon*, SPACE 88 Engineering, Construction, and Operations in Space, 1988, pp.362–374; P.Y. Chow, and T.Y. Lin:

Structural Engineer's Concept of Lunar Structures, Journal of Aerospace Engineering, Vol.2, No.1, January 1989.

45. G.T. Moore, J. Huebner-Moths, C.M. Brinlee, D.S. Erdmann, L.H. Matheson, W.A. McCambridge, S.M. Schmidt, and A.J. Wellings: *Domus I and Dymaxion: Two Concept Designs for Lunar Habitats*, Proceedings of the 9th Summer Conference, NASA/USRA Advanced Design Program, 1993.

46. K.J. Kennedy: *Inflatable Habitats Technology Development*, Proceedings 1999 Workshop on Mars Greenhouses: Concepts and Challenges, NASA/TM-2000-208577.

47. K.J. Kennedy, J. Raboin, G. Spexarth, and G. Valle: *Inflatable Habitats*, Chapter 21, **NASA JSC Inflatable Structures Technology Handbook**, JSC-CN-6300, 2000.

48. K.J. Kennedy: *Lessons from TransHab – An Architect's Experience*, AIAA Space Architecture Symposium, October 10-11, 2002, Houston, AIAA 2002-6105.

49. W.K. Belvin, J.J. Watson, and S.N. Singhal: *Structural Concepts and Materials for Lunar Exploration Habitats*, Space 2006, September 19-21, 2006, San Jose, CA, AIAA 2006-7338.

50. **Tensile Structures - Design, Structure, and Calculation of Buildings of Cables, Nets, and Membranes**, Frei Otto, Editor, MIT Press, Cambridge, Vol.1 1962, Vol.2 1966. Translation from German.

51. P. Gruber, S. Häuplik, B. Imhof, K. Özdemir, R. Waclavicek, and M.A. Perino: *Deployable structures for a human lunar base*, Acta Astronautica 61 (2007) pp.484-495.

52. J-S. Lew, L.G. Horta, and M.C. Reaves: *Uncertainty quantification of an inflatable/rigidizable torus*, Journal of Sound and Vibration 294 (2006) pp.615-623.

53. D. Cadogan, J. Stein, and M. Grahne: *Inflatable Composite Habitat Structures for Lunar and Mars Exploration*, 49th International Astronautical Congress, Sept.28-Oct.2, 1998, Melbourne. IAA-98-IAA.13.2.04.

54. D. Cadogan, J. Stein, and M. Grahne: *Inflatable Composite Habitat Structures for Lunar and Mars Exploration*, Acta Astronautica Vol.44, Nos.7-12, pp.399-406, 1999.

55. Derbès: *Case Studies in Inflatable Rigidizable Structural Concepts for Space Power*, 1999, AIAA-99-1089.

56. D.P. Cadogan, and S.E. Scarborough: *Rigidizable Materials for Use in Gossamer Space Inflatable Structures*, 42nd AIAA/ASME/ASCE/AHS/ASC Structures, Structural Dynamics, and Materials Conference & Exhibit, AIAA Gossamer Spacecraft Forum, April 16-19, 2001, AIAA 2001-1417.

57. A.E. Hoyt, L.A. Harrah, M.R. Sprouse, R.E. Allred, P.M. McElroy, S.E. Scarborough, and D.P. Cadogan: *Light Curing Resins for Rigidizing Inflatable Space Structures*, Proceedings 49th International Society for the Advancement of Material and Process Engineering Symposium and Exhibition, Longbeach, CA, May 16-20, 2004.

58. E.J. Brandon, M. Vozoff, E.A. Kolawa, G.F. Studir, F. Lyons, M.W. Keller, B. Beiermann, S.R. White, N.R. Sottos, M.A. Curry, D.L. Banks, R. Brocato, L. Zhou, S. Jung, T.N. Jackson, and K. Champaigne: *Structural health management technologies for inflatable/deployable structures: Integrating sensing and self-healing*, Acta Astronautica 68 (2011) pp.883-903.

5

Habitat studies

"How do we begin?"

There have been numerous studies on lunar bases and settlements that have summarized the respective current thinking on the challenges of returning to the Moon, with the purpose of long-term and permanent settlements. As might be expected with an endeavor that is always many years in the future and where much of the data is limited and not current, there is a bit of repetitiveness. There is a critical need for data. Fortunately, the ISS has gathered data on human life in microgravity and there have been autonomous missions to the Moon since Apollo that have gathered data to help scientists and engineers better understand and characterize the lunar surface and environment.

For historical interest, we mention two early lunar outpost studies: *Project Horizon*, and *LESA*. *Project Horizon* was an extensive study that considered all aspects of the process to create a habitable environment for the lunar surface, including design criteria, site location, medical requirements and surface transportation, as well as the space transportation system, communications, logistics, and R&D requirements and cost estimates.[1] A number of interesting and relevant considerations were mentioned. For example, due to the photoelectric effect and the radiation environment at the lunar surface, there will be an accumulation of charge on metals. Also, oxygen leakage from suits and other enclosed structures will be absorbed on the exterior of suits and structures, leading to the creation of ozone and ionic oxygen radicals by ultraviolet radiation. Retainment at joints may be a severe problem. Nuclear reactors were selected as the basic power sources and the construction and permanent outpost structures were sized for a dozen people. Figures 5.1 to 5.6 depict the concepts involved for the buried structure, the living quarters, and a construction vehicle.

The construction vehicle in Figure 5.4 is described as capable of moving lunar material, excavating lunar trenches, heavy cargo handling, and other prime mover functions. At the time of this report, the lunar electrostatic dust problem was not understood, and current concepts for construction equipment need to take those effects into consideration, as the Apollo astronauts discovered. Also suggested was the use of explosives to facilitate the excavation of trenches. The challenge here is that unless the explosives and the

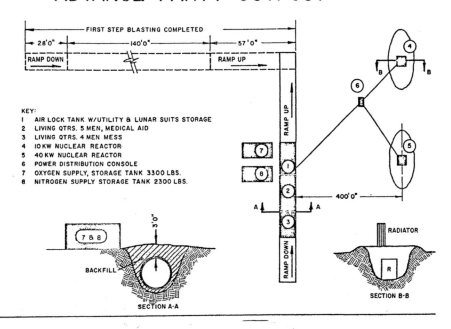

Fig. II-3. Layout of Initial Construction Camp

Figure 5.1. Project Horizon – Layout of Initial Construction Camp. (Courtesy NASA/U.S. Government)

surrounding material are contained, projectiles will be launched over a very large area, travelling at speeds that will undoubtedly cause damage.

In a discussion of biological issues, *Project Horizon* briefly mentioned using experimental animals to "ensure that no unsuspected problems are overlooked. ... Measurement evaluations of the biological effects of nuclear weapons on animals should be considered." There was also mention of the possible existence of extra-planetary life, with such life potentially indicated if certain combinations of carbon and nitrogen are found. "There is also the possibility of rudimentary life having, for example, silicon rather than carbon as the basic element."

The LESA study developed procedures, design equipment, and structural concepts for the deployment of exploration systems on the lunar surface.[2] A basic shelter was considered for six people who would stay on the Moon for six months. "Total free volume of this shelter concept is 2570 ft³, or about 430 ft³ per man. Free floor area is estimated at about 178 ft², or almost 30 ft² per man." This is a very small amount of free space: 30 ft² is approximately 5.5 ft. × 5.5 ft. Perhaps for the relatively short stay of six months this space will be sufficient. Figure 5.7 sketches through the process of gathering regolith and placing it atop the habitat structural cylinder.

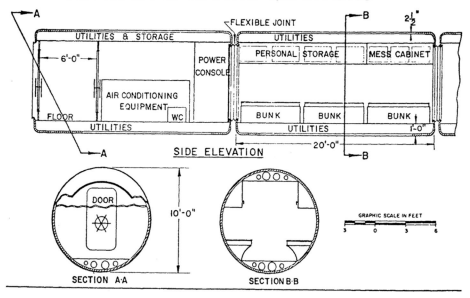

AIR LOCK & LIVING QUARTERS

SIDE ELEVATION

UTILITIES & STORAGE

UTILITIES

POWER CONSOLE

PERSONAL STORAGE

MESS CABINET

6'-0"

AIR CONDITIONING EQUIPMENT

FLOOR

WC

UTILITIES

BUNK

BUNK

BUNK

UTILITIES

FLEXIBLE JOINT

2½"

20'-0"

1'-0"

SECTION A-A

DOOR

10'-0"

SECTION B-B

GRAPHIC SCALE IN FEET

3   0   3   6

Fig. II-4. Cross-Section of Typical Outpost Compartments

Figure 5.2. Project Horizon – Cross-Section of Typical Outpost Compartments. (Courtesy NASA/U.S. Government)

Figure 5.3. Project Horizon – Overall View of Initial Construction Camp. (Courtesy NASA/ U.S. Government)

Figure 5.4. Project Horizon – Typical Lunar Construction Vehicle. (Courtesy NASA/U.S. Government)

An early and pioneering work on the design of structures for the lunar surface was by Johnson.[3] This was an exceptional effort and document, both for the breadth of the work and given that it was carried out at a very early point in time in the space race. The research and report focused on technical issues that would need to be considered by engineers designing the lunar structure. These include the characterization of the lunar soils and how to design foundations and embankments, the challenges of excavations, and the structural design itself given the environmental conditions: thermal, micrometeoroid, radiation, and local materials. An evolution of lunar base concepts from a very broad perspective included some ideas from science fiction and other literary writers.[4]

In 1984, the Lunar Base Working Group met in order to gather current thinking on the uses of the Moon (scientific, industrial, and human and life support systems development), to outline the technological and systems requirements for a lunar base (transportation, habitation, local resource utilization, energy), and the policy and legal assumptions and implications.[5] The primary workshop conclusions were that such a lunar base could make possible a wide range of scientific experiments, provide access to lunar resources that enable the development of new industrial products, and enhance human development. The group concluded that there were no scientific or engineering show stoppers, that there was a need for an evolution in space law, and that the costs were not prohibitive. There was not then, nor is there now, a legal regime for governing space development or private enterprise activity that utilizes local resources.

Figure 5.5. Project Horizon – Overall View of 12-Man Outpost. (Courtesy NASA/U.S. Government)

Figure 5.6. Project Horizon – Lunar Settlement Nearing Completion. (Courtesy NASA/U.S. Government)

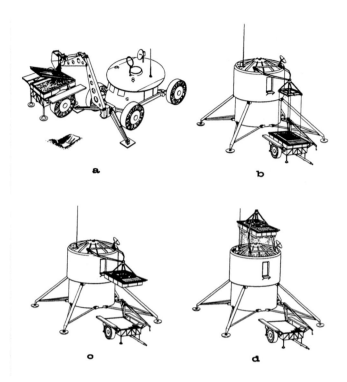

Figure 5.7. LESA – Moving Regolith to the top of the Lunar Habitat in the LESA Concept. (Courtesy NASA/U.S. Government)

The working group defined an evolutionary approach to the settlement of the Moon that would begin with an intermittently manned scientific and technological outpost, which would then evolve into a permanently inhabited facility. Finally, the facility would become a fully self-sufficient base with nascent manufacturing abilities, and a transportation hub to cislunar space, to Mars, and eventually the outer planets. A major part of this evolution was the design of a Closed Ecological Life-Support System (CELSS) of high reliability and reparability. It was assumed that the initial habitats would be derived from space station modules, with an initial crew size of 6 to 12 people.

During the period between the late 1980s and early 2000s, the number of space meetings, in particular about lunar bases and to a lesser extent Mars settlement, grew exponentially. With every new American administration, NASA received a new exploration direction, sometimes along a similar trajectory from the past, and sometimes in an orthogonal or worse direction. There was always a need for retreading. Apollo technical staff were retiring without new generations being hired and these gaps in generations meant that the critical continuity of knowledge transfer was lost. Much of that knowledge is not written, but mentored and nurtured. The space meetings did serve a purpose, through the development of a knowledge base along with knowledge transfer to the younger generations who were enthusiastic about space. There were the Space XX meetings sponsored by the American Society of Civil Engineers, organized by S.W. Johnson in Albuquerque, beginning in 1998 and held every two years into the early 2000s. One meeting was also held in Denver, organized by W.Z. Sadeh. There were proceedings for each meeting that are full of ideas and speculations. The early meetings were a lot of fun, and generated many ideas.

One NASA-sponsored conference resulted in a massive trove of papers (over 700 pages in two volumes), primarily on the technical aspects of creating a habitable environment on the Moon.[6] Many of the ideas introduced and discussed in that meeting formed the basis for engineering and scientific studies that continue three decades later. They included ideas on transportation, lunar site selection, lunar surface architecture and construction, scientific investigations, the utilization of lunar resources, life support, and operations and infrastructure.

The American Society of Civil Engineers was the home of the Task Committee on Lunar Base Structures, which provided a brief review of the study of lunar base structures as well as an overview of current (1992) ideas.[7] Lunar environmental characteristics were highlighted, and questions of structural functionality were categorized. In an appendix to the overview report, building systems proposed for lunar applications were categorized according to applications, application requirements, types of structures, material considerations, structures technology drivers, and requirement definitions.

A Jet Propulsion Laboratory design and trade-off study surveyed and considered various classes of lunar habitat designs with respect to the key environmental challenges: temperature, lack of atmosphere, radiation, soil properties, meteorites, and seismic activity.[8] Over twenty habitat designs were classified with respect to mission size, crew size, duration, modularity, environmental protection measures, and emplacement (above/below surface and equatorial/polar locations). Of these twenty designs, five were selected for future trade studies based on the criteria: post-Apollo design, design uniqueness, design layout, design dimensions, and structural material selection. These schemes can be loosely labeled logistical, delivered pre-made structures, underground structures, and structures made of local materials:

Logistical
Classified by mission type, crew size, duration of stay, environmental protection measures, and emplacement.

Delivered pre-made structures
Classified as prefabricated modules, pneumatic structures, prefabricated frame structures, and tent structures.

Underground structures
Those created by tunneling into the lunar surface to create large, safe volumes within which to place habitats, or by taking advantage of existing geologic structures and formations, such as craters or lava tubes, for innovative habitats.

Structures made of local materials
Also called ISRU-based structures, and may be composed of lunar regolith, lunar basalt, locally-derived composites, and locally processed ore.

The JPL study summarized the key challenges to engineering for lunar surface habitation. The lunar surface exhibits large temperature variations, between −280 to 224 °F. Heat flowing from either the subsurface or from surface structures radiates into space. Heat flow into the Moon during the day is higher than the heat flow out of the Moon during the night because of the increased efficiency of radiative heat transfer between particles at higher temperatures. Because of this, the temperature gradient within the top few cm of the subsurface is significant. Temperature fluctuations are virtually undetectable below 100 cm.

Based on the low thermal conductivity of the regolith, using it as a shielding material for surface structures may require only about 50–100 cm of depth, where the gathered regolith needs to be packed to the same density as the original untouched regolith. This last point is important to emphasize. If regolith is excavated, it becomes less dense, and any calculation based on the undisturbed density needs to be adjusted.

The temperature fluctuations can have severe design repercussions on exposed structures. Large strains and stresses can occur, and thermal cycling affects fatigue life. While the structure can be shielded after completion, that same structure may need to be temporarily shielded from the Sun during the construction operations; perhaps an umbrella-type system can provide temporary cover.

This study went into great detail on the meteorite and radiation threats, and we will discuss these threats in subsequent chapters, but an important conclusion stated that "if the habitat is protected for radiation, it will be adequately protected for micrometeorite damage." The chances of a micrometeorite striking an astronaut on the lunar surface are very small, much smaller than a similar habitat strike, but the outcome of such a strike would be fatal for the astronaut.

A special issue of *Applied Mechanics Reviews* introduced and discussed a variety of technical issues to be considered in the design and placement of a surface lunar structure.[9] In addition to a foreword to the volume by astronaut Charles "Pete" Conrad, Jr., there were chapters on the mechanics of materials of lunar base design, the lunar environment, the mechanics of lunar regolith, transport on the lunar surface, and the use of indigenous materials for lunar construction and structural concepts, in particular for lunar-based astronomy. Pete Conrad wrote: "I sincerely hope the politicians of this country can clear their vision to see the far reaching positive effects of space exploration."

In 1995, a special issue of the *Journal of the British Interplanetary Society* was devoted to the technical issues that need to be resolved in the design of a surface lunar structure.[10] Its considerations included lunar base system design, site selection, infrastructure, thermal control, inflatable structures, lunar soil mechanics, a mobile base, and a commercial lunar Helium-3 fusion power infrastructure.

An extensive reference on human spaceflight and planetary settlement was provided by Larson and Pranke.[11] Because it included spaceflight, topics such as trajectories, guidance, navigation, and propulsion were also discussed.

Two reviews of the literature provided an overview of the engineering, design, and construction of a lunar base.[12] [13] The environmental conditions were summarized, together with the mechanics of the regolith, the various loading conditions, and the classes of structures considered as possible lunar habitats. Discussions also introduced an overview on the creation of the infrastructure, the excavation of the very dense lunar soil, what could be expected with the use of explosives for this purpose, and the building of a transportation infrastructure. In order to avoid some of the regolith and dust issues, a robotic cable transport system was proposed. A cable-based robot shovel excavator was also explored.

The very detailed and extensive set of guidelines and capabilities provided by Allen *et al.* was a compilation and listing of the numerous design constraints that need to be considered across all the design parameters.[14] These include vehicle requirements, radiation risks, medical care, psychosocial issues, anthropometric design considerations, crew accommodations, habitat architecture and habitability, the crew environment, interfaces, tools, plant and animal factors and operations. A few interesting highlights are mentioned here.

For missions of 50–100 days (Moon) or 500–1000 days (Mars), it was assumed that all crewmembers would return to Earth alive and without serious injury or illness, while noting: "the program shall be designed so that the cumulative probability of safe crew return over the life of the program exceeds 0.99. This will be accomplished through the use of all available mechanisms including mission success, abort, safe haven, and crew escape." The question is by how much it should exceed 0.99, which is not a very high probability in this context. If applied to the Space Shuttle, for example, a probability of survival equal to 0.99 means that there is a probability of 0.01 of failure, or 1 of 100 Shuttle flights will not survive. In reality, two out of 135 failed catastrophically.

The study discussed artificial gravity possibilities for space travel involving long missions, for example using rotational systems and mechanisms. There is a high probability of significant illness or injury during a space mission. During a 2.5-year mission for a crew of six, it is almost definite that there will be a significant illness or injury.

For missions beyond LEO, a recommendation was made to allocate a minimum of 16.99 m³ (600 ft³) of usable space per crewmember. Individual private crew quarters should provide 1.50 m³ (53 ft³) for sleeping, 0.63 m³ (22 ft³) for stowage of operational and personal equipment, and 1.19 m³ (42 ft³) for donning and doffing clothing. As per Allen *et al.*: "These volumes do not include the crew quarters structures or the additional space needed for a desk, a computer and communications system, trash stowage, personal grooming, medical equipment and supplies required for convalescence, and an area for off-duty activities."

Hearing protection requirements are a challenge given the significant continuous noise environment. Protections and countermeasures are against peak and continuous levels to ensure safety and avoid stress, loss of sleep, distraction, an increase of errors in judgement, and decreased morale.

Pressurization and oxygen content are critical design parameters for crew health and also for structural design. Habitat pressures have ranged from 34.5 kPa (5.0 psia) to 101.4 kPa (14.7 psia). Space suit pressures are considerably lower in general. Cabin pressures are reduced prior to EVA operations, and the ability of crewmembers to move easily between spacecraft cabin, habitat and space suit atmospheric environments is critical to safety and mission success.

The comprehensive *The Lunar Base Handbook* provided an extensive overview of primarily the technical aspects of lunar base design and implementation, but also examined the physiological and psychological aspects of space travel and lunar settlement.[15]

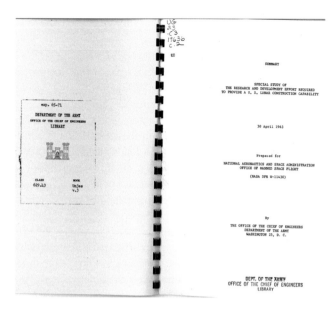

Figure 5.8. Cover page from the Army Corps of Engineers report: Special Study of the Research and Development Effort Required to Provide a United States Lunar Construction Capability. (Courtesy U.S. Government)

A more recent overview of lunar base structures provided a historical context, suggested a classification scheme for lunar structures, detailed key environmental challenges for designers, and summarized a variety of concepts:[16] Class 1 structures are pre-integrated, generally composite, for an exploration mission; Class 2 structures are prefabricated, generally hybrid and inflatable, for a settlement mission; and Class 3 structures are ISRU-derived and locally created – likely using layered manufacturing – for colonization. All three classes are surface structures requiring significant regolith shielding, but the overview recognized that lunar colonies and small cities would eventually mostly be buried, whether in lava tubes or in engineered subsurface volumes, in order to shield the inhabitants adequately and to provide them with large volumes for satisfactory living environments.

Badescu's *MOON* is an extensive collection of papers on the Moon, its resources and their recovery, and the challenges to the creation of facilities and habitats there, creating an off-world foothold in the Solar System.[17] In the foreword, there is a figure recounting the ILEWG technology roadmap for lunar exploration. Phase I (2000–2005) is for robotic exploration. Phase II (2005–2012) sees a permanent robotic presence and robotic construction. Phase III (2011–2018) sees the beginning of lunar resources utilization, robotic construction of manned systems such as habitats, food production systems, and energy production systems. Phase IV (2018 forward) sees crewed missions, manned systems operations, and human outposts. Clearly, the sequence makes sense. The reality, however, is phase lagged. It appears that, as of the writing of this book (summer 2017), Phase I will not begin until after 2020.

5.1   FRAMEWORKS

Given that we have not yet designed and placed an actual structure on the Moon, it seems appropriate and useful to understand the framework within which the process operates. On Earth, all aspects of engineering, especially those that deal with civil structures, are framed within design codes. Such codes are the procedures and rules for the design of structural components for a given purpose at a given location. The codes are a distillation of the engineering community's understanding, through theory and practice, of how structures need to be designed in order to meet the design loads (forces and other environmental aspects), while doing so safely and economically. Such codes exist for steel structures, concrete structures, for wind loads, and seismic loads. There are electrical codes, mechanical codes, and codes for the design of chemical plants. The codes reflect decades, even centuries, of a growing understanding of, and experience with, the physical world that exists and the engineered world that we create.

Going back to the Moon, and given our lack of experience, we wonder whether a "design code" is currently feasible. We have no practical knowledge base except for the science of the Moon; that is, its constituents and their characterizations. We understand the physical environment to an acceptable level and we have a grasp of the dangers of that environment to humans, machines, and structures. Our experience with the ISS is useful in many ways, although the regolith problem does not exist in LEO and the microgravity in space is significantly different in its effects than the 1/6 *g* we find on the Moon.

Good practice in structural design implies a cognizance of materials, structural behavior, environmental loadings, assumptions made in analysis and behavior, and the uncertainties inherent in all of these. The American Institute of Steel Construction's (AISC) *Manual of Steel Construction* is such a codification for the design and construction of steel structures. It includes information, some tabular and the rest in the form of specifications and commentaries, necessary to design and provide for the safe erection of steel-framed structures. The design equations are generally semi-empirical; that is, they are based on a mix of theoretical analysis, experimental data, and factors of safety. Each of these components has associated implicit assumptions.

Some of these assumptions have been explored in order to understand how – and if – the Earth-based design code could be used for the design of a lunar outpost.[18] Topics were discussed based on the AISC *Code of Standard Practice* and the commentaries, as well as issues such as the scaling of loads and strength in the 1/6 *g* lunar environment, thermal cycling effects and fatigue, and stiffening and buckling. Also discussed were impact loads due to launchings and landings, and accidents such as vehicular impacts.

> "Penetration of facilities caused by man-made projectiles resulting from explosions will be difficult, if not impossible, to design against. However, structural concepts must include some sort of compartmentalization to ensure that a breach of one portion of the facility does not lead to a domino effect, resulting in the failure of adjacent facilities. This may be achieved in part by a spacing criterion between modules. The second consideration is caused by natural impacts of meteoroids. It is estimated that one object of mass 10 g or greater will strike each 70,000 km^2 on the

lunar surface during a 24-hour cycle. During that same period of time, at least one micrometeoroid (0.001 mm diameter) will hit each square foot of the surface; 99 percent of the 40,000 kg of material deposited on the lunar surface per 24-hour. period is of this latter type. One can estimate the probabilities of such occurrences based on frequency information, thereby assisting in the definition of acceptable reliability and confidence levels.

"Such impact loads are of importance in the design of lunar structures, since any material is liable to penetration and rupture when impacted by such particles traveling at hypervelocities (order 10 km/s). Regolith shielding depths of 2–3 m will be able to absorb much of this rain of micrometeoroids. However, the rare larger particle will get through and the designed structure must be able to contain a possible explosive decompressive force."

It was noted that materials responding to such impacts and penetrations experience strain rates on the order $\dot{\varepsilon} \sim 10^8$, and since the regolith shielding is tightly packed, then even if it is capable of stopping the majority of projectiles, it may transmit stress waves to the underlying structure with the wave velocity:

$$c_p = \left(\frac{1}{\rho} \frac{d\sigma}{d\varepsilon} \right)^{1/2},$$

where ρ is the density of the packed regolith, ε is the engineering strain, and σ is the axial engineering stress. Displacements x in the direction u are governed by the wave equation:

$$\frac{\partial^2 u}{\partial t^2} - c_p \frac{\partial^2 u}{\partial x^2} = 0.$$

We cannot decouple the human from the machine and structure. A major goal of the structural design is to safeguard humans from the hazardous lunar environment. Therefore, we see numerous papers and conferences on the settlement of the Moon with that coupling intact. It makes no sense to speak of a structural design of a habitat without acknowledging the hazards to humans. Efforts have been initiated to create such a design framework for the engineering community.

Sherwood and Toups introduced and discussed technical, operational, and programmatic issues that are "fundamental to understanding the facts of life in this promising new design arena."[19] They emphasized that, of lunar base structures, only about 10–25 percent of the delivered mass was related to habitats, and about 90 percent of the mass used in the construction of the base would be for foundations, roads, and unpressured open structures. We will depend on Class I structures that are brought intact from Earth for the initial habitats. Organizing our efforts for the design and construction of lunar habitats and facilities requires us to define and understand our framework. Sherwood and Toups pointed to the design scope (what is to be included), scale (facility size), timing (how fast the project must be paced), and technology level (what technology level is defined and appropriate for the facility). It is also important to realize that the return to the Moon has been, and will continue to be, a slow process spanning generations. *A staircase cannot be built with only the upper steps; it must include the lower steps.* "Modest, solid steps appear to be both the quickest and surest way to grander lunar base structures." In the last several decades, the advent of layered manufacturing has opened up possibilities that were not recognized in the earlier studies of lunar bases. If we wait long enough before returning to the Moon, perhaps our first habitats will be locally manufactured on the lunar surface by autonomous systems utilizing the regolith.

Efforts to develop frameworks and logical approaches to the design of lunar structures and habitats have taken a number of forms. A decision analysis approach has been proposed, where a function of the decision variables is created that must then satisfy design constraints.[20]

We require an analytical tool that models structural behavior, given environmental and performance constraints, and is coupled to an economic cost model. Since there are many ways to build a lunar base, and that each way satisfies different constraints to differing degrees with different costs, an approach is needed to select one concept as the "best." Selecting the least expensive will not likely be the best choice since, for a relatively small additional sum, it may be possible to build a considerably more robust lunar base. Such a base, for example, may be more easily expandable, provide larger volumes within which to work, or be easier to maintain. In the long run, such a base will be less expensive, even with higher initial costs, while providing better service. We see the need, therefore, for a quantitative economic/design model of the lunar base. Such a model is generic in the sense that it represents all possible lunar base structures. It is incomplete until all the constraints are imposed upon it, and then the model with constraints is optimized with respect to multiple objectives. Such an approach, as just described, is one embodied in operations research and, specifically, decision theory.

In constructing a mathematical model of the lunar base system, if there are n quantifiable decisions to be made, the decision variables $x_1, x_2, \ldots, x_n$ must be representative of those decisions. The decision variables are the choices made in the specification, or design, of the structure. For example, materials, ISRU, geometry, mechanisms, sensors, implicit factors such as design life, and factors of safety, can be decision variables. A composite measure of the effectiveness E of the n decisions is given as an objective function:

$$E = f\left(x_1, x_2, \ldots, x_n\right).$$

Objective functions that need to be optimized (minimized or maximized) can be any of the following: minimize the total structure mass; minimize maximum deflection; minimize some aspect of the structural configuration (for example, to minimize complexity). We also want to maximize design life and minimize risk, while minimizing cost and time.

Typically, all designs are subject to constraints. (Life is subject to constraints, so why not structural designs?) In lunar structural systems, examples of constraints include the following among others. All parameters will use mass rather than weight representation:

- dust damage
- radiation dosage limits
- construction limitations, and ease

- 1/6 g
- internal air pressurization, and magnitude, and
- shielding levels.

All of the technical constraints must then satisfy cost constraints. Such constraints on the decision variables individually (and systematically on the objective function) are generally inequalities:

$$Ax \leq D, \qquad (5.1)$$

where A and D are property matrices particular to the problem being analyzed, that include all the above decision variables and constraint parameters. Choosing the decision variables, formulating the objective function, and identifying the constraints leads the analyst to a mathematical statement of the problem to be addressed, Equation 5.1.

There are numerous concepts for lunar structures, and, depending on their purpose, they all fare differently in addressing their design function, in part because there are many components affecting that function. One way to approach the evaluations of competing designs was discussed previously via decision analysis. We learn in such studies that there is never an optimal solution to a complex design problem; there are only sub-optimal solutions that satisfy the multiple constraints to varying degrees. The designer must make a choice about which constraint is the most important (to this designer, but not others), and move forward on that basis.

An approach adopted by a number of designers is to compare competing habitat designs by rating how each design accomplishes a long list of goals, using a subjective numerical scale applied consistently across all designs and all goals.

Drake and Richter recounted the ten NASA structural classifications for large volume habitats from the 1988 Exploration Studies Technical Report: prefabricated modules; pneumatic/inflatable structures (air-inflated, air-supported); prefabricated frame structures; tent structures; lunar-assembled canopies; craters (as shelters for habitats or as habitats); lava tubes (as shelters for habitats or as habitats); tunnels; structures constructed of lunar-derived materials (sintered blocks, ceramics, glasses, metals); and hybrid structures (combinations of the foregoing).[21] We note, again, that in the 25 years since this paper was published, the unanticipated development of advanced, layered manufacturing has likely altered the equation for extra-planetary construction. Perhaps by the time we truly move to return to the Moon, and begin to create an infrastructure, we will have such advanced layered manufacturing technologies that what we currently call Stage 3 development may become Stage 1 or 1.5.[1]

Drake and Richter selected eight concepts, summarized the concepts along with their respective assumptions, and then set them up against three sets of factors that were parametrized and numerically scaled against each other. The product of these three sets of numbers led to an evaluation rating number:

$$R = \sum EMT,$$

---

[1] This is one of the advantages of doing nothing. It is that as time goes by, not only is it easier to do nothing, it is also easier to do something better.

where R is the evaluation rating, E is the effectiveness factor (the degree to which the characteristic has a positive effect on habitat deployment, 1–5, with 5 representing most effective), M is the mission factor (the degree to which the characteristic is important to the mission of the outpost, 1–3, with 3 being very important), and T is the timing factor (quantifies the degree to which the performance characteristic is important and readily achievable at the current state of habitat development, 1–3, with 3 representing important and achievable). Four timing steps were used, based on the 1988 NASA exploration study:

Robotics phase:
The outpost site has been visited by autonomous hardware in a precursor mission.

Emplacement phase:
A first habitat has been placed on the Moon and scientific missions have been initiated.

Consolidation phase:
Habitats have been enhanced and scientific missions expanded.

Operations phase:
Operations were beyond the initial growth phase and are steady-state, with a further expansion of scientific missions.

Twenty-eight characteristics were defined including, for example, total weight, size limitations, size expandability, failure modes/safety, reliability and maintainability. For each of the three evaluation factors, each structure was assigned a numerical value signifying the success of each design concept.

As an example demonstration of the proposed methodology, a lunar observatory was proposed for the lunar consolidation phase, with the purpose of determining the sub-optimal structural habitat concept. A "hexagonal module" concept was identified as the best for this purpose. It would be constructed in a lunar precast concrete production facility that was assumed to be already in place by the consolidation phase. Drake and Richter emphasized "that the selected concepts, characteristics, missions, and developmental stages do not include all that may need to be considered in future decision-making processes. The methodology can easily accommodate future additions or deletions to those considered in this paper. Evaluation factors are proposed based on the experience of the writers." Others, however, can use this methodology, with their views reflected in the assigned weighting factors.

Design criteria have also been suggested to provide a framework for engineers who will eventually consider such structures.[22] Categorization of structures based on their type and their purpose helps designers anticipate the level of complexity of the potential design. The critical environmental conditions were outlined, as well as the need for a probabilistic approach to design. Load combinations were suggested: dead loads, live loads, internal pressure loads, and thermal loads. Occasional loads due to meteoroid impact and moonquakes can also be added. Since many uncertainties exist for all these lunar environmental loads, they can be interpreted to be random variables governed by their respective probability density functions. These density functions are also difficult to define because they rely on data, of which we have only a very limited amount for the Moon.

It is noted that there are likely to be epistemic uncertainties; that is, knowledge that we do not know that we are missing (we don't know what we don't know). These unforeseeable uncertainties might not only be technical, but also social and physiological factors.

They will reveal themselves once we are on the Moon and are living there. The key then is to make our habitats and machines robust in the face of such uncertainties. This is a very challenging task given cost limitations.

Another framework approach to the modeling of lunar bases is the *parametric* model. This name has customarily been attached to first-order magnitude estimates of masses required for all the aspects of the construction and maintenance of a lunar base. Mass, being a critical cost parameter, appears to be a useful unit of measure for estimating timing and costs. Eckart provided such first-order mass estimates as functions of crew size, lunar base location, environmental conditions, specific system masses, power requirements, and thermal loads.[23] The parametric model is for early lunar bases occupied by several tens of crew members. It is an integration of a dozen element system models, each of which is represented by an input-output, block diagram element. For each such element, inputs can be provided either by the user, or by other system element outputs. The resulting outputs are then fed to other coupled elements. The block in the block diagram represents the element properties. For example, it can represent a power supply model for lunar systems power requirements. The inputs and outputs couple power supply requirements as a function of other base operations. This model is not dynamic. It represents steady-state operations, so that in order to estimate changes in operations, the user of the model needs to vary inputs and then observe the changes in the outputs. Also, in this simplified form, this parametric model does not incorporate crew-time availability, maintenance requirements, operational aspects, or schedules. As a first-order planning model, it can inform preliminary lunar base conceptual models.

An early developer of parametric and comprehensive lunar base models was Koelle.[24] In this work, detailed proposals were made that included the rationale for a lunar base, base concepts, performance, schedules and costs. The path from a parametric analysis to a conceptual design was illustrated. Based on this model, a ten-year planning and development period leads to a 30-year operational phase for the lunar base, where final crew size approaches 100 astronauts. A series of flow charts detailed the numerous tasks and their flow. A representative base concept was utilized to demonstrate process flow.

The lunar construction code idea has been revisited, based on updated knowledge about the lunar surface.[25] Some strategic and other recommendations were made, many of which require precursor missions to the lunar surface. We can expect that design guidelines will be developed and tested as the concepts for the first lunar habitats are compared, and a single concept emerges.

5.2   STRUCTURES AND LIVING ON THE MOON

"Habitability"

The survival of living things in space and on extraterrestrial bodies is the key challenge to a spacefaring humanity. The first priority is living; the next priority is living with sufficient food and comfort; and the next priority after that is to live and enjoy life. Morphew provided an extensive overview and discussion on the spaceflight environment and how it induces numerous physiological, biomedical and environmental stressors to the flight crew.[26] Much of this discussion is relevant to those who would settle the Moon. In particular, "it is often the human element pertaining to poor human-technology interface design, team and interpersonal dynamics, spacecraft internal environmental conditions (habitability) and psychological factors that limit successful performance during space-flight, rather than the purely technological factors of the environment ... the most critical problems facing humans in long duration spaceflight, after the biomedical, are the psychological and psychosocial." Of particular interest is a table of stressors in that work. Five overlapping categories of stressors are listed, along with key examples:

Physiological/Physical:
Radiation risks, lack of natural time parameters, altered circadian rhythms, minimal exposure to sunlight, one-sixth Earth gravity, lack of natural sensory inputs, sleep cycle changes, space adaptation sickness.

Psychological:
Isolation and confinement, dangerous conditions due to hostile environment, lack of quick rescue possibilities, complex mission demands, sleep cycle changes, limitations on accustomed hygiene.

Psychosocial:
High level of teamwork demands, interpersonal tensions and enforced contact, family life disruptions and losses, cultural differences, crew differences – both physical and temperament, other potential social conflicts.

Human Factors:
Workload levels, limited off-world communications, limited facilities, equipment and supplies, dangers due to improperly functioning and damaged equipment, adaptation to an artificial environment, food limitations, limitations in technology interface designs, adaptation to lunar gravity, and the operation of equipment in that environment.

Habitability:
Limitations on accustomed hygiene, continuous noise and vibration, limited sleep facilities and privacy, lighting and illumination discomfort.

From the above stressors, we highlight a few. Isolation and confinement can lead to motivational decline, fatigue, somatic complaints, and social tensions. "The human body must physically adapt to the foreign microgravity [and lunar gravity] environment and, in doing so, undergo cardiovascular, muscular, and skeletal deconditioning as well as changes in the immune and nervous systems, and radiation exposure." Morphew provided an enlightening definition of space habitability in conjunction with the kind of support that NASA JSC specialists provide in the creation of such environments:

"Operational habitability refers to the design, integration and support of human, machine, mission, and environmental elements that promote optimal performance, physical and psychological health, and safety in long duration spaceflight. Habitability pertains to the qualities of a mission that enable people to live and work in a safe and productive manner. Habitability specialists at NASA JSC provide support in the following areas: architecture, acoustics, clothing, command structure, communications, crew interface/displays, dining, environmental conditions, emergency, exercise, EVA, fatigue, equipment, food and nutrition, group interaction, housekeeping, hygiene, lighting, maintenance, multicultural issues, psychological effects, privacy, recreation, restraints, supplies/provisioning, scheduling, sleep, stowage, translation/mobility, trash, training, and waste collection."

The Apollo program, and the Mercury and Gemini programs that preceded it, focused on the first priority of survival. When Skylab, Mir and the ISS were built, there was more volume for the inhabitants. Astronauts could move around – they even had a bit of privacy. Was it comfortable?

The design of the equipment and the human/machine interfaces began to look beyond survival and into ease of use. Concepts for space stations and lunar settlements take into account the appearances of the interiors – what colors are stress reducers and make people at ease, for example. With each new adventure, and each new generation of the adventure, designs will add comfort and take into account the full human being. Layout design facilitates ease of human motility. Improving the social and organizational aspects of life in the settlements will take a higher priority. Social and psychological issues – effects of stress, recreation and exercise, interpersonal dynamics in space, personal space, privacy, crowding, territoriality – are understood to be critical as we evolve into a lunar civilization.

Human physiology takes the front row, since being alive is a prerequisite for being comfortable and happy, but the serious consideration of human psychology is close behind. Just being alive will not be enough for the permanent inhabitants of the Moon. Soon after the first settlers occupy lunar habitats, people from all backgrounds will begin to populate the lunar settlements, not just engineers and scientists with years of training. Approaching the design of a lunar surface structure, or any extraterrestrial structure, for long-term habitation purely based on structural engineering considerations is a sure recipe for a viable structural design, but a failing human habitat. The complexities of creating a place for humans to live and thrive are challenging. When structures for habitation are designed for Earth, many aspects do not need to be considered because they are automatically satisfied by virtue of being on Earth. For example, the Earth habitat has an outdoors that is generally accessible and expansive, with many colors and vistas. People can visit the habitat; residents can leave and go on vacation; the environmental risks are generally not of a life-threatening nature.

On the surface of the Moon or Mars, all the above attributes of an Earth residence are generally missing. The outside is only accessible via a spacesuit, and for brief periods of time; a broad spectrum of color is nonexistent; there are no visitors and no vacations, and the environment is deadly. Except for the effects of gravity, we generally understand how to design a structure for extra-planetary use where the residents will be relatively safe, but more thought is necessary for the creation of an environment for long-term habitability where humans will thrive.

Early in the Apollo program, when there was a belief, if not a consensus, that our space program would not end with a few lunar landings, thought had already been given to the development of lunar concepts and designs. Habitability was already a concern. Celentano *et al.* developed a habitability index to assist engineers in the design of safe and thriving environments.[27] Criteria included duration of stay, the nature of the missions and assigned tasks, the number of people, and the distance from Earth. Habitability factors included the design of environmental control systems, nutrition and personal hygiene needs, living area and volume, crew work/rest cycles and fitness requirements, and accounting for the low gravity environment. These factors are in addition to those of a structural analysis that accounts for internal pressurization and external threats such as radiation, temperature extremes and micrometeoroids. Wise and Wise discussed the many aspects of colors and how small changes can lead to major advantages.[28] "There are demonstrable perceptual impressions of color applications that in turn can affect the experiences and performances of people in settings."

Human Psychology

In addition to physiological challenges, the psychological pressures of lunar habitation can be severe. The early settlers will be isolated in close quarters, but over the decades we will evolve and build larger facilities and residences. Eventually, we can expect to have as much interior space, both private and public, as the average Earth dweller. Of course, we will never have outside space, and we can expect that many of the public spaces will eventually be large atriums, with few hallways except as necessary, and with very high ceilings. Our public spaces, while interior, will be our outside space.

There is much to learn about the psychological aspects of long-duration space flights.[29] While travel to the Moon takes a relatively-short three days, there are significant psychological challenges to the almost year-long flight to Mars. The first people back on the Moon, but especially those who settle there permanently, will experience a profound isolation.[30] These will be true pioneers in the full sense of the word – enduring isolation, danger, close quarters – what a challenge to the human body and mind!

Mixed crews are expected for such long-duration missions. Couples may be considered, along with emotionally mature individuals with training in conflict resolution and an ability to work through issues such as jealousy. Many of these early problems are expected to disappear as the settlements grow larger and the populations evolve into the hundreds. Greenhouses will help many during the expected depressions. Not only will the plants provide all sorts of sustenance, but they will help residents keep their emotional balance and put feelings into perspective, especially during periods when living is one day at a time, when the main goal is to make it to the next day.

We have learned much about the psychological impact of space, even during the short spaceflights of the Apollo and Shuttle eras. One study explored how the value hierarchies of people in the space program changed after a trip to space.[31] "Transcendence[2] leapt from last place to second place among the men and from second to unrivalled first place among the women." Such changes did have a significant impact on the lives of the astronauts and their close associates. Perhaps among the future populations on the Moon and Mars, a higher purpose and spirituality will be very common. Of course, these personality traits could be predicted as being dominant for those who leave the comforts and beauty of Earth, as well as family and friends, for a life on planetary bodies that can never feel like home, in the case of the Moon, or will take centuries to have some semblance of Earth, in the case of Mars.

It has been well known from the beginning of human spaceflight that inhabitants require a certain area and volume for living, as well as an option for privacy in order to retain high morale and effectiveness. Humans in long-term situations require the possibility to move around, and to move between different locations. It is recognized that small

---

[2] Transcendence – a combination of spirituality (unity with nature, inner harmony, detachment from material desires) and universalism (protecting the environment, wisdom, peace).

group dynamics can be a critical factor affecting mission success and safety. In the habitability index of Celentano *et al.*, relative values were calculated, with weights to help prioritize factors, and then the sum of the factors was normalized and added so that a percentage was calculated. This percentage is the habitability index that can be used to compare different concepts, and the sensitivities of each concept to adjustments and changes. They provided charts on acceptable sound pressure levels, acceptable vibration amplitudes as functions of frequency, and area/volume minimums per person, as functions of mission duration. In these last curves, the area/volume requirements level off after about six months. For example, at six months, for a crew of five, area/volume = 100 ft^2/700 ft^3 per person.

In a more recent study about volume estimation, Simon *et al.* provided a historical review of such estimates and essentially arrived at similar optimal volume numbers as cited in Celentano.[32] The psychological and social dimensions of habitation in space were explored by Lockard.[33] Habitation implies long-term. The point was made that humans can endure short-term excursions into space but will need to thrive in long-term habitats. "The focus on mitigating the psychological and social stresses of spending long durations in space has been addressed almost exclusively from the perspective of crew selection and training, rather than on environmental factors. Integrating an architectural perspective could help remedy many psychosocial problems to a significant degree – especially those stemming from isolation and confinement." Part of the challenge is that well-being is a subjective concept and an individual reaction to an environment. Different individuals react differently.

Cohen and Haeuplik-Meusburger provided a deep discussion of habitability from a new perspective.[34] Rather than approaching habitats from an additive and minimalist – and usually unrealistic – perspective, as many do, they provided a subtractive view that began from the Earth habitat that we are familiar with, noting items that are lost in the transition to a space or extra-planetary location, in particular Mars and the trip to Mars. The discussion included images of classical and impressionist paintings and other images to highlight "normal" life on Earth, in order to identify what will be lost to the astronauts and settlers when they settle off-Earth. The various plans for Mars settlement were discussed, focusing in particular on key issues that are ignored by their proponents. Examples include adequate volume for living, reliability, and sufficiency of stores. "Critical Habitability emerges as most fundamental for the question of crew survival on extremely long duration missions ..., whether flyby, round trip, or one way. Critical Habitability arises as a leading concern, particularly in terms of what is lacking from the living and working environment."

Stressors leading to degraded performance and health, and countermeasures to alleviate these, were tabulated and discussed. Key stressors are: volume limitations; noise, inadequate housekeeping, lack of hygiene and cleanliness; thermal/humidity, closed atmosphere issues such as odors and bad air; confinement, isolation, separation from society, separation from nature; and artificial lighting. A key point proven is that minimalist and sub-minimalist plans cannot succeed for a variety of reasons. Cohen and Haeuplik-Meusburger concluded that:

> "... mission designers must be honest and open-minded about the risks and challenges that the crew will face over their long voyage and return or their permanent stay on the surface. What mission designers must do with the close involvement of well-qualified space architects is to take into account the human needs of the crew, finding ways either to satisfy them or to compensate for the loss of what they give up and leave behind to fly on these missions. These solutions include bioregenerative life

Figure 5.9.   Moon Base Precursor – Robotic Phase. (Courtesy M.A. Perino)

support systems that are sustainable to produce a breathable and healthy atmosphere, to grow food, and to bring essential features of the Earth's ecosystem with the crew to their new home world. These quality of life provisions will prove essential to sustain the crew for years. Human missions to Mars and beyond can succeed only if they take fully into account how best to accommodate and support the crew's needs across the broad spectrum of human experience."

Perino provided a very broad and deep discussion of all aspects related to a successful design of a surface lunar structure:[35]

"The return to the Moon will require a systematic exploration for the selection of the best site to establish a lunar outpost. In the following phases, the Moon Base will grow according to the selected scenario to include laboratories, production plants, additional habitat capabilities, power production facilities, etc., and it could be regarded as a staging base for more distant space travel."

The kind of base will depend on the mission characteristics. Generally, this can be a base for scientific research or a base for long-term habitation, with the development of an infrastructure and in-situ resource utilization (ISRU). It is unlikely that a manned base will be created purely for science, though it is likely that a habitat infrastructure will include a science component. Perino's estimate for volume requirements for a crew of nine, for a duration of greater than six months, was 216 ft^3. Psychological needs were mentioned as critical aspects of habitability, with the suggestion of habitat designs that do not resemble a tunnel; that is, the circular cylindrical concepts that are common. Triangular or square configurations were recommended, perhaps with inflatable dome structures. Sketches were provided of early and later bases, as well as possible interior configurations. Figure 5.9 depicts a precursor robotic phase. More advanced bases were anticipated to be either inflatable structures or those created from ISRU via automated construction.

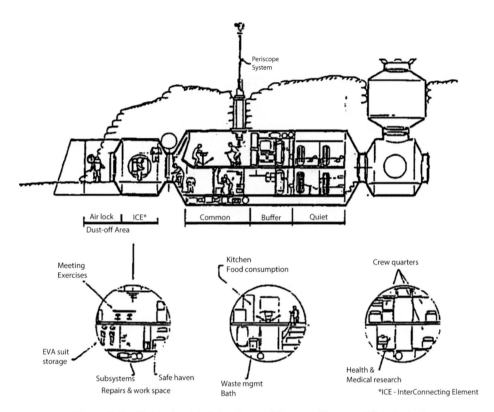

Figure 5.10.   Habitation Module – Internal Layout. (Courtesy M.A. Perino)

Figures 5.10 and 5.11 depict a second lunar base, showing a mostly buried facility taking advantage of the regolith shielding, along with an access port.

Figures 5.12 and 5.13 depict the advanced stages of lunar settlement with extensive facilities for possibly one hundred or more people.

Human Physiology

Evolutionary physiology will help us understand what to expect as we begin to colonize the Moon. Our environment causes our body and our mind to change. This change can be rapid, or take years and generations. We are always interested in the role that the environment plays – especially an extreme environment – in our adaptation and evolution. Gravity, or its lack thereof, shapes our lives. Earth gravity is directly related to what we have become physically and psychologically. Therefore, as we begin to move into space in the early 21st century, we are rightly concerned about how microgravity in space, and low gravity on the Moon and Mars, can impact our daily and long-term lives.

Some of the effects of changes in gravity are overcome by the body after a period of time. Some are temporary and others may become permanent. On shorter missions, astronauts report severe motion sickness. This is because "the vestibular system relies on gravity, so as the head tilts, hair cells in the inner ear are displaced, creating a signal to the brain

Figure 5.11. Second Moon Base – Manned Phase. (Courtesy M.A. Perino)

Figure 5.12. Third Moon Base – Manned Phase. (Courtesy M.A. Perino)

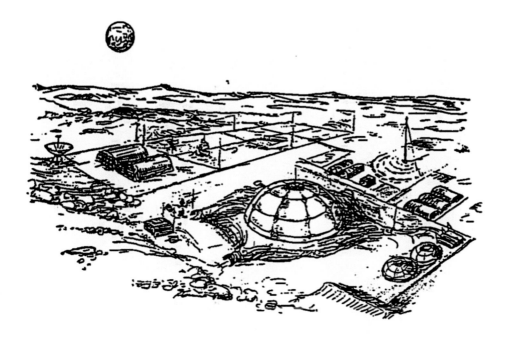

Figure 5.13. Fourth Moon Base – Manned Phase. (Courtesy M.A. Perino)

regarding the head position and balance. When these signals from the vestibular system and the brain are incongruent, an individual feels nausea."[36] Astronauts on longer missions can adapt to this incongruence.

Valeri Polyakov holds the distinction of spending the most consecutive time in space, 438 days on the Russian space station Mir. On longer missions, more serious effects are noticed. Due to the lower gravity, blood and body fluids shift, and there are muscle atrophy and bone density loss. In microgravity, fluids shift upward in the body – in Earth gravity we walk and the mechanical loads result in leg muscle contraction, arterial and venous constriction, and a resulting blood flow through one-way valves. "However, without gravitational opposition, blood pressure decreases causing the potential of cerebral ischemia [blood loss in the brain] and neuronal [brain cell] death."[37]

Additionally, the lack of gravitationally-induced signals leads to bone loss and slows biological growth. Microgravity results in an annual 1–2 percent bone loss in astronauts in weight-bearing areas such as the pelvic bones, lumbar vertebrae and femoral neck. As bones in the lower body atrophy due to lack of use, upper body skeletal regions grow in density.

Research has been performed on the utility of artificial hypergravity in countering these effects. Exercise will be a significant fraction of days spent in space and on extra-planetary surfaces. Pregnancies in a low gravity environment will be very challenging to women and their fetuses. But we may envision that, as we become "permanent" residents of the Moon, we may decide that we will let evolution take its course. Very few of us will likely visit Earth. Our lives will be lived on the Moon and beyond. We may need to become prepared to see subsequent generations with less development of their lower bodies.

Figure 5.14. Apollo 8 crew of Frank Borman, Bill Anders and Jim Lovell preparing for the first manned mission to the Moon, December 21–27, 1968. They made a Christmas Eve live broadcast to the people of Earth from lunar orbit. (Courtesy NASA/spacefacts.de)

However, some of the effects can be temporarily pleasing: "Former Astronaut Susan Helms liked how her body changed while she was on the International Space Station for six months. 'I got taller. I shed about 20 pounds. Your legs get very skinny. If you have varicose veins, they'll go away. Wrinkles go away because the fluid shifts. You're getting the idea that they could build a spa in space and there would be a lot of people paying money to go,' she joked. 'It's almost like a fountain of youth in a sense. Your body goes back 20 years and sheds the effects of gravity. It's pretty amazing. Unfortunately, it all reverses in a short time once you return to Earth'."[38]

Another effect of microgravity and low gravity is on our ability to judge size and distance.[39] Apollo astronauts reported difficulties judging distances and features. On the Moon, vision can be distorted, making it difficult to judge the speeds of objects, which we estimate by observing how an object changes in size.

Those who one day move to other planetary bodies with larger-than-Earth gravity fields will need to undergo hypergravity training to build up their ability to survive in gravity fields several times that of Earth.

Plants

Green has always been a part of the vision for space. Not money, that is, but plants. The role of plants in a manned facility on the Moon is multi-faceted. We have plants in our homes and offices because their presence makes us feel better. They make the rooms feel more in tune with our psychological needs. We like the colors of plants. They become a part of the atmosphere of the room in numerous ways.

Plants can become as crucial to human survival in space as oxygen. They are "complex eukaryotic[3] organisms that share fundamental metabolic and genetic processes with humans and all higher organisms, yet their sessile nature requires that plants deal with their environment by adaptation *in-situ*."[40] This is very different to the way humans deal with their environments, which is by creating a protective structure, by changing the environment, or by leaving.

Plants address their environment by adjusting their metabolism. In this way, plants can be viewed as biological sensors that can be monitored to report on their environments. They are sensitive to fluctuations in gravity, radiation, temperature and pressure – as are humans. "Truly insightful experiments that address fundamental questions about biological adaptation and responses to extreme extraterrestrial environments can be answered using plants. They are easy to transport in spaceflight in their seed, under complete vacuum and in extremely low temperatures, and are fundamental components of any long-term bioregenerative life support system."

Plants and animals have complementary existences, as "plants recycle human wastes and provide human nutrition, while humans recycle plant wastes and provide plant nutrients. As biosensors and due to their adaptation to the space environment via their ability to alter their gene expression, plants are viewed as an extremely useful indicator of how humans would evolve over the generations in space and on the planetary bodies."

We can expect to see tremendous greenhouses on the Moon at a well-developed lunar settlement. The interiors of our habitats will have plants distributed throughout. While we do not speak to our plants – that is, most of us don't – we communicate with them psychologically, and through sounds and music. Flowering plants will be provided appropriate light so that they can bloom. In the early years on the Moon, those blooms will give tremendous joy to our pioneering settlers.

Given the pace of genetic engineering, it is not unrealistic to expect that certain plants will be engineered to survive in the lunar environment via what may be called *lunaponics*. Recall hydroponics on Earth, where the cultivation of plants is accomplished by placing their roots in liquid nutrient solutions rather than in soil, leading to the soilless growth of plants. Lunaponics is similar – plants are coupled to bioengineered vines that provide the plant access to cyanobacteria, which can break down lunar rocks into constituents that the plants can use for food. These evolved plants will require minimal shielding on the lunar surface.

Marigolds can grow in crushed rock very much like that found on the lunar surface, with no need for plant food.[41] The marigolds are planted in crushed anorthosite – a rock very similar to much of the lunar surface – and, with added bacteria, the plants thrive.

---

[3] Eukaryote – any organism having as its fundamental structural unit a cell type that contains specialized organelles in the cytoplasm, a membrane-bound nucleus enclosing genetic material organized into chromosomes, and an elaborate system of division by mitosis or meiosis, a characteristic of all life forms except bacteria, blue-green algae, and other primitive microorganisms.

"The medicinal significance of marigolds is high, and a marigold may be recommended for antiulcerogenic action, for prevention of liver and kidney inflammation, and for onco-protection."[42] The first plants on the Moon may be utilized to create a fertile soil that could grow second-generation plants for human consumption. The pioneer plants may be engineered to survive on lunar regolith, be resistant to diseases, with periods of little light, low gravity, and all of the harsh conditions found on the Moon. The microbial community infused into the regolith will be critical in allowing the plants to survive.

Experiments show that cyanobacteria can grow in lunar soil, if provided with water, air and light. When cyanobacteria are placed in a container with water and simulated lunar soil, they produce acids that can break down tough minerals including ilmenite, which is relatively abundant on the Moon. The cyanobacteria use only sunlight for energy. "Cyanobacteria in growth chambers, where water, sunlight and lunar soil are provided, ... can be harvested and further processed to make use of the elements they extract from the lunar soil ... that can be used as fertilizer for food plants grown in hydroponic greenhouses ... [and] ... methane given off by the breakdown of the cyanobacteria could be used as rocket fuel."[43]

Of tremendous interest is the multifaceted value that plants and bacteria have in support of a lunar habitat. From medicinal value, to food, and in support of biotechnologies and material extractions, plants and bacteria are truly wondrous. As Shiga noted:

"The most challenging technologies for future lunar settlements are the extraction of elements (for example, iron, oxygen, and silicon) from local rocks for life support, industrial feedstock and the production of propellants. While such extraction can be accomplished by purely inorganic processes, the high energy requirements of such processes drive the search for alternative technologies with lower energy require-ments and sustainable efficiency. Well-developed terrestrial industrial biotechnolo-gies for metals extraction and conversion could therefore be the prototypes for extraterrestrial biometallurgy. Despite the hostility of the lunar environment to unprotected life, it seems possible to cultivate photosynthetic bacteria using closed bioreactors illuminated and heated by solar energy. Such reactors might be employed in critical processes such as air revitalization, element extraction, propellant (oxy-gen and methane) and food production."[44]

These pioneering studies may lead to an ability to utilize bacteria in all of the industrial and life support processes on the Moon. Plants and humans will form a partnership in space, creating a sustainable biosystem that can flourish.

Quotes

- "Magnificent desolation." Buzz Aldrin reflects on the view from the Moon after stepping off the Eagle landing module onto the lunar surface.
- "Hello Neil and Buzz. I'm talking to you by telephone from the Oval room at the White House, and this certainly has to be the most historic telephone call ever made. I just cannot tell you how proud we all are of what you have done. For every American, this has to be the proudest day of their lives." President Richard Nixon congratulates the astronauts on being the first men to walk on the Moon.

- "Fate has ordained that the men who went to the Moon to explore in peace will stay on the Moon to rest in peace. These brave men, Neil Armstrong and Edwin [Buzz] Aldrin, know that there is no hope for their recovery. But they also know that there is hope for mankind in their sacrifice." The opening lines of a speech, prepared by President Nixon's speechwriter William Safire, to be used in the event of a disaster that would maroon the astronauts on the Moon.
- "Here Men From The Planet Earth First Set Foot Upon the Moon, July 1969 A.D. We Came in Peace For All Mankind." The inscription on a plaque left behind.
- "As soon as somebody demonstrates the art of flying, settlers from our species of man will not be lacking [on the Moon and Jupiter]. ... Given ships or sails adapted to the breezes of heaven, there will be those who will not shrink from even that vast expanse." Johannes Kepler, letter to Galileo, 1610.
- "I think there's a supreme power behind the whole thing, an intelligence. Look at all of the instincts of nature, both animals and plants, the very ingenious ways they survive. If you cut yourself, you don't have to think about it." Clyde Tombaugh.
- "Space flights are merely an escape, a fleeing away from oneself, because it is easier to go to Mars or to the Moon than it is to penetrate one's own being." Carl Gustav Jung.
- "There is a theory which states that if ever anyone discovers exactly what the Universe is for and why it is here, it will instantly disappear and be replaced by something even more bizarre and inexplicable. There is another theory which states that this has already happened." Douglas Adams.

References

1. **Project Horizon**: *Volume I - Summary and Supporting Considerations; Volume II - Technical Consideration & Plans*, United States Army, March 20, 1959 (declassified September 21, 1961).
2. **Study of Deployment Procedures for Lunar Exploration Systems for Apollo (LESA)**, Lockheed LMSC-665606, Department of the Army and NASA, February 15, 1965. (Courtesy NASA/US Government)
3. **Criteria for the Design of Structures for a Permanent Lunar Base**, S.W. Johnson, Ph.D. Dissertation, Civil Engineering, University of Illinois, Urbana, 1964. (Advisors: Nathan Newmark and Joseph Murtha.)
4. S.W. Johnson and R.S. Leonard: *Evolution of Concepts for Lunar Bases*, pp.47–56, 1985 Lunar Base Conference, Lunar and Planetary Institute, Houston.
5. Report of the Lunar Base Working Group, April 23–27, 1984, NASA LALP 84–83.
6. The Second Conference on Lunar Bases and Space Activities of the 21st Century, W.W. Mendell, Editor, sponsored by the L.B. Johnson Space Center and the Lunar and Planetary Institute, Houston, April 5-7, 1988. NASA Conference Publication 3166, Vol.1 and Vol.2, 1992.
7. *Overview of Existing Lunar Base Structural Concepts*, H. Benaroya, TC Chairman, Journal of Aerospace Engineering Vol.5, No.2, April 1992.

8. G.B. Ganapathu, J. Ferrall, and P.K. Seshan: *Lunar Base Habitat Designs: Characterizing the Environment, and Selecting Habitat Designs for Future Trade-Offs*, JPL Publication 93–20, May 1993.

9. *Applied Mechanics of a Lunar Base*, ASME Applied Mechanics Reviews, H. Benaroya, Editor, Vol.46, No.6, June 1993.

10. Lunar Structures, Journal of the British Interplanetary Society, H. Benaroya, Editor, Vol.48, No.1, January 1995.

11. **Human Spaceflight: Mission Analysis and Design**, W.J. Larson and L.K. Pranke, Editors, McGraw-Hill, 1999.

12. H. Benaroya, L. Bernold, and K.M. Chua: *Engineering, Design, and Construction of Lunar Bases*, Journal of Aerospace Engineering, Vol.15, No.2, April 1, 2002, pp.33–45.

13. H. Benaroya, and L. Bernold: *Engineering of lunar bases (Review)*, Acta Astronautica 62 (2008) pp.277–299.

14. C.S. Allen, R. Burnett, J. Charles, F. Cucinotta, R. Fullerton, J.R. Goodman, A.D. Griffith, Sr., J.J. Kosmo, M. Perchonok, J. Railsback, S. Rajulu, D. Stilwell, G. Thomas, T. Tri, J. Joshi, R. Wheeler, M. Rudisill, J. Wilson, A. Mueller, and A. Simmons: **Guidelines and Capabilities for Designing Human Missions**, NASA/TM-2003-210785, January 2003.

15. **The Lunar Base Handbook, An Introduction to Lunar Base Design, Development, and Operations**, Second Edition, edited and written by P. Eckart, 2006, McGraw-Hill.

16. M.M. Cohen and H. Benaroya: *Lunar-base structures*, **Out of this World, the New Field of Space Architecture**, A.S. Howe and B. Sherwood, Editors, AIAA 2009.

17. **MOON - Preospective Energy and Material Resources**, V. Badescu, Editor, Springer 2012.

18. H. Benaroya, and M. Ettouney: *Design and Construction Considerations for a Lunar Outpost*, Journal of Aerospace Engineering, Vol.5, No.3, July, 1992, pp.261–273.

19. B. Sherwood, and L. Toups: *Technical Issues for Lunar Base Structures*, Journal of Aerospace Engineering, Vol.5, No.2, April, 1992, pp.175–186.

20. H. Benaroya, and M. Ettouney: *Framework for Evaluation of Lunar Base Structural Concepts*, Journal of Aerospace Engineering, Vol.5, No.2, April, 1992, pp.187–198.

21. R.M. Drake, and P.J. Richter: *Concept Evaluation Methodology for Extraterrestrial Habitats*, Journal of Aerospace Engineering, Vol.5, No.3, July, 1992, pp.282–296.

22. E.P. Steinberg, and W. Bulleit: *Considerations for Design Criteria for Lunar Structures*, Journal of Aerospace Engineering, Vol.7, No.2, April, 1994, pp.188–198.

23. P. Eckart: *Lunar Base Parametric Model*, Journal of Aerospace Engineering, Vol.10, No.2, April 1997, pp.80–90.

24. H-H. Koelle, and H-N. Mertens: **Conceptual Design of a Lunar Base**, Shaker Verlag, Aachen 2004.

25. A.M. Jablonski, and K.A. Ogden: *A Review of Technical Requirements for Lunar Structures - Present Status*, International Lunar Conference, 2005. Also, *Technical Requirements for Lunar Structures*, Journal of Aerospace Engineering, Vol.21, No.2, April 1, 2008, pp.72–90.

26. M.E. Morphew: *Psychological and Human Factors in Long Duration Spaceflight*, MJM 2001 6:74–80.

27. J.T. Celentano, D. Amorelli, and G.G. Freeman: *Establishing a Habitability Index for Space Stations and Planetary Bases,* AIAA/ASMA Manned Space Laboratory Conference, CA, May 2, 1963.

28. B.K. Wise, J.A. Wise: *The Human Factors of Color in Environmental Design*: *A Critical Review*, NASA Report N89-15532, August 1988.

29. D.A. Vakoch, Editor: *Psychology of Space Exploration, Contemporary Research in Historical Perspective*, The NASA History Series, NASA SP-2011-4411, 2011.

30. **Isolation: NASA Experiments in Closed-Environment Living - Advanced Human Life Support Enclosed System**, H.W. Lane, R.L. Sauer, and D.L. Feeback, Editors, American Astronautical Society, Vol. 104, Science and Technology Series, 2002.

31. P. Suedfeld: *Space Memoirs: Valuable Hierarchies Before and After Missions – A Pilot Study*, Acta Astronautica, Vol.58, 2006, pp.583–586.

32. M. Simon, M.R. Bobskill, and A. White: *Historical Volume Estimation and a Structured Method for Calculating Habitable Volume for In-space and Surface Habitats,* Acta Astronautica 80 (2012) pp.65–81.

33. E.S. Lockard: *Beyond Habitation in Space: The Need to Design for Human Adaptation*, Astrosociology Research Institute, 2015.

34. M.M. Cohen, S. Haeuplik-Meusburger: *What Do We Give Up and Leave Behind?*, 45th International Conference on Environmental Systems, July 2015.

35. M.A. Perino: *Moon Base Habitability Aspects*, Proceedings of the 4th European Symposium on Space Environmental and Control Systems, Florence, Italy, October 21–24, 1991.

36. S.E. Bell, and D.L. Strongin: *Evolutionary Psychology and its Implications for Humans in Space*, in **Beyond Earth: The Future of Humans in Space**, B. Krone, Editor, CG Publishing, Apogee Space Press, 2006, pp.78–83.

37. V.M. Aponte, D.S. Finch and D.M. Klaus: *Considerations for Non-invasive In-flight Monitoring of Astronaut Immune Status with Potential Use of MEMS and NEMS Devices*, Life Sciences, Vol.79, 2006, pp.1317–1333.

38. L.S. Woodmansee: **Sex in Space**, p.48, CG Publishing, 2006.

39. R. Courtland: *Zero-gravity May Make Astronauts Dangerous Drivers*, NewScientistSpace.com, September 22, 2008.

40. R.J. Ferl and A.-L. Paul: *Plants in Long Term Lunar. Exploration*, 2159.pdf, NLSI Lunar Science Conference (2008).

41. R. Black: *Plants 'Thrive' on Moon Rock Diet*, BBC News website, April 17, 2008.

42. N.O. Kozyrovska, T.L. Lutvynenko, O.S. Korniichuk, M.V. Kovalchuk, T.M. Voznyuk, O. Kononunchenko, I. Zaetz, I.S. Rogutskyy, O.V. Mytrokhyn, S.P. Mashkovska, B.H. Foing, and V.A. Kordyum: *Growing Pioneer Plants for a Lunar Base*, Advances in Space Research, Vol.37, 2006, pp.93–99.

43. D. Shiga: *Hardy Earth Bacteria Can Grow in Lunar Soil*, NewScientistSpace.com, March 14, 2008.

44. I.I. Brown, J.A. Jones, D. Garrison, D. Bayless, S. Sarkisova, C.C. Allen, and D.S. McKay: *Possible Applications of Photoautotrophic Biotechnologies at Lunar Settlements*, Rutgers Symposium on Lunar Settlements, June 3–8, 2007.

6

Lunar-based astronomy

"A terrific place to stargaze."

Space exploration and science go hand in hand. This is true for both robotic and manned missions. The advantage of missions that include astronauts is that they may be able to figure out how to repair something that stopped working, and they can modify the research itinerary if something unexpected of greater scientific interest is found. (It would be a shame if non-carbon-based life forms were sitting to the side of a robotic explorer that is looking for carbon-based life forms, and the robot's sensors did not notice these life forms because the sensors were programmed so narrowly!)

The objective of the lunar observatory case study depicted in Figure 6.1 was to understand the effort required to build and operate a long-duration, human-tended astronomical observatory on the Moon's far side. Some scientists feel that the lunar far side, which is quiet, seismically stable, and shielded from Earth's electronic noise, may be the Solar System's best location for such an observatory. The facility would consist of optical telescope arrays, stellar monitoring telescopes and radio telescopes, allowing almost complete coverage of the radio and optical spectra. The observatory would also serve as a base for geologic exploration and for a modest life sciences laboratory. In the left foreground of the image, a large fixed radio telescope is mounted on a crater. The telescope focuses signals into a centrally located collector, which is shown suspended above the crater. The lander in which the crew would live can be seen in the distance on the left. Two steerable radio telescopes are placed on the right; the instrument in the foreground is being serviced by scientists. The other astronaut is about to replace a small optical telescope that has been damaged by a micrometeorite. A very large baseline optical interferometer system can be seen in the right far background.[1]

Astrophysicists, astronomers, and space scientists have long viewed space and the Moon as ideal sites from which to observe and gather data about the universe.[1] Lester *et al.* suggested that an existing lunar site is an excellent and cost-effective location, but that orbital

---

[1] This description is from the original caption for the image.

© Springer International Publishing AG 2018
H. Benaroya, *Building Habitats on the Moon*, Springer Praxis Books,
https://doi.org/10.1007/978-3-319-68244-0_6

Figure 6.1. This is an artist's concept depicting a possible scene of an observatory on the far side of the Moon. The artwork was part of a NASA new initiatives study that surveyed possible future manned planetary and lunar expeditionary activity. (This painting was done by Mark Dowman and Doug McLeod. S89-25054 Courtesy NASA)

vantages are superior.[2] Even though interest in the Moon has revitalized ideas about lunar astronomical observatories, the dramatic improvement in capabilities for free-space observatories suggests a re-evaluation. The lunar surface does offer huge performance advantages for astronomy over terrestrial sites, but Earth orbit, or the Lagrange points, offer performance that is superior to what could be achieved on the Moon. "While astronomy from the Moon may be cost-effective once infrastructure is there, it is in many respects no longer clearly enabling compared with free space." A disadvantage of free space observatories is their susceptibility to micrometeoroids and radiation from all directions.

The Moon has been viewed as an "outstanding [platform for *X*-ray astronomy] for its stability, the availability of unlimited space for large instruments, and the ability to observe any object continuously for as much as 14 days."[3] "Infrared astronomical observations from the Moon [have] the most obvious apparent advantage [of] the lack of atmosphere, … and the relative ease with which beams from an array of telescopes can be interferometrically combined [due to] the vacuum environment with constant refractive index."[4] "An observatory on the Moon is an ideal place for low frequency, solar system radio astronomy."[5]

A theme through some scientific and policy circles, regarding the debate between purely robotic and manned-plus-robotic space science, is won by the latter. Some have suggested that the added cost of sending people into space should be rolled over into purely robotic exploration missions, but it has been noted that such a roll over would never happen, and the extra funds would instead be funneled into other government expenditures unrelated to science. But the interest in human activities on the Moon, with an eye to science, continues to this day, unsurprisingly.[6]

Studies and proposals have been made for the construction and operation of large lunar telescopes, and a radio observatory at the lunar South Pole or elsewhere on the lunar surface.[7–9]

Figure 6.2. Setting up a lunar telescope array. (Courtesy NASA)

The engineering issues for designing and placing telescopes on the lunar surface are challenging.[10] While telescopes are unmanned structures, except for the usual tending and maintenance, that they are unshielded while operational raises additional engineering design constraints: extreme temperatures and gradients, micrometeoroid impacts, and dust accumulation. As per Johnson *et al.*, design challenges include: "dust mitigation, passive thermal cooling, passive thermal self-compensation of the structure for stability, [and] tailoring of thermal expansion coefficients (feasible with some composite materials)," in addition to the challenges for habitat designs, such as "design for lower lunar gravity and relevant dynamic loads, minimization of structure mass consistent with a multiplicity of constraints, selection of construction techniques appropriate for remote automated/robotic deployment, and [finding some way to use] lunar materials for shielding against cosmic and solar radiation and micrometeoroids."

The special 1995 issue of the *Journal of the British Interplanetary Society* on lunar-based astronomy provided an overview of interest in such astronomy, suggested a potential site for a robotic lunar observatory, and discussed telescope reliability and the critical issue of lunar dust.[11] In a paper in that issue, it was suggested that optics may have to be covered during rocket landings and launches, where, it was estimated, the dust would return to the surface after about an hour.[12] In addition, a 10 km diameter circle surrounding the observatory could be stabilized, neutralizing the dust problem. Studies were cited that summarized the types of optical degradation that can be expected on mirrors. Additionally, a number of dust mitigation approaches have been suggested by Johnson *et al.*:

1. shielding and baffles to protect mirrors and sensitive equipment from dust, meteoroid impacts and stray light
2. restricted zones for operations, and stabilized zones, as well as berms, to help intercept ejected dust
3. periodic cleaning and restoring of mirrors, and use of segmented mirrors to allow for replacement of individual damaged segments

4. development of dust mitigation technologies, especially after a presence is established on the Moon and tests are possible locally, and
5. dustlocks to capture and dispose of dust before it is carried to sensitive areas.

Murphy *et al.* summarized the damage to the Apollo retroreflector arrays over decades
of usage.[13] The Apollo astronauts placed the first of several retroreflector arrays on the
lunar surface, but new laser ranging data has revealed that the efficiencies of the three
Apollo reflector arrays have diminished by a factor of ten at all lunar phases, and by an
additional factor of ten when the lunar phase is near full Moon. The rate of decay of imaging quality suggests that the lunar environment damages optical equipment on the times-
cale of decades. "Dust or abrasion on the front faces of the corner-cube prisms may be
responsible, reducing their reflectivity and degrading their thermal performance when
exposed to face-on sunlight at full Moon. These mechanisms can be tested using laboratory
simulations and must be understood before designing equipment destined for the Moon."

Optical degradation was noticed in the 1970s, within one decade of placement on the
lunar surface. Dust appears to be the most likely candidate for the observed degradation.
It is suggested that electrostatic charging, in conjunction with mechanical disturbances by
micrometeorite and impact ejecta activity, could free the grains. Some of the data on the
degradation might be skewed by the fact that a Surveyor 3 camera lens was near the Apollo

Figure 6.3. Artist's concept of a large Arecibo-like radio telescope on the Moon, which uses
a crater for structural support. In the background are two steerable radio telescopes. (Artwork
done for NASA by Pat Rawlings of SAIC. S95-01561, February 1995. Courtesy NASA)

12 landing site and might have been corrupted by ejecta. An important distinction to make is between permanent abrasion and removable dust. An understanding of this difference can help in the design of resistant optical devices and safeguards.

Quotes

- "We would like to give special thanks to all those Americans who built the spacecraft; who did the construction, design, the tests, and put their hearts and all their abilities into those craft. To those people tonight, we give a special thank you, and to all the other people that are listening and watching tonight, God bless you. Good night from Apollo 11." Neil Armstrong concludes the final television broadcast from Apollo 11 on the night before splashdown, July 23, 1969.
- "Whoopee! Man, that may have been a small one for Neil, but it's a long one for me!" Charles Conrad Jr., Commander of Apollo 12, on becoming the third man to walk on the Moon on November 20, 1969.
- "There's a need for accepting responsibility – for a person's life and making choices that are not just ones for immediate short-term comfort. You need to make an investment, and the investment is in health and education." Buzz Aldrin.
- "Globalization means many other countries are asserting themselves and trying to take over leadership. Please don't ask Americans to let others assume the leadership of human exploration. We can do wonderful science on the Moon, and wonderful commercial things. Then we can pack up and move on to Mars." Buzz Aldrin.
- "Buzz Aldrin doesn't think we need to go back to the Moon – that we should go straight on to Mars. I'm more on the side that says we should go back to the Moon. I think there's a lot we can utilize the Moon for scientifically." Charles Duke.
- "So most astronauts getting ready to lift off are excited and very anxious and worried about that explosion – because if something goes wrong in the first seconds of launch, there's not very much you can do." Sally Ride.
- "Yes, I did feel a special responsibility to be the first American woman in space." Sally Ride.
- "Basically, all my life I'd been told you can't do that because you're female. So I guess I just didn't pay any attention. I just went ahead and did what I could and then, when the stars aligned, I was ready." Shannon Lucid.
- "I think there are huge lessons there, for young people who are getting started in life, as well as other people. And that is, to take responsibility for your own life. Only you are responsible for the course you take from there." Story Musgrave.
- "If women can be railroad workers in Russia, why can't they fly in space?" Valentina Tereshkova.
- "I was told having a website would help me. I have yet to figure out why my life story needs to be on the web." Wally Schirra.

References

1. R.C. Haymes, A.G. Opp, and R.V. Stachnik: *Study of Astrophysics from the Lunar Outpost*, Advances in Space Research, Vol.11, No.2, pp.(2)301-(2)306, 1991; B. Foing: *Astronomy and Space Science from the Moon: Panel Discussion and Perspectives*, Advances in Space Research, Vol.14, No.6, pp.(6)283-(6)288, 1994; B. Foing: *The Moon as a Platform for Astronomy and Space Science*, Advances in Space Research, Vol.18, No.11, pp.(11)17-(11)23, 1994.
2. D.F. Lester, H.W. Yorke, and J.C. Mather: *Does the lunar surface still offer value as a site for astronomical observatories?*, Space Policy 20 (2004) pp.99-107.
3. P. Gorenstein: *X-Ray Astronomy from the Moon*, Advances in Space Research, Vol.14, No.6, pp.(6)61-(6)68, 1994.
4. M. Harwit: *Infrared Astronomy from the Moon*, Advances in Space Research, Vol.14, No.6, pp.(6)69-(6)76, 1994.
5. A. Lecacheux: *Solar System, Low Frequency Radio Astronomy from the Moon*, Advances in Space Research, Vol.14, No.6, pp.(6)193-(6)200, 1994.
6. I.A. Crawford: *The scientific case for renewed human activities on the Moon*, Space Policy 20 (2004) pp.91-97.
7. W.W. Mendell: *The roles of humans and robots in exploring the solar system*, Acta Astronautica 55 (2004) 149-155; I.A. Crawford, M. Anand, C.S. Cockell, H. Falcke, D.A. Green, R. Jaumann, and M.A. Wieczorek: *Back to the Moon: The scientific rationale for resuming lunar surface exploration*, Planetary and Space Science 74 (2012) pp.3-14.
8. P.J. Van Susante: *Study Towards Construction and Operations of Large Lunar Telescopes*, Advances in Space Research, Vol.31, No.11, pp.2479-2484, 2003.
9. Y.D. Takahashi: *A Concept for a Simple Radio Observatory at the Lunar South Pole*, Advances in Space Research, Vol.31, No.11, pp.2473-2478, 2003.
10. T.J.W. Lazio, R.J. MacDowall, J.O. Burns, D.L. Jones, K.W. Weiler, L. Demaio, A. Cohen, N.P. Dalal, E. Polisensky, K. Stewart, S. Bale, N. Gopalswamy, M. Kaiser, and J. Kasper: *The Radio Observatory on the Lunar Surface for Solar Studies*, Advances in Space Research, 48 (2011) pp.1942-1957.
11. S.W. Johnson, K.M. Chua, and J.P. Wetzel: *Engineering Issues for Early Lunar-Based Telescopes*, Journal of Aerospace Engineering, Vol.5, No.3, July, 1992; J.P. Wetzel, and S.W. Johnson: *Engineering Technologies for Lunar-Based Astronomy*, Advances in Space Research, Vol.14, No.6, pp.(6)253-(6)257, 1994.
12. S.W. Johnson, K.M. Chua, and J.O. Burns: *Lunar Dust, Lunar Observatories, and Other Operations on the Moon*, in Lunar-Based Astronomy, H. Benaroya, Editor, Journal of the British Interplanetary Society, Vol.48, No.2, February 1995, pp.87-92.
13. T.W Murphy Jr., E.G. Adelberger, J.B.R. Battat, C.D. Hoyle, R.J. McMillan, E.L. Michelsen, R.L. Samad, C.W. Stubbs, and H.E. Swanson: *Long-term degradation of optical devices on the Moon*, Icarus 208 (2010) 31-35.

7

Materials and ISRU

"Live off the land."

The severe lunar environment not only causes concerns about human survivability, but also about material degradation. Issues include thermal stability, fatigue resistance, ultraviolet radiation effects, solar flare protons, galactic cosmic rays, high charge and high-energy particles, vacuum stability, and interior humidity gradients.[1] Materials at particular risk are composites, those fabricated by merging a suite of materials, each of which addresses part of the spectrum of strength or durability needed in a modern structure.

Thermal gradients and cyclic thermal inputs, due to various operations and changes in the lunar environment, can lead to fatigue. Cyclic loads of all types on a structure lead to a weakening of the material due to the breaking of bonds at the atomic level. This is a fact in the design of all structures, whether for the Moon or the Earth. Engineers have considerable experience in designing composites for extreme environments. Some of this experience will be transferred to the design of lunar structures.

Radiation effects on composites, such as polymeric matrix materials, leads to rapid aging, resulting in embrittlement and the development of microcracking. Radiation shields will be utilized and resistant composites are beginning to be developed.

Vacuum stability, sometimes called outgassing, is the process by which some of the elements in a material emerge from the compound. This can lead to problems in sensitive equipment, such as optical and other instrumentation, and if the outgassing occurs into the living quarters it places an added burden on life support systems.

Humidity gradients within the structure can result in damage to certain composite materials. Cyclic variations in humidity levels can redistribute internal stresses.

Some of these concerns are characteristic of the interaction of the material and the lunar environment, so burying the bulk of the structure would avoid many of these problems. Perhaps by using regolith-derived structural components, we can avoid many such issues.

ISRU will be fundamental for human survival beyond Earth. In time, we expect to create an infrastructure for the fabrication of everything needed for survival. A significant expansion of the human presence on the Moon requires easily fabricated habitats from

H. Benaroya, *Building Habitats on the Moon*, Springer Praxis Books, https://doi.org/10.1007/978-3-319-68244-0_7

local materials. Beyond the first transported habitats from Earth, second-generation habitats may be fabricated with sulfur-based concrete.[2] As Pinni pointed out:[3]

"Sulfur-based concrete, using sulfur as the cement, is commonly used in harsh terrestrial environments. It has properties different from Portland cement, in that it doesn't have capillary porosity, and this makes it impermeable to fluids. The concrete would be composed of mineral aggregates cemented together with highly polymerized native sulfur, thereby making a durable concrete. The typical percentages of ingredients for the terrestrial sulfur concrete are around 80 percent of aggregates, 12 percent of sulfur, and 8 percent of fly ashes. It is ... used in some civil construction [on Earth] because of its properties, especially in chemically aggressive environments and in the presence of salts. It is capable of being operational within 24 hours from the time of casting and also has the possibility of being cast at temperatures well below 0°C."

Whenever lunar settlements are envisioned, roads between the different structures are also included. The regolith dust needs to be contained and suppressed so that it does not track into machines and habitats. Microwave sintering of the lunar soil is an attractive concept. A robotic paving machine can pass over the path to be sintered sending microwave energy into the regolith, melting and solidifying it to a prescribed depth depending on the purpose of the road.[4]

A side benefit of the process is the release and capture of most of the solar-wind particles within the soil – hydrogen, helium, carbon, and nitrogen.

"Sintering of pre-formed blocks of soil can be used to form solid bricks for various construction purposes, for example, igloos! An impact crater might be selected and smoothed out to a parabolic shape. Subsequent microwave treatment of the surface could produce an antenna dish, complete with a smooth glass surface. Or, this dish might be cut into sections for movement and reassembly to another location. ... Processing of pre-molded soil can produce strong structural components. Melting of the mare low-viscosity soil [can be used] for blowing glass wool or pulling of glass fibers."[5]

Figure 7.1 suggests a possible future on the Moon and beyond where autonomous fabrication, in this case via extrusion, will populate sites.[6] For the settlement of extraplanetary bodies, beginning with the Moon, in-situ resources must be used to fabricate habitats, laboratories, berms, garages, solar storm shelters and greenhouses, for example. It is simply infeasible to carry large masses (and volumes) from the Earth to the Moon, and later Mars and beyond. Examples of what is needed include: "extruded concrete and inflatable concrete dome technologies based on waterless and water-based concretes, development of regolith-based blocks with potential radiation shielding binders, including polyurethane and polyethylene, pressure regulation systems for inflatable structures, production of glass fibers and rebar derived from molten lunar regolith simulant, development of regolith bag structures, … [and of great importance are] automation design issues."

Fused regolith structures have been suggested where the proposed structures are small and many, and reside on the surface.[7] The Sun's energy can be used to fuse regolith into components. The advantages offered by numerous, smaller structures are safety and

Figure 7.1. ISRU-derived, extruded structures. Such autonomous creation of facilities will and must be the future for extraplanetary construction. (Courtesy NASA)

reliability. As long as the structures are independent in a probabilistic sense, they are more reliable as a group by virtue of redundancy. However, if they are all fabricated by the same machines using the same feedstock, then they are vulnerable as a group to some systemic flaw. Couplings such as this can lead to systemic failures, even though it appears that there is a redundancy.

Various related concepts have also been studied: a precast, prestressed concrete lunar base with a floating foundation in order to minimize differential settlement; utilizing unprocessed or minimally processed lunar materials for base structures as well as for shielding; the use of indigenous materials for the design of a tied-arch structure; construction using layered embankments of regolith and filmy materials (geotextiles) via robotic construction; and fabric-confined soil structures.[8–12]

Landis proposed a process sequence for refining oxygen, metals, silicon, and glass from lunar regolith, consisting of separating the required materials from lunar rock with fluorine, since their extraction cannot be done using the industrial production processes used on Earth.[13] These raw materials will be required for long-term habitation, and for the production of structural materials and solar arrays on the Moon. Figure 7.2 charts the lunar regolith composition, identifying the most abundant elements, all of which are fundamental to the creation of a lunar infrastructure. Landis suggested the following materials as useful feedstocks to be manufactured:

- Metals (structural)
- Composite structural materials: glass fiber and matrix
- Transparent materials: glass
- Electrical conductors (wiring): aluminum or calcium
- Semiconductors: silicon
- Semiconductor dopants: aluminum, phosphorus.

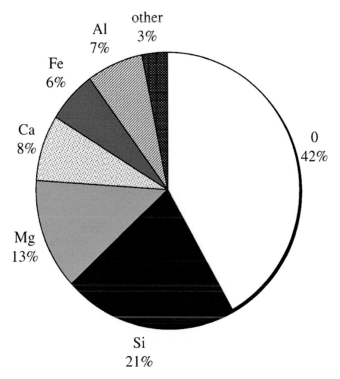

Figure 7.2. An averaged composition of the lunar regolith. (From Landis 2007, Courtesy Elsevier Publishers)

Landis concluded that "there seems to be no insurmountable barriers to producing from lunar materials the main raw materials required for basic manufacturing, including semiconductor-grade silicon and structural materials. ... Silicon, aluminum, and glass can be refined directly from the regolith using fluorine, and a multi-stage reduction process can separate out and purify the required materials. This refining process also produces oxygen, and possibly other refined materials of possible economic value."

A compilation of papers on ISRU in 2013 summarized the state-of-the-art.[14] In that journal, magnesium was considered to be a possible ISRU material useful for structures on the Moon.[15] Magnesium is one of the most pervasive metals in the lunar soil and has many characteristics that make it applicable to in-situ refining and production. This somewhat overlooked alkaline earth metal is easily cast, used, and recycled, characteristics that are required in the Moon's harsh environment. Moreover, alloys of this element have several properties fine-tuned for building shelters in a lunar environment, including several advantages over aluminum alloys. Magnesium alloys may prove to be the optimal choices for reinforcing lunar structures or manufacturing components as needed on the Moon. The following discussion is from Benaroya et al. (where citations can be found).

A particularly beneficial factor specific to high extraterrestrial magnesium alloy utilization is its high strength-to-weight ratio, especially in strong alloys such as AZ91, which even outperforms steel. Although aluminum alloys are typically proposed for such purposes, magnesium alloys have advanced greatly in recent decades and have been shown to have higher relative mechanical properties (property/density) and energy absorption than aluminum and steel. Because of these attributes, magnesium alloys have recently been used in the aerospace and automotive industries: many magnesium engine components are present in aircraft to reduce weight and retain strength.

Magnesium exhibits strong electromagnetic shielding, the effectiveness of which is established by its conductivity and permeability. For example, relative to copper, an excellent conductor, the relative conductivities of the following three metals show magnesium to be a good conductor: Mg (0.36), Al (0.61), and Fe (0.17). It has been considered, along with aluminum, as a radiation shield for space applications. Its neutron absorption cross-section value is given at 0.063 ± 0.004 barns/atom (1 barn = 10^{-24} cm^2).

Magnesium also has 30 times the vibrational damping of aluminum, and high impact resistance. These traits will be useful on the Moon for combating moonquakes and meteoroids respectively.

The low melting temperature of magnesium metal, 650 ± 2 °C is well within the range of the furnace heated by solar concentrators used on the ISS, and as such is doubtlessly achievable on the Moon. For comparison, aluminum's melting point is 660 °C, that of iron is 1535 °C, and that of titanium is a nearly unfeasible 1795 °C. Coupled with a latent heat of fusion that is two-thirds, and a specific heat that is three-quarters that of aluminum, these traits allow magnesium to be produced and recycled quickly and with little energy expenditure, when compared with other metals.

Magnesium is the most easily machined metal used in engineering designs and has excellent castability. Thus, magnesium replacement parts could be fabricated from regolith using a computer numerical control milling machine, leading to decreased dependence on Earth-based manufacturing.

Magnesium is a diamagnetic material, with a magnetic susceptibility of 1.2×10^{-5} that is close to aluminum's magnetic susceptibility of 2.2×10^{-5}. This dimensionless parameter equals zero if the material does not respond to any magnetization. Having a slight magnetization can be useful in refining or processing.

One of the main deterrents to the widespread use of magnesium is its reactivity with water and oxygen. However, because the surface pressure on the Moon is about 101.325×10^{-12} kPa, and there is essentially no atmosphere, this should not be a significant problem if the magnesium is only used externally on structures. Also, environmental corrosion in new nickel- and iron-free alloys is reportedly lower than that of steel. These alloys may minimize environmental corrosion to within acceptable levels.

Although its relative properties are advantageous, magnesium has a lower ultimate tensile strength and fatigue strength than aluminum and most structural metals. This is ultimately made less important by the Moon's low gravity, which is one-sixth of the terrestrial value. Furthermore, these minor strength deficiencies can be effectively mitigated by the use of strong alloys and well-designed casts with surface features, cross sections, and ribbing.

Additionally, magnesium is viewed as a safety risk because of its ability to ignite in the presence of oxygen. Most of this danger is dissipated by the Moon's nearly oxygen-free hard vacuum, yet some risk may linger as a result of oxygen's presence during some stages of regolith refining. It may be possible to take advantage of magnesium's thermal conductivity to quench a localized fire. Furthermore, the National Institute of Advanced Industrial Science and Technology of Japan has shown that the ignition points of magnesium alloys can be raised 200–300 °C by adding calcium in its alloys. Copper conductor rods may be packed in magnesium oxide to act as a seal and fireproof insulator.

Energy expenditure for the magnesium-refining process is largely caused by the difficulty of separating MgO from the regolith. Some proposed methods exist, but researchers and practitioners may favor processes that produce multiple desired products from the lunar soil for increased efficiency. Meanwhile, no study has focused directly on magnesium refining on the Moon, as it is considered refractory.

Although die-casting magnesium mandates costlier materials and more complex machinery, with higher clamping pressures and ram speeds than those for aluminum, the average magnesium die life is two (or more) times that of aluminum dies. As the current shipping cost to the Moon is about $10,000/kg using current rocket fuel, the additional costs and the higher power requirements of using magnesium dies are possibly less than the long-term costs of shipments of building materials from Earth. The goal is to forgo as many shipments from Earth as possible.

Of course, the use of magnesium requires that it be available in sufficient quantities and that an adequate infrastructure exists on the Moon for its extraction and processing.

As a recent example of using local materials for structural applications, Indyk has shown that minimally processed regolith can be used and is stronger than many other materials considered for the construction of a surface structure.[16] Other technologies and processes have also been suggested.[17] [18] Crawford provided an overview of our current knowledge about the resources available to us once we begin to inhabit the Moon.[19] While there is no single lunar resource that, on its own, can motivate an accelerated pace of lunar development, there are abundant raw materials that can push the creation and growth of lunar industrial capacity, benefitting cis-lunar space, the Earth, and, of course, the Moon. "Indeed, it is the fact that the lunar surface has the potential to support a growing scientific and industrial infrastructure, in a way that asteroidal surfaces do not, which is likely to make the Moon the linchpin of humanity's future utilization of the Solar System."

7.1   INTERVIEW WITH DONALD RAPP

This interview with Donald Rapp was conducted via email beginning on June 23, 2017 and concluded on July 5, 2017. We begin with a brief professional biography, and then continue with the interview.

Brief Biography
Donald Rapp is a chemical engineer, with degrees from The Cooper Union (BS), Princeton University (MS), and the University of California at Berkeley (PhD). His over 50-year career as a scientist and engineer included positions at Lockheed Palo Alto Research Laboratory, Polytechnic Institute of New York, University of Texas at Dallas, and the Jet Propulsion Laboratory. He has done fundamental work in chemistry and physics for many

years, and has worked on more applied problems, particularly in space technology and space mission design. He is a recognized leader in ISRU technology, and is an expert in requirements, architectures and transportation systems for space missions, with particular emphasis on the impact of in-situ resource utilization and water resources. Rapp has surveyed the wide field of global climate change and its relationship to energy consumption and is familiar with the entire literature of climatology. He is also knowledgeable on energy systems and thermodynamics. He is known within the broad NASA community for his abilities to plan, organize and lead studies of broad technical systems and was proposal manager on two major winning spacecraft proposals (Genesis and Deep Impact) that brought in $500 million. He has more than 70 publications in refereed journals, plus numerous informal in-depth reports, and has written the following books:

- *Quantum Mechanics*, 672 pages, published 1971 by Holt, Rinehart and Winston, self-published in 2013.
- *Statistical Mechanics*, 330 pages, published in 1972 by Holt, Rinehart and Winston, self-published in 2012.
- *Solar Energy*, 516 pages, published in 1981 by Prentice-Hall.
- *Human Missions to Mars: Enabling Technologies for Exploring the Red Planet*, Springer/Praxis, 2007; 552 pages, two 8-page color sections; 2nd edition 2015.
- *Assessing Climate Change – Temperatures, Solar Radiation and Heat Balance*, Springer Praxis Books – Environmental Sciences, third edition, 815 pages, 315 illus., Hardcover, 2014.
- *Ice Ages and Interglacials*, Springer/Praxis, 2nd edition, 405 pages, 188 illus., 2012.
- *Use of Extraterrestrial Resources for Human Space Missions to Moon or Mars*, Springer/Praxis, 185 pages, 28 illus., 2013.

Our Interview

Lunar structures have been grouped into three categories: I – habitats brought from Earth, II – habitats with components from Earth, but also components made from lunar resources, and III – habitats primarily made from lunar resources, ISRU, perhaps 3D printed structures. The longer it takes for us to get back to the Moon to stay, the more certain technologies advance, for example 3D printing. Would you keep the above categories, or modify them in some way?

I think that covers it well.

Given that we have been waiting almost 50 years to return to the Moon, do you have a feeling as to when we might return, with people, for an extended stay?

I think that the question of return to the Moon is very complicated and would require a lot of discussion – more than we have room for here. Since the glory days of Apollo, NASA has been searching for a target for human venture into space. The Shuttle proved to be faulty and expensive. The Space Station has not generated much and is also expensive. When Griffin was NASA Administrator, he wanted a specific target, and selected the Moon, probably because it seemed most feasible, affordable and could use mostly existing technology. As it turned out, it was more expensive than had been hoped, and they really couldn't figure out what good things there were to do on the Moon. Somehow, Congressional support waffled, Griffin left NASA, and return to the Moon fell back, amidst the battle cry of go to Mars. I simply don't have any idea whether or when NASA will return humans to the Moon. Right now, it seems improbable.

In your mind, what is a likely scenario and timeline for manned space?

The mindset seems to be to go to Mars. Overly optimistic projections claim that will happen in the 2030s. Elon Musk has fantastic dreams that make no sense to me. The problem as I see it is that NASA will need to spend a good part of its budget doing technology development and demonstration at Mars for more than 20 years just to have the capability to mount a human mission to Mars. When you look at the NASA budget, about 2/3 is so fully committed that the money is simply not there for this development. The Evolvable Mars Campaign seems to be mostly glossy viewgraphs that don't make sense. Maybe the 2060s?

Do you agree with those who have made a case for settling the Moon first and then using it as a base for travel to Mars?

No. Certainly not. I don't see how the Moon would help us get to Mars in proportion to the investment, since the lunar project will suck up more than $100 billion. Yes, there are connections and we can learn from lunar first, but not enough to justify the cost.

Is ISRU oversold for near-term lunar development?

Yes. As I have described in great detail in my books, lunar ISRU mostly does not make sense. The last lunar mission design under Griffin had space-storable ascent propellants brought from Earth as the most efficient approach, and the descending lander had to carry ascent propellants with it in case of abort to orbit. So there is no need to produce ascent propellants on the Moon, and furthermore the schemes to produce oxygen on the Moon all seem hopelessly complicated, expensive and technically challenging. On the other hand, lunar structures from lunar materials would make much more sense.

In your recent book on the use of extraterrestrial resources you provided very detailed cost/benefit analyses and found that ISRU is not the holy grail of space settlement, at least not for the first several generations of settlements.[20] Knowing that there are numerous detailed analyses behind your conclusions, can you summarize for us what are the key elements of your analysis and basis for your conclusions?

I don't know how to do this in the space available. ISRU on the Moon makes little sense because the cost/benefit ratio is poor, the processes are very difficult to implement, and the expedient mission design end runs ISRU. By contrast, Mars ISRU is technically feasible, has a very good cost/benefit ratio, and is essentially enabling. But even Mars ISRU requires a great deal of up-front investment, especially if it will include indigenous water from Mars.

You also analyze and compare lunar and Martian ISRU, and conclude that the latter is much more feasible and useful. Given the order of magnitude greater costs (and difficulties) of a manned mission to Mars as compared to the Moon, does it make sense to go to the Moon first and set up facilities there that can make the Martian trip easier and safer?

> *I briefly discussed this above. The problem is that they are both expensive (Moon and Mars) and while going to the Moon provides some help for Mars, not in proportion to the investment and the factor of 20- to 30-year time delay. Specifically for ISRU, going to the Moon provides little help.*

Key challenges to engineers and scientists who will be tasked to develop the lunar mission and structural design are the knowledge gaps that cannot be filled before being on site. For example, the lunar 1/6 *g* environmental effects on biology, and systems reliability in the lunar dust environment. Also of great concern is our understanding of the very complex systems that will not be integrated before they are placed on the Moon. Even if the behavior of all the components are fully understood, the complete integrated system will generally have unanticipated characteristics. Is there a way of dealing with these knowledge gaps before being there?

> *I am not very knowledgeable on this but I suspect that this can be done. Just a gut feel.*

Dust mitigation is a critical issue generally on the Moon. How far are we from getting a handle on that problem, both the biological risks and the engineered systems risk?

> *We are presently working on dust for a Mars experiment to fly on a 2020 rover of which I am a CoI. I am not aware of bio risks, but I do think dust can be filtered. Yet, no dust prevention scheme can be 100 percent. So I don't really know enough to answer you.*

The isolated environment that astronauts will face on the Moon, and even more isolated on Mars, makes critical the need for high levels of reliability. Part of that implies a self-healing capability. Such technologies are being studied and developed but are far from being usable. Do you see self-healing as a critical technology?

> *If a problem develops on the Moon, the crew can return in a few days. If a problem develops on Mars, the crew must stay for the entire mission sequence (6 months getting there, 18 months on Mars, 6 months returning). There is a vast difference in requirements for reliability on Mars vs. the Moon. Is self-healing the answer? Maybe. But testing, testing, testing is needed.*

Manned space settlement has been promoted as a justified expenditure, financed by taxpayers, for many reasons: economic development, scientific advances, SETI, manifest destiny, building bridges between nations, limitless energy production, for example. What is your case for manned space settlement? Do you have an overarching reason for supporting such an endeavor?

> *I am not an enthusiast for humans in space, nor am I opposed. I think that most of reasons given above are rather far-fetched. I don't have a self-generated wild enthusiasm for it. Ultimately, I think we provide artificial justifications for things we simply want to do out of a spirit of adventure and imagination.*

It appears that missions to settle the Moon or Mars have two underlying potential show stoppers: (1) assuring human survival and good health, and (2) assuring very high reliability for all engineered components. While we have shown with the ISS and Apollo that we can manage short forays to the Moon, or longer ones to LEO, satisfying (1) and (2) in limited ways, the bar is an order of magnitude higher for long-duration lunar stays, and an order of magnitude higher than that for Mars human missions. Are guaranteeing (with high probability) human and machine survival and high operability on the horizon, or beyond it?

As I show in my books, there are many technologies that need to be demonstrated at Mars on a large scale, and that entails an extended test verification program spread over at least two decades. Can the reliability be made high enough to justify such a mission? Depends on where you set the bar. I doubt that a human mission to Mars could be better than maybe 92 percent reliable. Risk will always be present.

Who would you consider as your key influence in your pursuit of space studies? Who inspired you?

The writers of DRM-1 around 1997 who wrote the first comprehensive Mars design reference mission.

All of us working on space recognize that much of what we do and are interested in promoting will not happen in our lifetimes. Of course, we have no choice but to accept this reality, but how do you view this?

Our job is to try to get NASA pointed in a good direction toward achievable goals. Preparing early roadmaps for future generations is OK with me.

Thank you.

7.2 LAVA TUBES

"Underground is the place to be."

Lava tubes were assumed to exist on the Moon and Mars long before they were actually located. Ideas regarding the utility of constructing the first outposts under the lunar surface have been long-proposed. Preliminary assessments recommended subselene development for the most effective evolutionary potential for settlement.[21] Based on recent photographs and analyses, there is hope that efforts can be made to identify a lava tube for exploration, and future habitation, after a human presence is initiated on the lunar surface.

Some point to the possibility of accessing larger lava tubes for the creation of underground settlements.[22] The advantage of utilizing caves for siting habitable bases is that the underground location offers natural protection against some of the most severe environmental problems, namely radiation, large temperature gradients and micrometeorites. Such caves may need to be shored up structurally, sealed with a polymeric material, and then pressurized. Once pressurized, structures can be built or placed inside the cave – these interior structures would likely not have to be pressurized and can therefore be of a much simpler design.

Summaries and speculations about the size and extent of lunar lava tubes, now known to exist, are of interest as potential sites for habitats, other structures and storage.[23] [24]

Lava tubes, formed at high extrusion rates when active low viscosity lava flows, develop a continuous and hard crust due to radiative cooling of their outermost parts. This thickens and forms a solid roof above the still-flowing lava. Once completely extruded, the molten magma is no more, leaving an empty flow channel that is a cylindrically-shaped tunnel below the surface. Lava tubes are generally correlated to "sinuous rilles." (See Figure 7.5.) Lava tubes exist on Earth, but those on the Moon can be much larger, by at least an order of magnitude in each dimension, due to the reduced gravity field. There is evidence of collapsed lava tubes, however. If their roofs are insufficiently thick, they are prone to collapse as a result of impacts that spall the inner roof structure, eventually reducing the cross-section to the point where the local basaltic tensile strength is exceeded. Meteorites and seismic activity are the sources of energy that can lead to such collapses. The meteorites can pierce the lava tube roofs, or just become embedded. In both cases, the impact results in stress waves that can create spalling.

The interiors of lava tubes provide a relatively constant-temperature environment, estimated at −20 °C, that is shielded from radiation. The interior is also relatively dust free. Recently observed lava tubes have large volumes, perhaps too large for early settlements. Depending on their orientation and angle of incidence with the surface, they can be difficult to enter.

Radiation safety is one of the major benefits of being within a lava tube.[25] Below a depth of 6 m, simulations show that there are no radiation effects due to, or induced by, GCRs. Below 1 m, there are no radiation effects due to, or induced by, SPE particles.

The first lava tubes were discovered from data gathered by the JAXA Kaguya mission.[1] A Japanese team analyzed this data and found holes in the lunar surface, called "skylights." One was found in the Marius Hills region of the Earth-facing side of the Moon, shown in Figure 7.3. The skylight was approximated to be 60 m wide, and its discovery caused quite a stir among proponents of lava tube bases.[26]

Haruyama *et al.*[27] [28] discovered a vertical hole on the Moon, a possible lava tube skylight, based on the Japanese lunar orbiter Selenological and Engineering Explorer (SELENE) data from the Terrain Camera and the Multiband Imager. "The hole is nearly circular, 65 m in diameter, and located in a sinuous rille at the Marius Hills [MHH] region, a volcanic province on the lunar nearside. ... Its depth [is estimated] to be 48 m." Two additional very large holes were also found and measured:

Mare Tranquillitatis Hole
with a near-circular opening of 98 m × 84 m, a long-axis angle of 165 degrees, and a depth of 107 m, and

Mare Igenii Hole
with an opening of 118 m × 68 m, a long-axis angle of 36 degrees, and a depth of 45 m.

It is interesting to note that the Tranquillitatis hole is not in or near any sinuous rilles and has no exposed surface bulges, which are often seen with terrestrial lava tubes. The MHH is strongly believed to have lava tube caverns (extended voids) at the bottom of the hole. A dramatic off-center view of this lava tube is seen in Figure 7.4.

---

[1] The Japanese SELENE (Selenological and Engineering Explorer) mission, also nicknamed Kaguya, was launched on September 14, 2007, and was intentionally crashed into the lunar surface at the end of its mission on June 10, 2009.

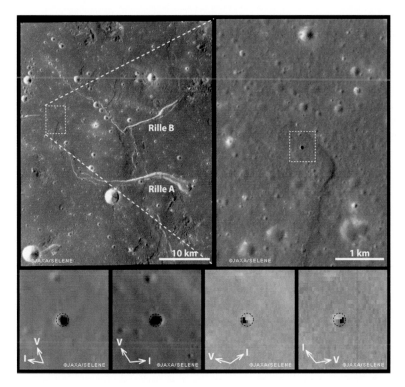

Figure 7.3. Images of the Marius Hills pit as observed under different solar illumination conditions by the SELENE/Kaguya Terrain Camera and Multiband Imager. (Courtesy JAXA/SELENE)

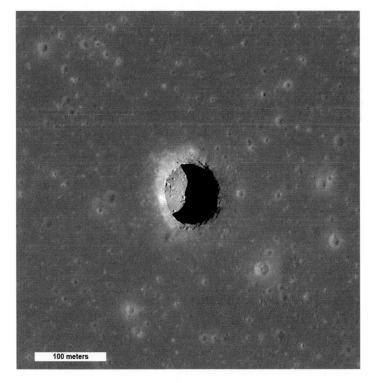

Figure 7.4. Spectacular high Sun view of the Mare Tranquillitatis pit crater revealing boulders on an otherwise smooth floor. Image is 400 meters wide, north is at the top, NAC M126710873R. (Courtesy NASA/GSFC/Arizona State University)

It is believed that important geological and mineralogical processes can be uniquely observed and studied inside lava tube holes, and they are therefore viewed as resources for scientific investigations as well as locations for habitats. In addition to protection from radiation and micrometeoroids, benefits for habitats include flat floors, sealed inner walls, and a relatively dust-free environment. It is suggested that, due to the geologic ages of most lava tubes, they are inherently stable in a structural sense. This needs to be proven empirically.

Ashley *et al.* confirmed that:[29]

"Lunar Reconnaissance Orbiter Camera (LROC) Narrow Angle Camera (NAC) images [reveal] potential opportunities for … exploration and habitation beneath the lunar surface at a variety of locations and geologic environments across the Moon. NAC imaging currently confirms the existence of sublunarean voids associated with two of three pits in mare deposits, and has revealed more than 140 negative relief features formed in impact melt deposits, some of which are likely to be the result of collapse into subsurface void spaces. The possibility for additional spaces beneath intact ceilings in both types of settings is plausible, if not likely."

Mission concepts for lava tube reconnaissance and for a scientific survey have been outlined.[30] [31] Concepts and assessments greatly depend on the lava tube geometry, especially its inclination with the surface, and the area material content. Vertical, or almost vertical tubes certainly pose a challenge for access and operations. Tubes at more modest inclines with respect to the lunar surface are more attractive.

A critical concern regarding inhabiting lava tubes is their structural stability.[32] Blair *et al.* performed a thermo-elastic finite element structural analysis that calculated the stresses and strains around a lava tube of a given shape and size. The lava tube's width, roof shape and roof thickness were varied and the analysis was carried out in two dimensions, employing a plane strain assumption. Their models accounted for thermal and gravitational stresses, and it was assumed that the [lava tube] material would fail when principal stresses in the roof exceeded 10 MPa in tension ($\sigma_1 > 10$ MPa) or 200 MPa in compression ($\sigma_3 < -200$ MPa), or when the von Mises stress exceeded 10 MPa. The roof and walls of the tube comprised a free surface, so confining pressure was taken to be zero. Their failure values were slightly conservative (lower in magnitude) since other stress sources, such as seismic shaking from meteorite bombardment and tidal forces, were not included in the modeling. Furthermore, as per Blair *et al.*:

"… Results show that the lava tubes inferred from GRAIL[2] data may in fact be structurally stable at widths in excess of 1.6 km, given sufficient burial by subsequent lava flows [and] provided thermal stresses are low. There are several reasons why this might be the case, or why [these models] may overestimate the importance of

---

[2]GRAIL was the NASA Gravity Recovery and Interior Laboratory spacecraft used to calculate the Moon's gravitational field to high precision. Two spacecraft, Grail A and Grail B – later renamed in a contest to Ebb and Flow – orbited the Moon for almost a year, and were able to map the lunar gravitational field using data from detected changes in their relative positions. The spacecraft took off within a few hours of each other on September 10, 2011, went into lunar orbit on December 31, 2011 and January 1, 2012, and remained there until both targeted the Moon on December 17, 2012.

thermal stresses. First, [these] models are entirely elastic; the addition of plasticity may allow the large thermal stresses to be accommodated by deformation, keeping strains small and the tube structurally sound. This idea is supported by the observation that even small lava tubes as exist on Earth are subject to large thermal stresses, but remain standing. Second, the thermal history of a real lava tube is substantially more complex than the models presented here, ... the possible cooling of the tube prior to burial [is not accounted for], nor ... the fact that surrounding areas may be cooler than the lava tube itself."

However, this is a valuable assessment of lava tube stability. Haruyama *et al.* provided recent additional information on lava tubes identified using the SELENE Terrain Camera.[33]

Figure 7.5. Rima Krieger winds its way through Oceanus Procellarum. Image width is ~ 3.5 km. Taken from LROC NAC image M1152172510R (Courtesy NASA/GSFC/Arizona State University)

A structural engineer who is tasked to build or place a structure within a lava tube needs to gather the following information:

1. lava tube location on the lunar surface
2. local site and interior material (soil) properties
3. detailed lava tube dimensions and topology, orientation with respect to the lunar surface, the existence of interior volumes and caverns, and
4. a mapping of the temperatures within the lava tube.

Figure 7.6.   Schematic of how a structure placed inside a lava tube might look. (Courtesy Ana Benaroya)

Once such information is gathered, the tube topology and orientation determines the possibility of placing structures within the tube, perhaps on its floor. If the incline is not precipitous, then it is possible to roll or lower the structure down the tube. Otherwise, a complex lowering winching may be needed. Another possibility is to land an inflatable structure inside the tube for inflation. ISRU and other possibilities exist. A conceptualization in a final setting is shown in Figure 7.6.

The lava tube provides shelter from the hazardous space environment, but structural engineers may be concerned about the structural stability of the tube. Even if it is structurally stable, pieces may become dislodged and fall. This can be due to meteoroid strikes on the surface, rocket landings, surface activity, or seismic activity. Also, even though the lunar ground is very dense, we may want to seal the lava tube to make it airtight.

In order to satisfy both these concerns, we propose a hybrid, deployable-inflatable structure. This is a deployable structural framework that extends into the lava tube, exerting pressure on the lava tube walls if the tube is narrow enough, or just rolling down the side of the tube if it has a large diameter. A small distance inside the framework there is a membrane that also deploys with the framework, and can be inflated or pressurized once the structure is fully deployed. The membrane can be rigidized, or not, depending on whether the framework is intended to contain a smaller pressurized structure. This concept is under study, and shown schematically in Figure 7.7.

Figure 7.8 is an advanced concept for a lava tube city, showing profile and plan views.

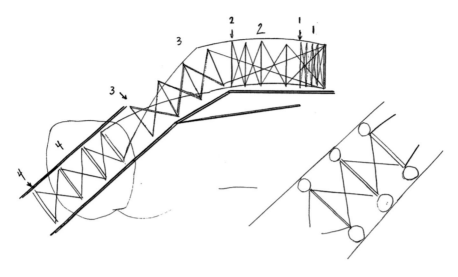

Figure 7.7. A schematic of an inflatable-deployable structure being deployed into a small diameter lava tube with a mild slope. If the lava tube has a large diameter, then the deployment may be one that only needs to be rolled down the slope.

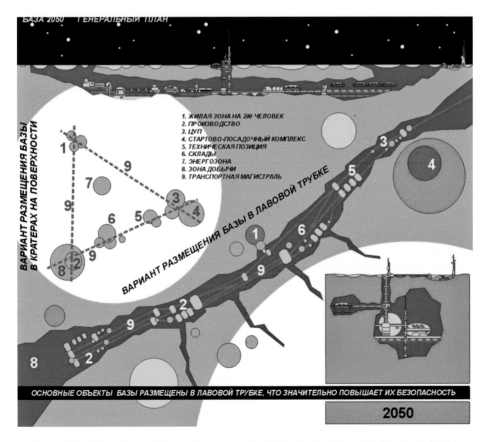

Figure 7.8. This advanced lava tube contains the following facilities: (**1**). Residential (200 people); (**2**). Manufacturing; (**3**). Control; (**4**). Launch complex; (**5**). Technical; (**6**). Warehouses; (**7**). Power systems; (**8**). Shelters; (**9**). Roads. (Courtesy V. Shevchenko)

Quotes

- "We shape our buildings; thereafter they shape us." Winston Churchill.
- "Buildings, too, are children of Earth and Sun." Frank Lloyd Wright.
- "Architecture aims at Eternity." Christopher Wren.
- "Architecture is a visual art, and the buildings speak for themselves." Julia Morgan.
- "I see Earth! It is so beautiful!" Yuri Gagarin.
- "The greatest advances of civilization, whether in architecture or painting, in science and literature, in industry or agriculture, have never come from centralized government." Milton Friedman.
- "I must study politics and war, that my sons may have the liberty to study mathematics and philosophy, natural history and naval architecture, in order to give their children a right to study painting, poetry, music, architecture, tapestry, and porcelain." John Adams.

References

1. D.W. Radford, W.Z. Sadeh, and B.C. Cheng: *Material Issues for Lunar/Martian Structures*, IAF-91-302, 42nd Congress of the International Astronautical Federation, October 5–11, 1991, Montreal.
2. V. Gracia, and I. Casanova: *Sulfur Concrete: A Viable Alternative for Lunar Construction*, SPACE 98 Engineering, Construction, and Operations in Space, 1998, pp.585–591; W.N. Agosto, J.H. Wickman, and E. James: *Lunar Cement/Concrete for Orbital Structures*, SPACE 88 Engineering, Construction, and Operations in Space, 1988, pp.157–168; R.S. Leonard, and S.W. Johnson: *Sulfur-Based Construction Materials for Lunar Construction*, SPACE 88 Engineering, Construction, and Operations in Space, 1988, pp.1295–1307; H. Namba, T. Yoshida, S. Matsumoto, K. Sugihara, and Y. Kai: *Concrete Habitable Structure on the Moon*, SPACE 88 Engineering, Construction, and Operations in Space, 1988, pp.178–189; H. Namba, N. Ishikawa, H. Kanamori, and T. Okada: *Concrete Production Method for Construction of Lunar Bases*, SPACE 88 Engineering, Construction, and Operations in Space, 1988, pp.169–177; D. Strenski, S. Yankee, R. Holasek, B. Pletka, and A. Hellawell: *Brick Design for the Lunar Surface*, SPACE 90 Engineering, Construction, and Operations in Space, 1990, pp.458–467.
3. M. Pinni: *Lunar Concrete*, in **Lunar Settlements**, H. Benaroya, Editor, CRC Press, 2010.
4. T.T. Meek, F.H. Cocks, D.T. Vaniman, and R.A. Wright: *Microwave Processing of Lunar Materials: Potential Applications*, Los Alamos Report LA-UR-85-312, also Lunar Bases and Space Activities of the 21st Century.
5. L.A. Taylor, and T.T. Meek: *Microwave Sintering of Lunar Soil: Properties, Theory, and Practice*, Journal of Aerospace Engineering, Vol.18, No.3, July 2005.
6. M.P. Bodiford, K.H. Burks, M.R. Perry, R.W. Cooper, and M.R. Fiske: Lunar In-Situ Materials-Based Habitat Technology Development Efforts at NASA/MSFC, 2006.
7. E.W. Cliffton: *A Fused Regolith Structure*, SPACE 90 Engineering, Construction, and Operations in Space, 1990, pp.541–550; R.S. Crockett, B.D. Fabes, T. Nakamura, and C.L. Senior: *Construction of Large Lunar Structures by Fusion Welding of Sintered*

Regolith, SPACE 94 Engineering, Construction, and Operations in Space, 1994, pp.1116–1127.

8. T.D. Lin: *Concrete for Lunar Base Construction*, Concrete International (ACI), Vol.9, No.7, 1987; T.D. Lin, J.A. Senseney, L.D. Arp, and C. Lindbergh: *Concrete Lunar Base Investigation*, Journal of Aerospace Engineering, Vol.2, No.1, January 1989.

9. E.N. Khalili: *Lunar Structures Generated and Shielded with On-Site Materials*, Journal of Aerospace Engineering, Vol.2, No.3, July 1989.

10. J.A. Happel: *The Design of Lunar Structures Using Indigenous Construction Materials*. A Thesis submitted to the Faculty of the Graduate School of the University of Colorado, in partial fulfillment of the Master of Science in Civil Engineering, 1992; J.A. Happel: *Prototype Lunar Base Construction Using Indigenous Materials*, SPACE 92 Engineering, Construction, and Operations in Space, 1992, pp.112–122.

11. M. Okumura, Y. Ohashi, T. Ueno, S. Motoyui, and K. Murakawa: *Lunar Base Construction Using the Reinforced Earth Method with Geotextiles*, SPACE 94 Engineering, Construction, and Operations in Space, 1994, pp.1106–1115.

12. R.A. Harrison: *Cylindrical Fabric-Confined Soil Structures*, SPACE 92 Engineering, Construction, and Operations in Space, 1992, pp.123–134.

13. G.A. Landis: *Materials refining on the Moon*, Acta Astronautica 60 (2007) pp.906–915.

14. Special Issue: *In-situ Resource Utilization*, H. Benaroya, P. Metzger, and A. Muscatello, Editors, Journal of Aerospace Engineering, January 2013, Vol.26, No.1.

15. H. Benaroya, S. Mottaghi, and Z. Porter: *Magnesium as an ISRU-Derived Resource for Lunar Structures*, Journal of Aerospace Engineering, Vol.26, No.1, January 1, 2013, pp.152–159.

16. S.J. Indyk, and H. Benaroya: *A Structural Assessment of Unrefined Sintered Lunar Regolith Simulant*, Acta Astronautica 140 (2017), pp.517–536; S.J. Indyk: *Structural Members Produced From Unrefined Lunar Regolith, A Structural Assessment*, Masters Thesis, Rutgers University, October 2015, submitted in partial fulfillment for the MS Degree.

17. G. Cesaretti, E. Dini, X. De Kestelier, V. Colla, and L. Pambaguian: *Building components for an outpost on the Lunar soil by means of a novel 3D printing technology*, Acta Astronautica 93 (2014) 430–450; B. Khoshnevis: *Automated construction by contour crafting-related robotics and information technologies*, Automation in Construction, 13 (2004) pp.5–19.

18. R.J. Soileux, and D. Osborne: *The in-situ construction, from vitrified lunar regolith, of large structures including habitats in artificial lava caves*, Journal of the British Interplanetary Society, Vol.68, Nos.9–10, September–October 2015, pp.268–274; R.J. Soileux, and K.I. Roy: *Safe comfortable habitats on the Moon, Mars, and Mercury using soil vitrification*, Journal of the British Interplanetary Society, Vol.68, Nos.9–10, September–October 2015, pp.275–281; N.J. Werkheser, M.R. Fiske, J.E. Edmunson, and B. Khoshnevis: *Development of additive construction technologies for application to development of lunar/Martian surface structures using in-situ materials*, CAMX 2015 - 2nd Annual Composites and Advanced Materials Expo, Dallas, October 26–29, 2015, pp.2395–2402.

19. I.A. Crawford: *Lunar resources: A review*, Progress in Physical Geography, 2015, Vol.39(2) pp.137–167.

20. **Use of Extraterrestrial Resources for Human Space Missions to Moon or Mars**, Donald Rapp, Springer-Praxis, 2013.

21. A.W. Daga, M.A. Daga, and W.R. Wendell: *A Preliminary Assessment of the Potential of Lava Tube-Situated Lunar Base Architecture*, SPACE 90 Engineering, Construction, and Operations in Space, 1990, pp.568–577.

22. R.R. Britt: *Lunar Caves: The Ultimate Cool, Dry Place*, posted on Space.com, 1pm ET, March 21, 2000.

23. F. Hörz: *Lava Tubes: Potential Shelters for Habitats*, 1985 Lunar Bases and Space Activities of the 21st Century, W.W. Mendell, Editor, Lunar and Planetary Institute, Houston, pp.405–412.

24. C.R. Coombs, and B.R. Hawke: *A Search for Intact Lava Tubes on the Moon: Possible Lunar Base Habitats*, 2nd Conference on Lunar Bases and Space Activities of the 21st Century, Houston, April 5–7, 1988, pp.219–229.

25. G. de Angelis, J.W. Wilson, M.S. Clowdsley, J.E. Nealy, D.H. Humes, and J.M. Clem: *Lunar Lava Tube Radiation Safety Analysis*, Journal of Radiation Research, 43: Suppl. S41-S45 (2002)

26. I. O'Neill: *Living in Lunar Lava Tubes*, Discovery Channel, posted online October 27, 2009.

27. J. Haruyama, K. Hioki, M. Shirao, T. Morota, H. Hiesinger, C.H. van der Bogert, H. Miyamoto, A. Iwasaki, Y. Yokota, M. Ohtake, T. Matsunaga, S. Hara, S. Nakanotani, and C.M. Pieters: *Possible lunar lava tube skylight observed by SELENE cameras*, Geophysical Research Letters, Vol.36, L21206, 2009.

28. J. Haruyama, T. Morota, S.K. Kobayashi, S. Sawai, P.G. Lucey, M. Shirao, and M.N. Nishino: *Lunar Holes and Lava Tubes as Resources for Lunar Science and Exploration*, **MOON - Prospective Energy and Material Resources**, Chapter 6, V. Badescu, Editor, Springer 2012.

29. J.W. Ashley, M.S. Robinson, B.R. Hawke, A.K. Boyd, R.V. Wagner, E.J. Speyerer, H. Hiesinger, and C.H. van der Bogert: *Lunar Caves in Mare Deposits Imaged by the LROC Narrow Angle Cameras*, First International Planetary Cave Research Workshop (2011) 8008.pdf.

30. S.W. Ximenes, J.O. Elliott, and O. Bannova: *Defining a Mission Architecture and Technologies for Lunar Lava Tube Reconnaissance*, 13th ASCE Conference on Engineering, Construction, and Operations in Challenging Environments, Pasadena, April 15–18, 2012.

31. A. Dage: *Lunar and Martian Lava Tube Exploration as Part of an Overall Scientific Survey*, A White Paper submitted to the Planetary Sciences Decadal Survey 2013-2022

32. D.M. Blair, L. Chappaz, R. Sood, C. Milbury, H.J. Melosh, K.C. Howell, and A.M. Freed: *Determining the Structural Stability of Lunar Lava Tubes*, 46th Lunar and Planetary Science Conference (2015) 2174.pdf.

33. J. Haruyama, T. Kaku, R. Shinoda, W. Miyake, A. Kumamoto, K. Ishiyama, T. Nishibori, K. Yamamoto, K. Kurosawa, A.I. Suzuki, S.T. Crites, T. Michikami, Y. Yokota, R. Sood, H.J. Melosh, L. Chappaz, and K.C. Howell: *Detection of Lunar Lava Tubes by Lunar Radar Sounder Onboard SELENE (Kaguya)*, Lunar and Planetary Science XLVIII (2017) 1711.pdf.

8

Structural design of a lunar habitat

"An actual design."

This chapter is a summary of the work of Ruess, and Ruess *et al.*[1][2] A method that assigns numerical values to the level of success of a structural concept was used to decide on one habitat for a structural analysis. Environmental conditions, structural requirements, materials including local resources, and ease of construction were the criteria groupings. Transportation issues were also considered, since it is anticipated that the first- and second-generation lunar bases will be brought to the Moon from Earth. It may happen that 3D printing and layered manufacturing will continue to revolutionize our advanced manufacturing abilities, and it may be that the first habitat on the lunar surface is created in this way. In this case, we will still have to transport the fabrication equipment to the Moon.

8.1 HABITAT GEOMETRY AND LOADS

Five habitat concepts were considered: two inflatable concepts, a lunar crater base, a three-hinged arch structure that resembles an igloo, and an underground structure. Based on transportation, construction, experience base, foundation requirements, and excavation requirements, the three-hinged arch was considered to be slightly more optimal than the inflatable habitat, and was chosen for a design. It needs to be noted that the grading system had some subjective aspects and therefore, given the closeness of the grading, another analyst might conclude that the inflatable structural configuration is a better choice. Such a weighting procedure is time dependent, since it is a function of the available and anticipated technologies.

The three-hinged arch concept is shown in Figure 8.1, and with dimensions and interior partitions in Figure 8.2. The concept of this arch structure is attractive due to its relative simplicity, and that its components can be brought to the Moon without the need for an excessively large launch system. Hinges were chosen rather than fixed connections because they tend to be easier to construct. The three-hinged arch is a statically determinate

© Springer International Publishing AG 2018
H. Benaroya, *Building Habitats on the Moon*, Springer Praxis Books,
https://doi.org/10.1007/978-3-319-68244-0_8

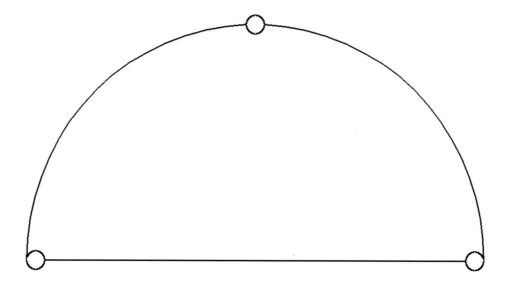

Figure 8.1. Three-hinged concept. Two arches are connected by a base tie beam. The interior is pressurized once endplates cap the structure. (All the images in this chapter were created by F. Ruess, unless stated otherwise. Courtesy F. Ruess)

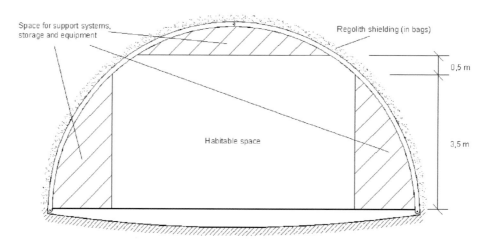

Figure 8.2. Three-hinged arch chosen for the design, showing the shielding and space for storage and equipment. (Courtesy F. Ruess)

structure. Therefore, temperature loading during construction does not introduce stresses in the members, only deflections. The two arches are held in configuration by the floor tie beam. The arches and tie beam are all designed to withstand tensile and bending loads.

The cross-section is semi-circular with a radius of 5 m, and 4 m headroom, allowing for 1/6 g bouncing-walking. Figure 8.3 shows the internal pressurization, as well as the ground reaction. This design takes advantage of the strength of arched structures in tension.

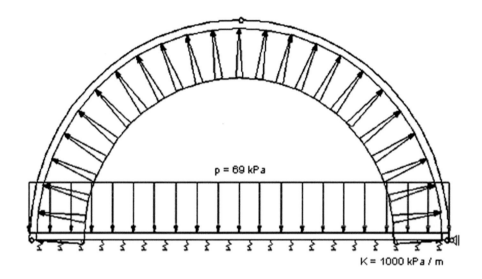

Figure 8.3. Internal pressurization and ground reaction. The internal pressurization is the governing static load. It is only partially balanced by the external regolith cover load. (Courtesy F. Ruess)

Figure 8.4 depicts the regolith cover loads acting on the structure after it is completely covered. Most of the loads on the structure may act at the same time after completion. But there are also a number of different design scenarios requiring consideration: starting with the construction stages when the structure is not yet complete and therefore not shielded; and when it is initially pressurized with the regolith partially placed on top of it. We then have the regular operational mode with all loads acting, and finally, the structure needs to withstand a planned or accidental decompression.

The maximum loading on the structure has to be found using a number of load combinations. For each scenario, only the loads that increase the stresses in the structure are included. Self-weight (dead load) is always present. Four combinations are used to find the maximum stresses in the members:

1. internal pressure plus floor loads
2. regolith cover plus installation loads
3. all loads, and
4. one-half of the regolith cover during construction.

Figure 8.5 shows the partial load on the structure during the process when regolith is placed on it, demonstrating that the structure must withstand the partial loads experienced

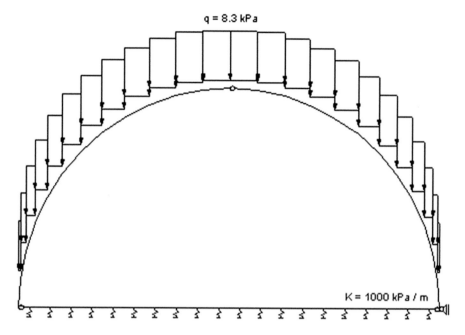

Figure 8.4. Structure subjected to regolith shielding load. The structure sits on the ground, which has the stiffness constant K shown in the figure and is represented by the line of springs. (Courtesy F. Ruess)

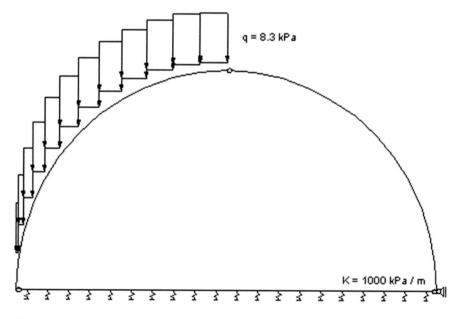

Figure 8.5. Partial regolith cover placement loads. In this configuration, at an intermediate stage in the placement of regolith atop the structure, the arches are subject to the largest bending moments. (Courtesy F. Ruess)

during construction, as well as in the final configuration. This loading case, where the structure is not yet pressurized, results in the maximum stresses in the arch segments.

The load case depicted in Figure 8.3, where the structure is pressurized but not yet shielded by a regolith layer, results in the highest stresses for the tie beam. Therefore, the structural design (stresses and deflections) must be carried out for these loading cases.

The construction material chosen was high strength aluminum, the members being pre-fabricated on Earth and transported to the Moon. Composite materials that could further save mass are possible, but care needs to be given to the behavior of these in a vacuum environment. So, all calculations were based on a high strength aluminum alloy with a yield strength of 500 N/mm²

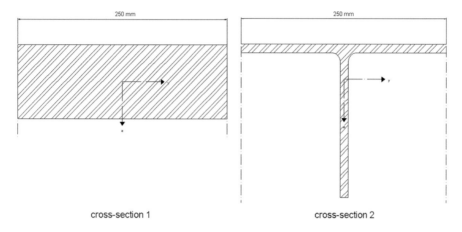

Figure 8.6.   Possible cross-sections of the arch structure. Four possibilities were considered. (Courtesy F. Ruess)

8.2   HABITAT DETAILS

Four possible cross sections were considered for the design of the arch, shown in Figures 8.6 and 8.7. Of these, the double I-beam was selected for the design due to its weight-strength advantage. Two design scenarios were carried out, one with a factor of safety of 4, and another with a factor of safety of 5. The goal is to show that even with such high safety factors, the resulting design is still a feasible structure. Therefore, the structure is significantly over designed. Once we have more experience on the Moon, these designs can be scaled back considerably.

We point to Figures 8.8 and 8.9 that depict the connections between the arch panels, and the arch with the foundation tie. All internally pressurized lunar structures need to contain their atmosphere reliably. The design of the structural and other connections is critical, especially when considering that many of these connections will be under shielding or within walls after completion, with access for repairs being difficult. These designs provide a mating mechanism, as well as a large contact area, minimizing leakage. Not discussed here are the welds and taped fabric at interfaces and joints. Details are in the original documents.

Two main conclusions result from the structural analysis. First, the arch segments can have a uniform cross section. It is possible, but not necessary, to adjust the arch cross

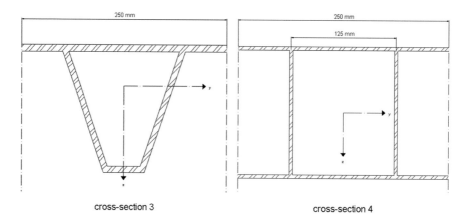

cross-section 3   cross-section 4

Figure 8.7. Possible arch cross-sections. The fourth section – the double I-beam – was selected for analysis and design. (Courtesy F. Ruess)

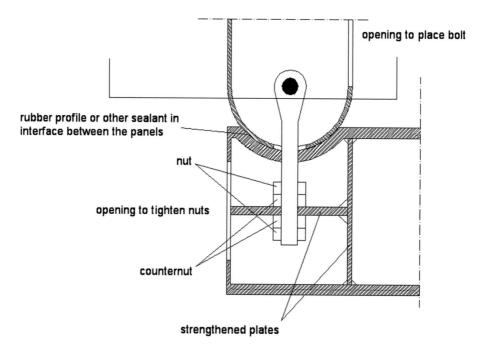

Figure 8.8. Hinge-floor connection, designed for ease of linking and to prevent leaks of the internal atmosphere. (Courtesy J. Schänzlin)

section to the distribution of internal forces since these are almost uniform. Second, in order to get an efficient cross section for the tie, its shape was adjusted to the distribution of internal forces. The bending moment has the shape of a parabola, suggesting the tie should have a similar shape. The depth of the tie cross section varies, with the smallest

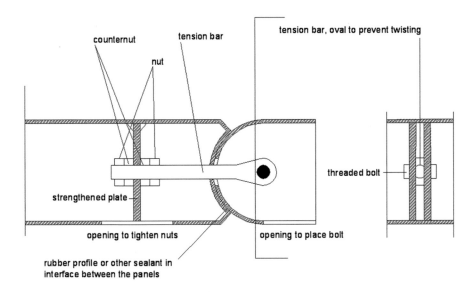

Figure 8.9. Panel connection, designed for ease of linking and to prevent leaks of internal atmosphere. (Courtesy J. Schänzlin)

depth at the ends and the largest in the middle. Figure 8.10 shows the principal shape of the tie/floor/foundation, identified as element 1.

The habitat dimensions in the longitudinal direction depend mainly on the crew size. The transportable length for structural members is limited by the transportation system, so dividing the members into different segments for transportation is necessary. A reasonable segment length is 2.5 m. Ten segments put together would then provide enough space for a crew of six astronauts. Connecting the segments has to ensure that all the loads can be transferred, and that the pressure is sealed. It is believed that welding the full section is the best solution to this problem. Welding in space has been performed (on the Russian Soyuz T12 mission, for example).

8.3 HABITAT ERECTION AND APPEARANCE

Erection of the structure can be done step-by-step, as depicted in Figure 8.10. It is a linear process. First, the floor is constructed. The 2.5 m-long panels are laid out on the site and joined using tongue and groove joints. Cables may help to pull all the panels together. Welding all the panels provides sealing and guarantees that all forces between structural sections can be transferred. The arch panels are erected one after the other, by placing one arch segment on a temporary construction scaffolding and bolting it to the floor at the bottom. The second arch segment is placed in the same way. Finally, the two arch segments are bolted together. The temporary scaffolding is lowered and transferred to the next section, where the placement of the next arch segments can begin. The panels then have to be welded together. All welding is best done at night to minimize inherent deflections due to temperature-induced deformations. If construction is performed during the lunar day, unprotected members are exposed to the Sun and temperatures on the member surface climb up to 150°C. However, the member's other side will be in the shadow, and therefore at a temperature of –100°C. This causes the member to deflect. These deflections are in the range of 0.5 m. However, they do not play a role in the arch construction. A three-hinged arch can always be joined together, for example, by lowering the scaffolding until the pieces fit. Figure 8.10 shows the structure after one arch is put in place.

When the desired structure length is achieved, the end walls can be slid in at the arch ends. The connections between arch and walls have to be established next. The structure is then sealed with membrane fabric strips glued to the structure along the structural connections where necessary. After completing all necessary seals, the habitat can be initially pressurized to test the seals, and finally the regolith cover can be put into place.

For the safety factor equal to 5, the floor segments have a mass of 2945 kg, so they weigh approximately 5 kN, while the arch segments have a mass of 605 kg and weigh about 1 kN each. Equipment will certainly be needed to move the parts from their landing site to the construction site and into place there, but no heavy equipment is needed to move 5 kN. A light crane is a good solution to put the segments in place onsite.

A solid rendering of a habitat module for a crew of three is illustrated in Figure 8.11. The lines on the surface are to enhance curvature and three-dimensional appearance.

Figure 8.12 is a sketch of the appearance of the structure with 3 m of regolith on top for shielding, while Figure 8.13 depicts an expanded settlement based on the structure designed here. There are solar energy panels in the background. Figure 8.14 suggests other alternative configurations using this structural module.

This design study demonstrates how standard Earth-based design methodologies can be applied to lunar surface structures.

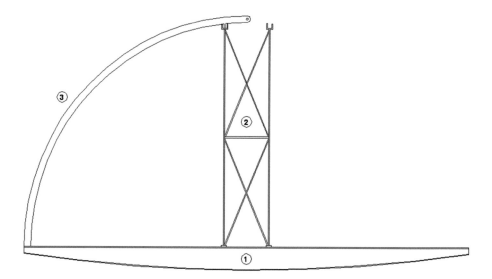

Figure 8.10. Erection sequence using temporary support. The tie, element 1, is placed on the smoothed lunar surface. Temporary scaffolding is next placed atop the tie. One half-arch is then attached to the left end of the tie and is supported by the scaffolding. The next step is to attach the right half-arch, connecting to the tie and left half-arch, utilizing the connectors discussed herein. (Courtesy F. Ruess)

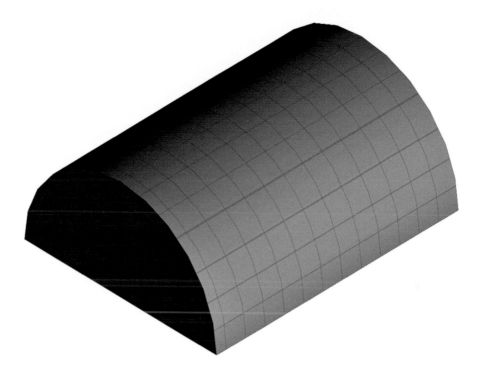

Figure 8.11.   A rendering of the structure. (Courtesy F. Ruess)

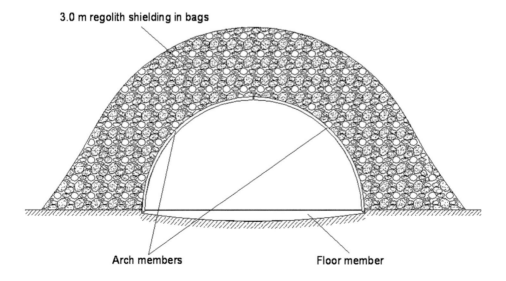

Figure 8.12.   Structure under 3 m of regolith shielding. (Courtesy F. Ruess)

Figure 8.13. Full base concept, including regolith shielding, access ports, and solar panels. (Courtesy Andre Malok)

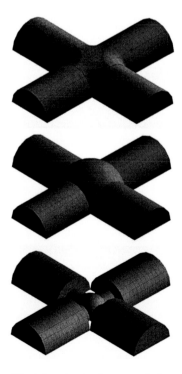

Figure 8.14. Realizations of the igloo structural concept in three possible expanded configurations. (Courtesy F. Ruess)

References

1. *Structural Analysis of a Lunar Base*, Diploma Thesis, F. Ruess, Advisor: H. Benaroya, University of Stuttgart, Rutgers University, April 2004.
2. F. Ruess, J. Schaenzlin, and H. Benaroya: *Structural Design of a Lunar Habitat*, <u>Journal of Aerospace Engineering</u>, Vol.19, No.3, July 1, 2006, pp.133-157.

9

Thermal design

*"It's hot for two weeks.
It's cold for two weeks."*

The extreme thermal environment has been outlined earlier. The extreme temperatures (roughly 100 K – 400 K) affect the design of the structure, its material components, its electrical and electronic systems, and its control systems, and will require systems to maintain nominal temperatures inside the habitats. A parallel challenge is energy generation during the long lunar night. Possibilities include the use of nuclear power. Also explored is the storage of heat in the lunar soil during the day for extraction during the night. We outline the challenges and possibilities here briefly, and close with a thermal structural analysis.

Temperature and humidity control are critical systems for habitat livability, and for the proper operations of equipment.[1] A lunar habitat is a Controlled Ecological Life Support System (CELSS), requiring air temperature and humidity control. Various systems exist for such control. During the lunar day, when temperatures are at their peak, dissipation of habitat heat into the environment is challenging due to the adverse temperature gradient.

A 2012 review of cooling systems for lunar surface habitats provided an extensive discussion of the technologies as well as tradeoffs.[2] The two primary objectives of a thermal control system are to remove excess waste heat from electrical components – and humans – in the habitat, and to keep a thermal energy balance with the outside environment. Of the three mechanisms of heat transfer: convective (through an intermediate medium such as an atmosphere), conductive (through direct contact), and radiative (by electromagnetic radiation), only the last is a significant option for the lunar surface. Radiative heat transfer includes the absorption of solar radiation, and the emission of thermal radiation. As per Fraser, for a spacecraft:

> "On the Moon, radiative heat transfer primarily occurs between the surface of an object and the lunar surface, the Sun, the albedo of the Earth or any other object, including the IR shine of the Earth and other objects. The net sum of this radiative heat transfer determines whether heat is accumulated or rejected by the object. The waste heat

H. Benaroya, *Building Habitats on the Moon*, Springer Praxis Books,
https://doi.org/10.1007/978-3-319-68244-0_9

released within the spacecraft and the net radiative heat flux therefore have to be carefully balanced in order to keep the spacecraft's interior and temperature-sensitive components within their safe operating temperature ranges. This can be achieved by applying a wide range of different active and passive systems. The combination of these systems is referred to as the thermal control system."

Temperatures are adjusted by the transfer of heat, or heat flux. Thermal control systems are generally grouped as passive or active. Passive systems follow a load profile defined as a function of the operating conditions. For example, the thermal energy removed by a specific radiator is a function of the radiative energy balance at the radiator. Radiator temperature, environmental conditions and energy dissipation rates balance each other so that environmental conditions remain constant. "Passive thermal control systems do not require a power input and do not have any (electro)mechanical components or mechanisms. Thus, they are very reliable, normally do not require any maintenance, and impose low mass, volume and cost requirements. The operational lifetimes of passive systems are often limited by degradation of the thermo-optical properties." Passive thermal control systems include a special surface coating, thermal insulation, Sun shields, or radiation fins. Heat pipes can be used to achieve an efficient heat transfer between the internal waste heat source and the radiator, for example.[3]

Active thermal control systems allow variations in the heating or cooling load by applying a higher driving voltage, or by changing the frequency of intermittent operating periods, for example. As per Fraser, "Active thermal control systems normally require a power input and often apply (electro)mechanical components or mechanisms. Reliability and maintenance issues, therefore, have to be considered. Mass, volume, and cost requirements often have to be evaluated more carefully than with passive systems." Examples for active thermal control systems are heat pumps, heaters, coolers, shutters/louvers, or fluid loops. Active thermal control systems can be further broken down into systems with and without temperature lift. A system without temperature lift can only transfer source heat to a lower temperature heat sink. Systems providing temperature lift consume more energy and are more complex than those that do not provide temperature lift, thus impacting maintenance and reliability. On the lunar surface, during the high temperature (400 K) daytime, an active system with temperature lift may be a necessity. Heat pumps of various cycles are considered as possible solutions to the challenging thermal cooling problem for lunar habitats.[4] The heat pumps provide temperature lift and a radiator that dissipates waste heat.

Thermal storage in deep lunar soil is put forward as a heat storage and dissipation mechanism.[5] Thermal wadis (dry valleys) using modified regolith are proposed as engineered sources of stored solar energy for the nighttime operation of lunar rovers, and in a preliminary concept to run a heat engine during the lunar night.[6,7]

The design of a surface or other lunar structure, whether for habitat or other purposes, requires an analysis of the thermal balance between the structure and its environment. Malla and Brown modeled surface and subsurface temperatures and their variations for the purpose of designing a surface habitat.[8] That habitat was a rectangular configuration supported by an external frame, and sealed via a membrane. Given that the frame was an external structure, it would be exposed to the lunar thermal cycles and undergo expansions and contractions that would alter the static, dynamic, and frequency response of the structure.

9.1   A STATIC ANALYSIS

This section is based on the work of Mottaghi, and Mottaghi and Benaroya, and is a summary of that work.[9] [10] The images herein have been drawn by Mottaghi, and appear here by permission.

An igloo-shaped structure, made of a magnesium alloy, supported on a sintered regolith foundation, and covered with sandbags of regolith shielding, was selected as a generic habitable structure for the lunar surface. Its location was taken to be on a mountain in the lunar south polar region at 0°0'0" W, 86°0'0" S. This location was selected due to the fact that the Sun never completely sets there, and that the Earth is visible 100 percent of the time. This igloo-shaped structure has been revisited periodically in this book. For a static analysis, it can be considered to be a pressure vessel.

Static analysis
Cylindrical or spherical thin-walled pressure vessels are those with ratios of inner radius, r, to wall thickness, t, greater than 10. Such pressure vessels can be analyzed as a plane stress problem (membrane analysis). Several assumptions can be made:

1. Plane sections remain plane.
2. The ratio $t/r \leq 0.1$, with t being uniform and constant.
3. The material is elastic, isotropic, and homogenous.
4. The resultant stresses are less than the proportional limit.
5. The applied pressure, p, is the gauge pressure, defined as the relative pressure of inner pressure with respect to outer pressure.
6. Stress distributions throughout the wall thickness are constant and equal to the average stress.
7. The cross section is selected so that it is far from geometric discontinuities.
8. The fluid/gas contained has negligible weight compared to the pressure.

If these conditions can be satisfied, the stress distributions can be found by considering the free body diagram of an appropriate cut section, and by the equations of static equilibrium.

The regolith shielding layer should be thick enough to protect the structure from solar flare radiation, extreme temperature fluctuations and meteoroid impacts. Various thicknesses have been proposed in the literature, from well below one meter to over 3 m. This analysis is based on a 3 m layer of regolith.

Habitat dimensioning
Pressure vessel equations can be used as initial estimates for the wall thicknesses. Due to discontinuities and sharp curves in the geometry of the structure, and regions around openings, stress concentrations need to be considered. To reduce these effects, fillets are included. Slightly thicker walls can compensate for bending stresses that are not accounted for in the simplified pressure vessel equations.

Writing the static equilibrium equation for the free-body diagram of a section of the structure, we obtain:

$$pA = A_t \sigma_{allow},$$

where A is the cross-sectional area under air pressure, A_t is area of the wall of uniform thickness t, p is the gauge pressure, and σ_{allow} is the allowable stress, defined as the ultimate tensile strength of the material (σ_u) divided by an appropriate factor of safety (FS):

$$\sigma_{\text{allow}} = \sigma_u / FS.$$

In order to account for stress concentrations, the allowable stress must be corrected by a stress-concentration factor (K), defined as the ratio of maximum stress to the average stress computed in the section of discontinuity:

$$K = \frac{\sigma_{\text{max}}}{\sigma_{\text{allow}}}.$$

Then:

$$pA = A_t \frac{1}{K} \frac{\sigma_u}{FS}. \tag{9.1}$$

Using readily available curves for stress concentrations, K is estimated to be 2.65. The low gravitational acceleration of the Moon allows us to neglect the weight of the sandbags by simply considering the gauge pressure, p, equal to the internal pressure of 1 atm (1.01325×10^5 Pa). Based on the chosen geometry of the structure, $A = 150$ m² and $A_t = 50t$ m². The factor of safety can be chosen to be very high, say 5, even though such a structure can be adequately designed with a much smaller factor of safety. Substituting these values

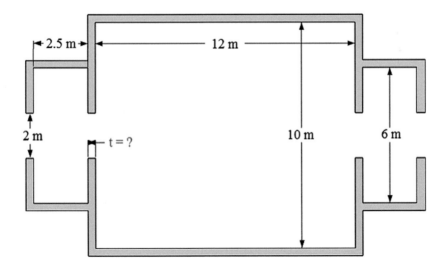

Figure 9.1. Floor plan for the structure.

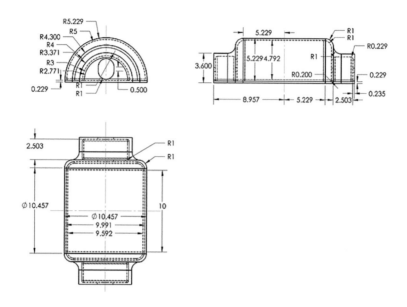

Figure 9.2.   Engineering drawing of lunar structure under consideration (using Solidworks).

in Equation 9.1, where the ultimate tensile strength of pure magnesium is 21 MPa, the required thickness of the shell is found to be 0.192 m. This likely underestimates the required thickness, since bending stresses are also induced by the regolith transverse loads. Furthermore, the stress concentration factor assumes that the openings are far from any geometrical discontinuities. Therefore, as a preliminary guess, this study assumes $t = 0.229$ m (9 in). The results of a stress analysis will show whether the chosen thickness is adequate. The high factor of safety assures us that we are safe.

It can be observed in Figure 9.2, and the CAD model Figure 9.3, that the sharp edges have been smoothed.

The structure is considered to be fixed to a foundation. A uniform pressure of 1 atm (1.01325 x 105 Pa) is acting on all interior surfaces. In order to account for the forces applied by the regolith shielding, the bulk density of the regolith is estimated to be 1650 kg/m³, which is within the range of about 1500 to 1740 kg/m³ for the upper 1 m layer of the Moon. The weight results in a uniform downward pressure of 8096 Pa (density × thickness). The physical properties of the magnesium structure are considered to be the same as for sand cast magnesium; that is, Young's modulus = 40 GPa, tensile (compressive) yield strength = 21 MPa, Poisson's ratio = 0.35, and density = 1738 kg/m³. Also, the lunar gravitational acceleration is taken as one sixth terrestrial, 1.64 m/s².

Figure 9.3. CAD model of the lunar structure under consideration (drawn in Solidworks).

Static computational results

Figure 9.4 shows that different parts of the structure have different factors of safety. This is generally the case in any design, since it is almost impossible to optimize a design with respect to all the parameters that govern that design. There are geometrical, material, con-structability, fatigue, and human factors constraints, to mention a few. Designers will often rank the constraints and then work towards a sub-optimal solution, meaning that the design will be optimal with respect to the top one or two constraints flagged as most critical. As expected, the range of factors of safety in this static analysis is high, ranging from around 7 to around 15.

Naturally, we expect to have higher factors of safety for most parts of the structure, because the chosen thickness for the structure is based on the stress concentration about the doorway. A structure with thinner walls, modified by adding few extra supports, can easily achieve higher safety factors. For example, adding a few ribs on the walls con-taining the exterior doors will increase the section properties by lessening the stress concentrations, and will result in a safety factor equal to 6 or higher. We did not modify or optimize the structural design, since the purpose of this work is to show the feasibil-ity of considering this type of structure for a lunar base, utilizing magnesium as a structural material.

Also, we have selected the physical properties of sand-cast pure magnesium, which possesses the lowest strength among unalloyed processed magnesium (using other meth-ods, for example, extrusion, hard rolled sheets, and annealed sheets). This is due to the fact

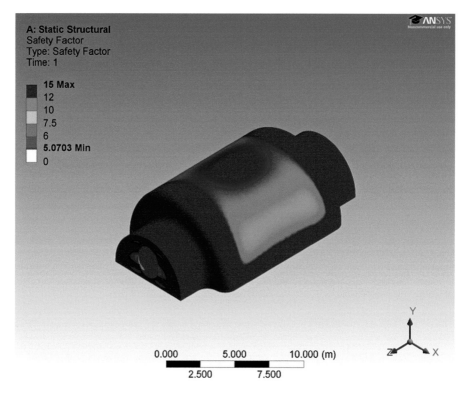

Figure 9.4.   Factor of safety isobars for a static analysis.

that rapid prototyping techniques generally result in structures with less strength when compared with other methods of manufacturing, and also that the mechanical properties of rapid prototyped magnesium could not be found.

Naturally, when designing such structures, an alloy will be selected. Most magnesium alloys possess an ultimate tensile strength of about five times larger than the sand-cast magnesium considered here. This affords us the possibility of designing much thinner-walled structures, saving material and time.

Figure 9.5 depicts the maximum principle stresses on the structural shell due to the static loads. We see, as expected, that the stresses are well below material strength. As discussed earlier, the high factors of safety can be reduced significantly without much risk.

Figure 9.6 depicts the maximum principal stresses on an end due to the static loads. We can see small regions of high stress at the lower wall near the opening. This region can be stiffened to reduce these stresses in a final design.

Figure 9.7 depicts the total static deformation. We note that the largest stresses (in red) are at locations where the bending moments in the shell are greatest. It is possible to place periodic circumferential stiffeners in the arches, thereby significantly reducing these deflections.

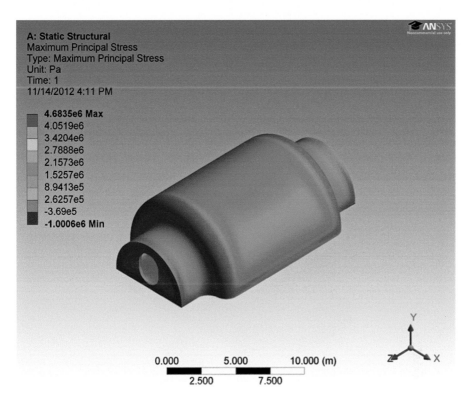

Figure 9.5. Maximum principle stresses due to a static analysis. This includes only gravity, internal pressure and regolith loads.

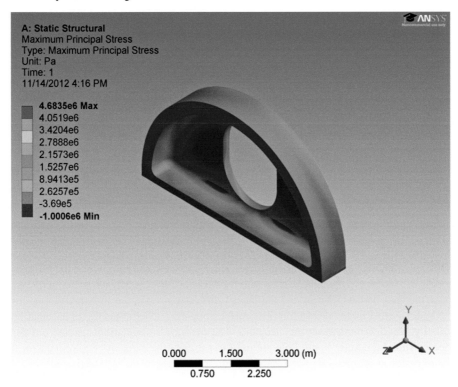

Figure 9.6. Maximum principle stresses on the end of the structure due only to static loads. This includes only gravity, internal pressure and regolith shielding.

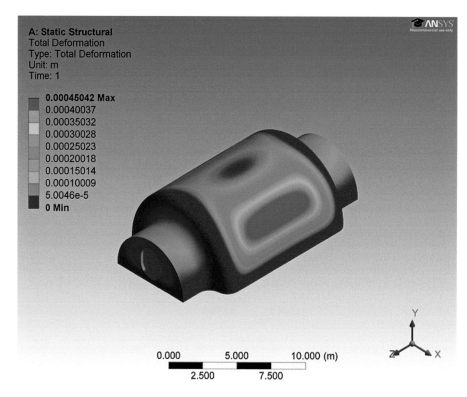

Figure 9.7. Total static deformation. Due only to gravity, internal pressure and regolith shielding.

9.2   A THERMAL ANALYSIS

A thermal analysis of a lunar structure requires knowledge of the temperatures on the lunar surface and the thermal properties of the regolith; in particular, its conductivity, specific heat, the lunar albedo, and the emissivity of the regolith. Thermal control of a lunar structure is a challenging problem. The temperature of the lunar surface swings in a large range during the long lunar day and night, and differently from location to location. Having essentially no atmosphere on the Moon, the heat exchanges between the base and the lunar environment are in the form of radiation and conduction. The outer surface of the base exchanges radiative heat with the lunar surface, space, and the Sun, while the structural foundation exchanges conductive heat with the Moon. Convective heat transfer takes place inside the base between the air and the inner walls, and also via the air conditioning unit.

The analytical model derived in this work can be used to estimate the radiational heat loads by the environment on the structure. The numerical simulations carried out for a selected site at the lunar South Pole are used to estimate the capacity requirement for the heat rejecting system based on the estimated thermal loads.

Protecting the structure from the radioactive environment, and from meteoroid impacts, requires a thick regolith shield, but the thicker shield acts as a thermal insulator. It prevents the rejection of heat generated inside by astronauts and instruments. Therefore, heat rejection radiators, of the kind found on the ISS, are needed to maintain the base temperature in a habitable range. In order to design these, we need to know the heat gain/loss of the base and the heat loads, as well as the environment in which the radiators are placed. The thermal analysis of the base, located at the lunar South Pole (latitude 88°), informs the designers of such radiators.

It is assumed that cooling after sunset happens very rapidly, and then remains constant for the remainder of the night. This assumption is based on data from the Apollo experiments. The temperature equation by Simonsen *et al.* is used:[11]

$$T_m = 373.9\left(\cos\phi\right)^{0.25}\left(\sin\theta\right)^{0.167},$$

where T_m is in K, ϕ is the latitude (deg) of the habitat location, and θ is the Sun's angle above the lunar horizon (deg). Using this equation, the temperature at the south polar location becomes:

$$T_m = \begin{cases} 161.607\left[\sin\left(11.4786d + 9.6495\right)\right]^{1/6}, & 0 \leq d \leq 14 \\ 120 & 14 \leq d \leq 28, \end{cases}$$

where d is the number of Earth days past lunar sunrise.

Most of our knowledge of the regolith's thermal properties comes from the Apollo era. Heat flow experiments were carried out during the Apollo 15, 16, and 17 missions, in addition to experiments on the collected lunar samples. Additional tests have been carried out on lunar simulants. Based on conductivity studies, Simonsen *et al.* state that:

> "The temperatures were found to reach constant values within the first meter of the surface that were higher than the mean surface temperatures. At a depth of 83 cm at the Apollo 15 site, the temperature was 45 K higher, and at a depth of 130 cm at the Apollo 17 site, the temperature was 40 K higher. Conclusions were made that a strong temperature-dependent conductive layer should be present in the upper few centimeters. In the other words, the lunar surface absorbs more heat during the lunar day than it rejects during the night. This requires the upper few centimeters to have very low density that increases with depth. This causes the upper few centimeters to have a large temperature gradient, which dies down at deeper levels where the temperature reaches a steady-state at some point less than a meter below the surface."

Based on Hemingway *et al.*, temperature-specific values of specific heat C_p, and thermal conductivity, were estimated from Apollo data.[12] For temperature ranges of 100 K to 350 K, the specific heat range is $275.7 \leq C_p \leq 848.9$ (J/kg K), and the thermal conductivity ranges between 0.0007 and 0.0017 (W/m K).

The lunar albedo has an average value of 0.09. On Earth, the range is 0.1 to 0.4 with an average of 0.3. The albedo is a measure of the planetary body's reflectivity, and is defined as the fraction of light or electromagnetic radiation reflected by a body or surface.

The emissivity is defined as the ratio of the energy radiated from a material surface to that radiated from a perfect emitter (blackbody) at the same temperature and wavelength, and under the same viewing conditions. It is a dimensionless number between 0 for a perfect reflector, and 1 for a perfect emitter. Lunar emissivity is in the range 0.90 to 0.95. Earth's emissivity is about half that value.

Given the slow variation of lunar temperature, and with the purpose of specifying the capacity of the thermal control system, the extreme conditions at lunar night and lunar noon are evaluated. The surface temperatures for these cases are 120 K and 161.61 K (−243.7°F and −168.8°F) respectively, for a site at 88° latitude at the south polar region.

For numerical simulations, a finite section of the lunar surface and depth are needed. Figure 9.8 is a sketch of the area around (10.5 m longitudinally and 11 m radially), and the depth of regolith beneath (3 m). The magnesium structure is assumed to be in thermal equilibrium; that is, there is no heat flux through the cut sections. The surface is modeled as isotropic, with a thermal conductivity of 0.009 W/mK (5.2011 x 10^{-3} Btu/ft h°F), and the radiative property of 0.93 is assumed for both absorptivity and emissivity.

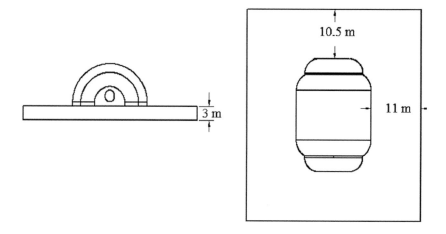

Figure 9.8.   Area around, and depth beneath, the structure, considered to be isolated from the rest of the Moon for purposes of thermal computation.

The structure is on a 1 m-thick foundation of sintered regolith. Although some experiments have been carried out on the sintering of regolith, no value is reported for its thermal conductivity. Therefore, two values for the conductivity of sintered regolith are considered in this study; Case 1: 0.009 W/mK (5.2002 x 10^{-3} Btu/ft h°F), and Case 2: 1.6 W/mK (0.9246 Btu/ft h°F). The first case is the same value as that for the lunar surface, and the second is a value in the range of the conductivity coefficients for cast basalt ceramic, where an assumption is made that the thermal conductivity of the sintered regolith should be close to that of the cast basalt ceramic, which ranges from 1.1 W/mK (0.6357 Btu/ft h°F) to 1.6 W/mK (0.9246 Btu/ft h°F).[13] We only show Case 1 results.

It is assumed that the sandbags used to contain the regolith shielding have the same thermal conductivity as the lunar regolith. The sandbags are assumed to be in perfect contact (no gaps) with the magnesium structure. Also, the emissivity and absorptivity of the shield is considered to be the same as that of the lunar surface, 0.93, since the shield will be covered by regolith dust over time due to dust migration. Therefore, it is probable that its radiative properties, regardless of the material used for the bags, will be very close to those of the lunar surface.

Based on Mackay, assuming uniform lunar temperatures in the vicinity of the structure, the heat flux per unit area between a thermal shield and its environs at the lunar surface is governed by:[14]

$$\frac{q}{A} = \left(1 - F_{bm}\right)\epsilon_b \sigma T_b^4 + F_{bm}\epsilon_m \sigma \left(T_b^4 - T_m^4\right) - S_c \alpha_s \cos\theta_s,$$

where q/A is the heat flux per unit area; F_{bm} is the configuration factor; ϵ_b and ϵ_m are non-dimensional emissivity factors; σ is the Stephan-Boltzmann constant; T_b is the temperature of the surface outside the shield; T_m is the lunar surface temperature; S_c is the solar constant (heat from the Sun); α_s is the absorptivity of the outside surface of the shield; and θ_s is the angle between the incident solar rays and the normal to the outside surface of the shield. The incident solar radiation reflected from the lunar surface and hitting the structure is neglected due to very low lunar albedo. The first term on the right-hand side of the Mackay equation is the heat per unit area that radiates from the structure into space. The second term represents the heat per unit area exchanged between the structure and the lunar surface. The third term is the heat per unit area absorbed by the structure from the Sun.

The configuration factor F_{bm} only depends on the shape of the structure and its surrounding environment, and it represents the fraction of heat radiating from the structure that impacts the lunar surface. In other words, $1 - F_{bm}$ is the fraction of the heat radiating from the structure that escapes into space. The factor is generally complicated to evaluate and is computationally intensive. Here, the values of this factor are estimated by utilizing the configuration factors of a few known geometries.

As mentioned earlier, it can be assumed that the outside surface of the structure has the same emissivity as the lunar surface ($\epsilon_b = \epsilon_m$) due to dust migration. Therefore, the above equation can be simplified, resulting in:

$$\frac{q}{A} = \epsilon_b \sigma T_b^4 - F_{bm}\epsilon_b \sigma T_m^4 - S_c \alpha_s \cos\theta_s$$

$$= \epsilon_b \sigma \left(T_b^4 - F_{bm} T_m^4 - \frac{\alpha_s}{\epsilon_b \sigma} S_c \cos\theta_s \right).$$

In order to reduce the computational requirements, this equation is solved for the ambient temperature, $T_a = T_b$, with $q = 0$, resulting in the relation:

$$T_a = \left(F_{bm} T_m^4 + \frac{\alpha_s}{\epsilon_b \sigma} S_c \cos\theta_s \right)^{1/4}. \tag{9.2}$$

Thermal computational results

Figures 9.9 and 9.10 depict the ambient temperatures on various parts of the lunar structure. The calculations are based on Equation 9.2, from which we can see that during the lunar day the ambient temperature is affected by the relative position of the Sun to the surface by the value of θ_s. The orientation of the base on the lunar surface is of importance

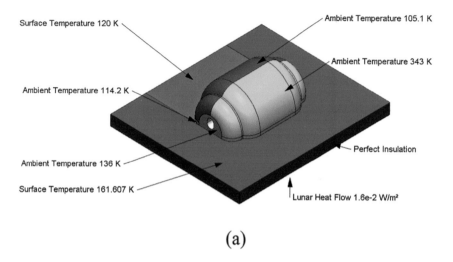

(a)

Figure 9.9. Temperatures at lunar noon at the South Pole. The lack of atmosphere results in sharp temperature transitions.

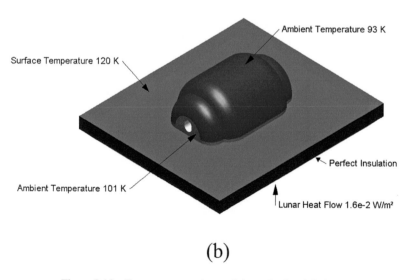

(b)

Figure 9.10. Temperatures at lunar night at the South Pole.

for a thermal analysis, the extreme case being an orientation that is east to west; that is, a large area of the structure facing east with the remainder facing west. This configuration allows the base to receive maximum heat from the Sun at lunar noon.

Considering the case with a foundation conductivity of 0.009 W/m K (Case 1), the rate of heat loss is estimated to be 142 W at lunar noon, and 331 W at lunar night. Figures 9.11 and 9.12 show the temperature distributions of a structural cross-section, at lunar noon and lunar night at the South Pole respectively.

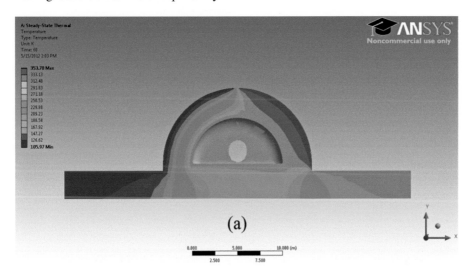

Figure 9.11. Temperature distribution at lunar noon at the South Pole, Case 1.

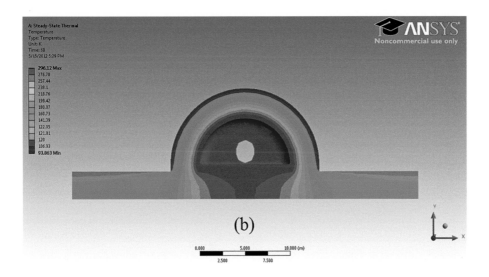

Figure 9.12. Temperature distribution at lunar night at the South Pole, Case 1.

Figures 9.13 and 9.14 depict the heat flux at lunar noon and lunar night at the South Pole respectively. The foundation conductivity is assumed to be 0.009 W/m K.

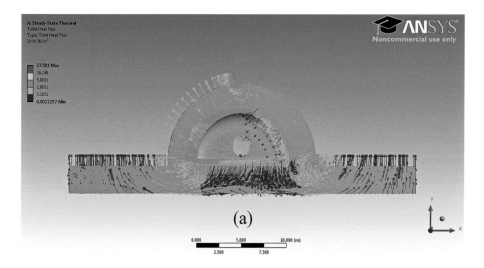

Figure 9.13.   Heat flux at lunar noon at the South Pole, Case 1

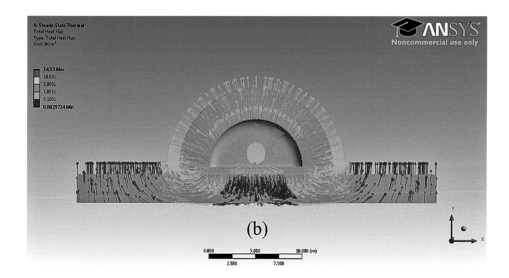

Figure 9.14.   Heat flux at lunar night at the South Pole, Case 1.

Based on these computations, the heat loads on the habitats, laboratories, nodes, and airlocks are used to estimate the heat rejection capacities of radiators to be designed. The 3 m regolith thickness seems to be sufficient to stabilize the thermal variations of the lunar day and night. Studies such as these can be used for trade studies that vary factor of safety, structural materials, and site preparations such as fusing the regolith around the structure, partially burying the structure, or placing it in a suitable crater, for example.

References

1. N. Izutani, N. Kobayashi, T. Ogura, I. Nomura, M. Kawazoe, and H. Yamamoto: *Temperature and Humidity Control System in a Lunar Base*, <u>Advances in Space Research</u>, Vol.12, No.5, pp.(5)41–(5)44, 1992.
2. S.D. Fraser: *Theory and Applications of Cooling Systems in Lunar Surface Exploration*, Chapter 18, **MOON - Prospective Energy and Materials Resources**, V. Badescu, Editor, Springer 2012.
3. A. Faghri: *Heat pipes: Review, Opportunities and Challenges*, <u>Frontiers in Heat Pipes</u> 5, 1(2014).
4. K.R. Sridhar, and M. Gottmann: *Lunar Base Thermal Control Systems Using Heat Pumps*, <u>Acta Astronautica</u> Vol.39, No.5, 1996, pp.381–394.
5. W.A. Marra: *Heat Flow Through Soils and Effects on Thermal Storage Cycle in High-Mass Structures, Technical Note*, <u>Journal of Aerospace Engineering</u>, Vol.19, No.1, January 1, 2006, pp.55-58.
6. R. Balasubramaniam, R. Wegeng, S. Gokoglu, N. Suzuki, and K. Sacksteder: *Analysis of Solar-Heated Thermal Wadis to Support Extended-Duration Lunar Exploration*, 47th AIAA Aerospace Sciences Meeting, January 5-8, 2009, Orlando, AIAA 2009–1339.
7. B. Climent, O. Torroba, R. Gonzalez-Cinca, N. Ramachandran, and M.D. Griffin: *Heat storage and electricity generation in the Moon during the lunar night*, <u>Acta Astronautica</u> 93 (2014) pp.352–358.
8. R. Malla, and K.M. Brown: *Determination of temperature variation on lunar surface and subsurface for habitat analysis and design*, <u>Acta Astronautica</u> 107 (2015) pp.196–207.
9. S. Mottaghi: *Design of a Lunar Surface Structure*, thesis submitted in partial fulfill-ment of the M.S. Degree, Rutgers University, October 2013.
10. S. Mottaghi, and H. Benaroya: *Design of a Lunar Surface Structure I - Design Configuration and Thermal Analysis*, <u>Journal of Aerospace Engineering</u>, 2014, https://doi.org/10.1061/(ASCE)AS.1943-5525.0000382
11. L.C., Simonsen, M.J. Debarro, and J.T. Farmer: *Conceptual design of a lunar base thermal control system*, The Second Conference on Lunar Bases and Space Activities of the 21st Century, W.W. Mendell, Editor, NASA Conference Publication 3166, Houston, TX, 579–591, 1992.
12. B.S. Hemingway, R.A. Robie, and W.H. Wilson: *Specific heats of lunar soils, basalt, and breccias from the Apollo 14, 15, and 16 landing sites, between 90 and 350 K*, Lunar and Planetary Science Conference Proceedings, Vol.4, SAO/NASA Astrophysics Data System, 1973, pp.2481–2487.
13. ITW Densit: *Data Sheet*, Densit.com (2012). (http://www.densit.com, accessed March 26, 2013)
14. **Design of Space Powerplants**, D.B. Mackay, Prentice-Hall, Englewood Cliffs, NJ, 1963.

10

Probability theory and seismic design

"More data will help,
but there is a limit to our ability to be precise."

Uncertainties always exist in engineering analysis and design. Even when they can be ignored, designers need to account for them by using safety factors. Uncertainty is the difference between what is known and what is reality. Of course, reality is only approximated by designers via tests and experiments. Uncertainty can be *aleatory* (random) and *epistemic* (subjective). Aleatory uncertainty refers to the inherent variability of a physical system, which can be modeled as a random variable or a random process, as described below. When data are scarce, probability distributions cannot be obtained, and epistemic uncertainty approaches are used. Epistemic uncertainty is defined as the lack of knowledge or information about some aspect of the system or model. Epistemic uncertainty can be reduced with the collection of more information. Sometimes, physical processes are so complex that, even with additional data, they are modeled as though they are inherently probabilistic. Examples of this include aerodynamic forces and seismic forces. In such cases, probability is used as an organizing principle. A review of reliability under epistemic uncertainty was provided by Kang *et al.*[1]

This chapter introduces elementary probability theory in order to discuss seismic design and reliability. There are numerous books and papers on probability, both applied and theoretical. Probabilistic analysis of dynamical mechanical systems can be found in detail in Benaroya *et al.*[2]

The main goal for studying dynamic systems under random loading is to predict the response (output) statistics given the loading (input) statistics. *Statistics* is the discipline that organizes data in a form that is meaningful and useful.

For example, the governing equation for a linear second-order system with loading, that is, a forced oscillator, is given by:

$$\ddot{x} + 2\zeta\omega_n\dot{x} + \omega_n^2 x = \frac{1}{m}F(t), \tag{10.1}$$

where the force $F(t)$ is a random function of time.

© Springer International Publishing AG 2018
H. Benaroya, *Building Habitats on the Moon*, Springer Praxis Books,
https://doi.org/10.1007/978-3-319-68244-0_10

Suppose the force oscillates in a complex manner, as shown in Figure 10.1. One possibility is to carry out many experiments and gather data on $F(t)$ in the form of time histories. Then, the time history with the largest amplitude can be used for the deterministic analysis and design. But if the largest amplitude force occurs only infrequently, the system would be over-designed. In other words, the system would be designed to be stronger than it needs to be, making it uneconomical. In this book, the structural designs for the Moon use very high factors of safety because of concerns about the unknown unknowns.

We distinguish between inherently random molecular forces, such as the Brownian motion experienced by atoms on a molecular scale, and the environmental forces of concern here. Environmental forces may not be inherently random, although they undergo very complex cycles. Natural randomness is called *aleatory variability*, and scientific uncertainty in the model of a process due to limited data or knowledge is called *epistemic uncertainty*. Such complex phenomena can only be modeled probabilistically.

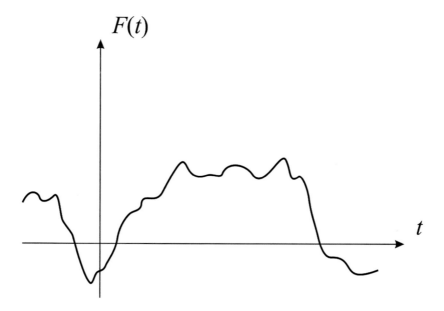

Figure 10.1. A sample random oscillation, or time history, can be representative of a turbulent force, or a seismic force.

10.1 A DEFINITION FOR PROBABILITY

Probability can be viewed as a measure of uncertainty, either as a measure of belief, or as a measure based on testing. One difficulty is that it is often not possible to perform enough experiments to arrive at that rigorous probability, and then judgment must be used. This is the situation we find ourselves in regarding the design of a structure for the lunar surface. While much data has been gathered since the days of Apollo, there are gaps in our knowledge about the eventual lunar site for a habitat. These gaps must be filled to inform engineers, as well as specialists in physiology and psychology.

Consider a lunar structure subjected to seismic loads or impact forces due to a meteoroid, resulting in a random structural response. In principle, the response amplitudes and frequencies can only be characterized probabilistically. In an analysis of this structure, we need to estimate the probability that the structural displacement amplitude A is greater than a specific value A_0. Using probability notation for this question, we have:

$$\Pr\{A > A_0\} = ?$$

This probability is related to the fraction of time the structural displacement has an amplitude greater than A_0, implying a *fraction* or *frequency* interpretation for probability. We can look at a lengthy time history of the oscillation and determine the amount of time that the amplitude A is greater than A_0. That *excursion frequency* is the probability estimate:

$$\Pr\{A > A_0\} = \frac{\text{amount of time } A > A_0}{\text{total time history}}.$$

A *histogram* can be used to estimate probabilities using frequencies of occurrence, which is the most common approach to estimating probabilities and the one we use here.

Randomness is possible for constants as well as functions (of time or space). A constant with a scatter of possible values is called a *random variable*. A function of time (or space) with scatter is generally called a *random process (or field)*.[1] Random variables are those which can be prescribed only to a specific level of certainty. An example is the material yield stress that defines the transition from elastic to plastic behavior. *Random processes* are time-dependent (or space-dependent, or both) phenomena that, with repeated observation under essentially identical conditions, do not show the same time histories.

Our motivations for such engineering studies are generally random vibration, fatigue life, random forces such as ocean waves and wind, and material properties. Lunar structural reliability and risk, fatigue, and responses to seismic and meteoroid forces, all require probabilistic tools for analysis. Even human factors, and human susceptibility to injury and disease, are characterized using probability. This chapter provides a brief introduction and background so that we can discuss these applications to the lunar habitat.

10.2 RANDOM VARIABLES

Probability affords us a framework for defining and utilizing random variables in the models developed for engineering analysis and design. Mathematical models of physical phenomena are essentially relationships between variables. Where some of these variables have associated uncertainties, there are a multiplicity of possible values for each random variable. An example is the set of possible values of Young's moduli, determined from a series of experiments on 'identical' test specimens. This multiplicity is represented by the

---

[1] From the Greek we also have the *stochastic (στοκος)* process.

probability distribution function, introduced in the following section. A random variable may be *discrete, continuous,* or *mixed.* If a parameter is a random variable, then its probability distribution function provides a complete description of its variability.

In the following discussion, we adopt the notation that random variables are represented as capital letters. We use lower case letters to denote *realizations,* or specific values, of random variables.

Probability distribution

The likelihood that a random variable takes on a particular range of values is defined by its *cumulative distribution function, $F_X(x)$*:

$$F_X(x) = \Pr\{X \leq x\},$$

where $\Pr\{X \leq x\}$ is the probability that the random variable X is less than or equal to a particular value x, which is called the *realization* of X. A *realization* is one of many possible values of a random variable. This probability is a function of the particular value x. (We will omit the subscript in $F_X(x)$ where clarity is not an issue.) For example, such a function can characterize the probability that a meteoroid will have a mass less than particular value m, $\Pr\{M \leq m\}$, or a velocity less than particular value v, $\Pr\{V \leq v\}$.

Based on the axioms of probability,[2] it can be shown that $F(x)$ is an increasing function of x, and is bound by 0 and 1. An *impossible* event has a zero probability, and a *certain* event has a probability of one. In particular:

$$F(-\infty) = 0,$$

since $\Pr\{X < -\infty\} = 0$. All realizations of the random variable must be greater than negative infinity. Similarly:

$$F(+\infty) = 1,$$

since $\Pr\{X < +\infty\} = 1$. All realizations of the random variable must be less than positive infinity. Thus, bounds on $F(x)$ are $0 \leq F(x) \leq 1$, and for $x_1 < x_2$:

$$F(x_1) \leq F(x_2),$$

since $\Pr\{X \leq x_1\} \leq \Pr\{X \leq x_2\}$. The probability distribution function is *non-decreasing.*

The cumulative distribution function is one way to describe a random variable probabilistically. The probability density function is another equivalent way.

---

[2]An axiom is a rule that is assumed to be true, and upon which further rules and facts are deduced. For engineering, the deduced facts must conform to reality. An excellent book on the basics of probabilistic modeling is Probability, Random Variables, and Stochastic Processes, A. Papoulis, McGraw-Hill, 1965.

Probability density function

The *probability density function* contains the same information as the probability distribution function. Assuming continuity of the distribution, the probability density function $f_X(x)$ is defined as:

$$f_X(x) = \frac{dF_X(x)}{dx},$$

and is sketched in Figure 10.2. Alternately, by integrating both sides and rearranging:

$$F(x) = \Pr\{X \le x\} = \int_{-\infty}^{x} f(\xi)d\xi, \tag{10.2}$$

where we have omitted the subscripts.

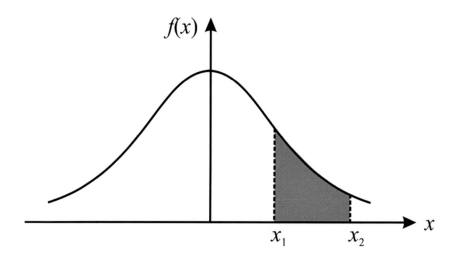

Figure 10.2. Probability density function $f(x)$. The probability that X is between x_1 and x_2 equals the area of the shaded region.

Equation 10.2 provides a useful interpretation of the density function: the probability that a continuous random variable X has a value less than or equal to the number x is equal to the area under the density function for values less than or equal to x. Similarly, for arbitrary x_1 and x_2, the probability that $x_1 < X \le x_2$ is equal to:

$$\Pr\{x_1 < X \le x_2\} = \int_{x_1}^{x_2} f(x)dx. \tag{10.3}$$

It is customary to write $\Pr\{x_1 < X \le x_2\}$ with $<$ rather than $\le$ on the left side so that there is no overlap between probability ranges, that is, between $\Pr\{x_0 < X \le x_1\}$ and $\Pr\{x_1 < X \le x_2\}$. For example, this function can be used to estimate the probability that a seismic force F will be within a certain range of values, $\Pr\{f_1 < F \le f_2\}$.

Note the important *normalization* property:

$$\int_{-\infty}^{+\infty} f(x)\,dx = 1, \tag{10.4}$$

signifying that the density function represents all possible outcomes or realizations of the random variable. Therefore, the area under the density function needs to be *normalized* to 1. This is an important point that is addressed further in our discussion of reliability design. There, we emphasize the importance of knowing the complete design, or sample space. Probabilities do not make sense unless they are with respect to the complete sample space of possibilities.

Since probability is numerically in the range 0 to 1, the density function must be a positive semi-definite function: $f(x) \ge 0$. For a continuous random variable, the probability $\Pr\{X = x\} = 0$ since there is no area under the density function at a point x.

Reliability theory and design, where both the loading and the strength of the material are random, can be understood schematically with Figure 10.3. While such curves are common to reliability, we can see similar curves characterizing any situation where there is a design and an environment. For example, the design of the regolith shielding is able to resist a certain level of radiation. But the environment sometimes has radiation levels that

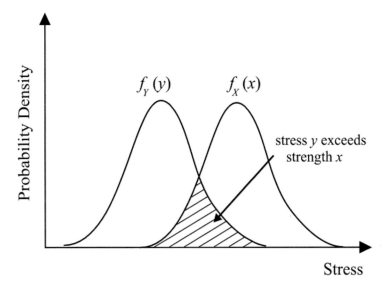

Figure 10.3. In this schematic, the strength is X and the stress is Y. Each is a random variable, signifying that there are uncertainties about both, and here they are modeled using normal probability density functions. When the stress exceeds the strength, the system has failed. A measure of the reliability (probability of failure) is given by the overlapped area, shown hatched.

exceed the design values and in those instances, the shielding is not adequate and the probability of that inadequacy is measured by the hatched area in Figure 10.3, when it is drawn for radiation variables.

The random variable is a static property – the shape of the density function does not change with time. Where the density function is time-dependent, the variable is called a *random*, or *stochastic*, *process*.

Before examining some commonly used densities, we define an averaging procedure known as the mathematical expectation for probabilistic variables.

10.3   MATHEMATICAL EXPECTATION

The single most important descriptor of a random variable is its *mean* or *expected value*. This defines the most likely value of a variable. However, random variables may have the same mean, but their *spread of possible values*, or their *variance*, can be considerably different. This explains the need for the variance in addition to the mean value when characterizing a random variable.

The mean and variance of a random variable are statistical averages, and are evaluated using the concept of the *mathematical expectation of a function of random variable X*, defined as:

$$E\{g(X)\} = \int_{-\infty}^{\infty} g(x) f(x) dx. \tag{10.5}$$

The *expected* or *mean value* is defined, using Equation 10.5, as:

$$\mu = E\{X\} = \int_{-\infty}^{\infty} x f(x) dx, \tag{10.6}$$

where x is a realization of X. The integral is over all possible realizations with $f(x)$ acting as a weighting function. The expected value is a constant, *first-order* statistic, and is also known as the *first moment* because the variable X appears to the first power.

The density function, acting as a probabilistic 'weighting' function, is a larger factor in the integral for more probable values of the random variable. From the definition of expectation, the expected value of a constant is that constant. In the same way that we calculated the mean value, we can derive an equation that provides a measure of the scatter about the mean value.

When a deterministic analysis is performed, it is assumed that all parameters are known exactly. In that case, we use the mean values of all the parameters, since the variances are all zero.

Variance
The *variance* is a *second-order moment*. It is defined as:

$$\operatorname{Var}\{X\} = E\left\{\left(X - E\{X\}\right)^2\right\}$$
$$= \int_{-\infty}^{\infty} (x - \mu)^2 f(x) dx.$$

Expanding the squared term and integrating term by term, we find the variance equal to:

$$\mathrm{Var}\{X\} = E\{X^2\} - (E\{X\})^2,$$ (10.7)

which is the difference between the *mean square value* and the *mean value squared*. Here, the second moment analogy is with the *mass moment of inertia*.

In order that the measure of dispersion has the same dimensions as the random variable, the *standard deviation* is defined as the positive square root of the variance:

$$\sigma = +\sqrt{\mathrm{Var}\{X\}}.$$ (10.8)

This parameter is related to the factor of safety of a design.

A related and important dimensionless parameter is the *coefficient of variation:*

$$\delta = \sigma/\mu.$$ (10.9)

It is used as a non-dimensional measure of the degree of uncertainty in a parameter; that is, the scatter of its data. In engineering practice, one expects a δ value of between 0.05 to 0.15, or 5 to 15 percent. Values larger than this imply a serious lack of knowledge about the system itself and its underlying physics. If this is the case, then experiments are warranted before one can consider the analysis and design of such a system.

To engineer a product such as a structure or a machine, we need to understand the behavior of materials, the dynamic characteristics of the system, and the external forces. Usually, the largest uncertainties are with the loading. Even so, in practice we expect probability densities to have most of their area about the mean value – that is, with a small variance.

Sometimes in engineering, the uncertainty is such that we only know the *high* and *low* values of a variable. In this instance, it is assumed that all intermediate values are equally probable. This leads us to the use of the uniform probability density. Other times, our experience tells us that parameter values significantly different from the mean can happen, even if these are unlikely. A possible density is the Gaussian density. What we see is that data from testing and design are crucial to the decision regarding the choice of the most physically realistic probability density function.

10.4   USEFUL PROBABILITY DENSITIES

It turns out that a handful of density functions are sufficient for probabilistic modeling in many engineering applications. Important densities are the uniform, exponential, normal (or Gaussian), lognormal, and the Rayleigh densities.

Normal (Gaussian) density
Many physical variables are assumed to be governed by the normal, or Gaussian, density. There are two reasons for this: the Gaussian density is mathematically tractable and tabulated, and the *central limit theorem* is broadly applicable. The central limit theorem states that under very general conditions, as the number of variables in a sum becomes large, the density of the sum of random variables approaches the Gaussian density in the limit, regardless of the individual densities.

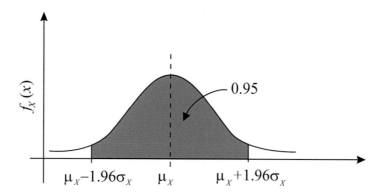

Figure 10.4. The Gaussian or normal probability density function has the property that 95 percent of the area under the curve is between the bounds $\mu - 1.96\sigma < X \le \mu + 1.96\sigma$.

Many naturally occurring physical processes approach a Gaussian density, shown in Figure 10.4. The Gaussian density function is given by:

$$f(x) = \frac{1}{\sigma\sqrt{2\pi}} \exp\left\{-\frac{1}{2}\left(\frac{x-\mu}{\sigma}\right)^2\right\}, \tag{10.10}$$

where $-\infty < x < \infty$, the mean value is μ, and the standard deviation is σ.

Note that the Gaussian density extends from $-\infty$ to ∞, and therefore cannot exactly represent any physical variable. It is an *approximate* representation since there are no physical parameters that can take on all possible values on the real number line. The Gaussian approximation is valid in many instances, but it is also invalid in many instances and cannot be used blindly without experimental verification.

10.5   RANDOM PROCESSES

A *random process* may be understood to be a time-dependent random variable. The random function of time $X(t)$ with probability density $f(x;t)$ is representative of many possible time histories, also known as a *sample population*. Theoretically, there are an infinite number of samples $X_i(t)$ with statistical properties governed by the density function $f(x;t)$. For any time t_1 there exists an infinite number of possible values: $x_1(t_1), x_2(t_1), x_3(t_1), \dots$. To account for all these possible values, we average them to obtain the most likely value of the function at time t_1. This process is known as *ensemble averaging,* since the group of samples is known as an *ensemble*.

The averaging procedure uses the same mathematical expectation of Equation 10.6. The second-order average is called the *correlation function*, and is given the shorthand notation $R_{XX}(t_1, t_2)$:

$$R_{XX}(t_1, t_2) = E\{X(t_1)X(t_2)\},\qquad (10.11)$$

where the end result of the double integral is a function of t_1 and t_2.

For a normal random process, for example, the time-dependent probability density function is:

$$f_X(x;t) = \frac{1}{\sqrt{2\pi C_{XX}(t,t)}}\exp\left[-\frac{(x-\mu_X(t))^2}{2C_{XX}(t,t)}\right].$$

Generally, specific correlation functions $R_{XX}(t_1, t_2)$ are derived experimentally. In practice, preliminary analyses assume a reasonable function for $R_{XX}(t_1, t_2)$ that is representative of the physical process under study. Predictions using this model are then verified with experimental data.

For example, seismic forces are random processes that vary in time in a way that can only be mathematically modeled probabilistically. In other words, any predictions about the characteristics of seismic waves cannot be accomplished deterministically. Similarly, the sizes and speeds of meteoroids that hit the lunar surface are random, and therefore so are the forces with which they strike the surface. These uncertainties change with time.

The correlation function is a measure of the relation between, and the uncertainties about, random behavior at two locations. For example, if a meteoroid strikes the lunar habitat at a particular location, how that impact affects the global response of the structure is characterized, in part, by a correlation function.

Stationarity

For steady-state random processes, it is common to make the assumption of *stationarity*. If the statistical properties of a random process are invariant under translation in time, the process is called *stationary*. While this assumption may appear to limit applicability, stationarity is a viable assumption for numerous practical applications. The assumption of stationarity implies steady-state behavior in the statistical sense.

For a stationary random process, the mean value becomes a constant (as for random variables):

$$E\{X(t)\} = \mu_X,$$

and the correlation function becomes a function of time difference:

$$R_{XX}(t_1, t_2) = E\{X(t)X(t+\tau)\} = R_{XX}(\tau),$$

$\tau = t_2 - t_1$, rather than a function of the specific times t_1 and t_2. For physical processes, the correlation is an even function, $R_{XX}(\tau) = R_{XX}(-\tau)$. Similarly, for two stationary processes, the *cross-correlation* function is $R_{XY}(\tau)$.

For $\tau=0$, the *mean-square* value of $X(t)$ is:

$$R_{XX}\left(\tau = 0\right) = E\left\{X^2\left(t\right)\right\} = \sigma_X^2 + \mu_X^2,$$

using Equation 10.7. The mean-square value has the following physical interpretation. If $X(t)$ is a displacement, then $E\{X^2(t)\}$ is a measure of strain energy, $V(t) \sim kX^2(t)$, where k is a stiffness parameter. If $X(t)$ is a velocity, then $E\{X^2(t)\}$ is a measure of kinetic energy, $T(t) \sim mX^2(t)$, where m is a mass parameter. The average energy of a stationary process is independent of time and equals the autocorrelation at $\tau=0$. This knowledge is useful for interpreting the meaning of the *spectral density* in Section 10.6 'Power Spectrum'. In physical processes, as $\tau \to \infty$, the correlation function approaches the mean-square value:

$$R_{XX}\left(\tau \to \infty\right) \to \mu_X^2.$$

As the time difference τ becomes larger, less correlation exists between the two respective values of the process, and the limit of the correlation becomes the square of the mean value.

While seismic forces are random processes, they are not stationary because they are not steady-state random processes. They 'turn on' when the seismic events strike; they are transient.

Ergodicity

As a practical matter, we rarely have the benefit of numerous experiments to generate statistics, but usually must make the best use of *one* trial. This is especially true for expensive testing environments, such as for space or for the ocean. Utilizing one trial requires the introduction of the concept of *ergodicity*. A stationary random process is said to be an *ergodic* process if the *time average* of a *single* record is approximately equal to the ensemble average. It is thus possible to average over a long *single* time history, rather than trying to obtain numerous records over which to perform an ensemble average.

The mean value is then given by:

$$\mu_X \simeq \overline{X} = \lim_{T \to \infty} \frac{1}{2T} \int_{-T}^{T} X(t)dt, \tag{10.12}$$

where it is assumed that $X(t)$ is one particular realization $x(t)$ of the random process. Such an average makes sense only if μ_X is constant. Otherwise μ_X will be a function of T, and the initial assumption is no longer valid. An ergodic process is always stationary; the opposite is not always true.

The corresponding ergodic definition for the autocorrelation function is:

$$R_{XX}\left(\tau\right) \simeq \lim_{T \to \infty} \frac{1}{2T} \int_{-T}^{T} X(t)X(t+\tau)dt. \tag{10.13}$$

The importance of ergodicity as a framework for gathering data cannot be overestimated. Unless we have a long history of tests of similar equipment, or we are able to perform dozens of tests on a prototype, we have no way to gather the needed data from which we can perform a statistical analysis to estimate the reliability of that piece of equipment or prototype empirically. Since the field of space engineering is relatively young, there is no history for similar components, except perhaps rockets. As we look forward to sending habitats to the Moon for the first time, we have zero data. We are in essence designing structures for an environment that we cannot test on Earth. We are piecing together the reliabilities of components as our estimate for the reliability of the complete structure. As we know, and as we will discuss in more detail subsequently, the complete structure can behave in ways that are unpredictable, if based only on the behavior of its components. Our reliability estimates are guesses about the future, not extrapolations from past data. More on this later.

Now that we have an understanding of the autocorrelation, we proceed to study its Fourier transform, the spectral density.

10.6   POWER SPECTRUM

A measure of the 'energy' of the stochastic process $X(t)$ is given by its *power spectrum*, or *spectral density*, $S_{XX}(\omega)$, which is the *Fourier transform* of its autocorrelation function:

$$S_{XX}(\omega) = \frac{1}{2\pi} \int_{-\infty}^{\infty} R_{XX}(\tau) \exp(-i\omega\tau) d\tau,$$

and thus:

$$R_{XX}(\tau) = \int_{-\infty}^{\infty} S_{XX}(\omega) \exp(i\omega\tau) d\omega. \tag{10.14}$$

These equations are known as the *Wiener-Khintchine* formulas.

Since $R_{XX}(-\tau) = R_{XX}(\tau)$, $S_{XX}(\omega)$ is not a complex function but a real and even function. For $\tau = 0$:

$$\int_{-\infty}^{\infty} S_{XX}(\omega) d\omega = R_{XX}(0) = E\{X(t)^2\} \geq 0,$$

where $S_{XX}(\omega) \geq 0$ since, as a measure of energy, it must be positive semi-definite. The integral of the power spectrum equals the 'average or mean square power' of the process $X(t)$, confirming our opening statement that it is an energy measure. Where there is no chance of confusion, the subscripts used above can be omitted.

The power spectrum is a function that can be time-dependent for nonstationary processes, such as seismic events, and is representative of the energy distribution of a random process as a function of frequency. It can be representative of the kinetic energy, or radiation energy, for example. The power spectrum for the seismic force on a surface lunar structure is used to estimate its response. This information is used to refine the structural design.

White-noise process

A *white-noise* process is an idealization that is used to simplify the algebra of an analysis, as we will see below. The term *white* is adopted from optics to signify that all frequencies are part of such a process, much like white light is composed of the whole color spectrum. For a white-noise process, the power spectrum is a constant:

$$S_{XX}(\omega) = \frac{1}{2\pi} \int_{-\infty}^{\infty} 2\pi S_0 \delta(\tau) e^{-i\omega\tau} \, d\tau = S_0.$$

This is the flat spectrum shown in Figure 10.5. The figure includes the equivalent one-sided spectrum.

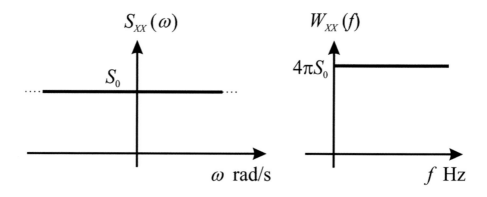

Figure 10.5. The white-noise power spectral density is a useful approximation for structural systems subjected to random loading. Here, it is shown in units of rad/s and Hz.

For the lunar site, spectral densities of the seismic energy, or of the radiation energy, are used to characterize the threats these natural phenomena pose to structures and the inhabitants inside.

10.7   RANDOM VIBRATION

Single-degree-of-freedom systems, when modeled as linear, are governed by the second-order differential equation:

$$\ddot{X}(t) + 2\zeta\omega_n \dot{X}(t) + \omega_n^2 X(t) = F(t)/m. \tag{10.15}$$

Here, we take the input force to be the stationary random process $F(t)$, with mean μ_F and power spectrum $S_{FF}(\omega)$, and the output displacement to be the random process $X(t)$. The stationarity assumption for the forcing means that transient dynamic behavior cannot be considered here. The initial loading transients of earthquakes, lunar quakes, wind gusts, or extreme ocean waves, cannot be considered as stationary. If the character of the loading does not change, steady-state behavior can be assumed to be statistically stationary.

We define the frequency-response function as:

$$G(i\omega) = \frac{1}{m\left[(i\omega)^2 + i2\zeta\omega_n\omega + \omega_n^2\right]}, \tag{10.16}$$

where the impulse response function $g(t)$ and $G(i\omega)$ are Fourier transform pairs:

$$g(t) = \frac{1}{2\pi}\int_{-\infty}^{\infty} G(i\omega)\exp(i\omega t)\,d\omega \tag{10.17}$$

$$G(i\omega) = \int_{-\infty}^{\infty} g(t)\exp(-i\omega t)\,dt, \tag{10.18}$$

where $1/2\pi$ appears in the Fourier, instead of inverse Fourier, transform formula.

Response mean and variance
Suppose that an analysis resulted in the statistics μ_F and σ_F, where stationarity has been assumed. Consider how the response mean and variance (or standard deviation) can be useful in a design procedure. The designer needs both the mean value *and* the variance to establish bounds on the possible response. Example bounds are $\mu_F \pm \sigma_F$, $\mu_F \pm 2\sigma_F$, or $\mu_F \pm 3\sigma_F$, where the larger the *sigma bounds*, the more likely that all possible responses are covered. Along with a higher probability comes this broader band with its vagueness. There is no way around this *uncertainty-type* principle. These upper and lower bounds are used to define the least and most likely range of responses. If designing for strength, then the upper sigma bound can be used to size the structural components.

How wide or narrow the sigma bounds are depends on the underlying density function. For parameters governed by the Gaussian probability density, there is a probability of 0.6827 of being within the one-sigma bounds, and a probability of 0.9545 of being within the two-sigma bounds. Different densities have different probabilities for their sigma bounds.

There is no easy or clear-cut answer regarding how many sigma bounds to use in a design. The designer must study the data in order to better understand the underlying density. As a practical matter, by retaining larger sigma bounds in the design, it becomes more conservative, which leads to a costlier structure or product.

Response spectral density
The *fundamental* result for random vibration and linear systems theory is the relation between the output spectral density and the input spectral density:

$$S_{XX}(\omega) = |G(i\omega)|^2 S_{FF}(\omega). \tag{10.19}$$

Figure 10.6 is a graphical image of this equation. A version of this equation is used to estimate the seismic response of a surface lunar structure later in this chapter.

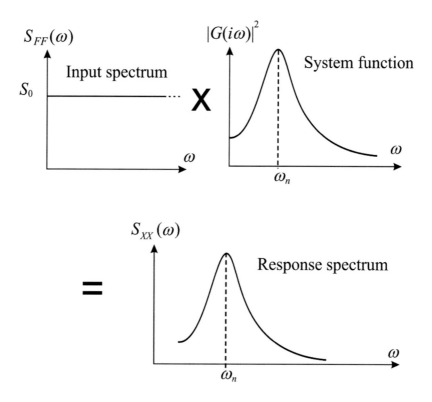

Figure 10.6. With input spectrum $S_{FF}(\omega)$, transfer function $|G(i\omega)|^2$, and output spectrum $S_{XX}(\omega)$, we see how the transfer function acts as a filter, allowing primarily input energy around the natural frequency of the system to pass through, resulting in the response spectrum with energy primarily focused around the system's natural frequency.

10.8　SEISMIC ACTIVITY

"The ground is shaking, and we can't ignore it."

While the Moon has a molten core like Earth's, it is susceptible to seismic activity due to tidal energy releases – the Moon is pulled on by Earth's gravity field, causing strains and stresses. Repetitive cycles of such loading leads to relatively minor quakes. Data mining of the Apollo lunar seismic data set has revealed that there had been 7245 deep moonquakes while the experiments were active, from 77 regions within the Moon.[3]

This analysis invigorated further research of the structure of the Moon, but also had implications for the designers of habitable and other structures. The largest recorded moonquake in the Apollo data was 2–3 on the Richter scale, with most magnitudes in the range 1–2.[4] "The deep moonquakes showed some periodic regularity depending on the point when the Moon's orbit is closest to Earth. They occurred at these centers at opposite phases of the tidal pull, so that the most active periods are 14 days apart. ... Much of the data collected in the Apollo seismic experiments were due to meteoroid impact, demonstrating that meteoroids will be a much more significant concern than moonquakes with an original source from the Moon interior."

Moonquakes, while rarely a serious issue for surface structures, could be potentially more serious as we begin to build underground. Buried structures may require a variety of vibration isolation devices, much like the ones used in tall buildings and bridges in seismic regions on Earth. Structures placed within lava tubes may require some shielding against debris, and the tubes themselves may require bracing if seismic activity is a concern.

10.9   A SEISMIC DESIGN

This section provides a particular seismic design of a lunar surface structure based on the work of Mottaghi, and Mottaghi and Benaroya, and is a summary of the work therein. [5] [6]

The analysis and design of structures in seismic environments is challenging. Uncertainties abound; for example, we do not know when a seismic event will occur – neither its location (epicenter) nor its magnitude. Furthermore, such an event is nonstationary given that the input turns on at a given time, making a random vibration analysis that much more difficult. On Earth, after extensive experience, we have been able to design and retrofit structures, underground pipelines and power stations so that they can survive certain levels of seismic activity with minimal damage. For the Moon, however, so much less is known about lunar seismicity that any studies performed without a significant presence on site are likely to be, at best, only order of magnitude accurate.

Seismic waves are divided into two classical types, P or primary waves (also called compressional waves), and S or shear waves (also called secondary waves). P waves are the first arriving waves, having traveled a direct path from the focus of the seismic disturbance to the structure (or seismometer). They cause the medium to move back and forth in the direction of propagation. Afterwards, a pulse pP that traveled from the focus gets reflected off the surface and arrives at the structure. Second to P waves are S waves, which cause the medium to move perpendicular to the direction of propagation. Analogous to the pulses due to pP waves, there are also reflected shear waves sS.

P and S waves, including their subcategories, are known as body waves since they travel through the Earth's, or the Moon's, interior. P and S waves are divided into more categories than discussed here.

After the arrival of the body waves, for terrestrial cases, seismograms are dominated by larger, longer-period waves. These waves are called surface waves, as these are trapped near the surface of the Earth, or Moon. In other words, their energy is concentrated near the body's surface. These are subcategorized into two types: Love waves and Rayleigh waves. Love waves result from SH waves (shear waves polarized in the horizontal plane) trapped near the surface, while Rayleigh waves are a combination of P and SV waves (shear waves polarized in the vertical plane) that can exist at the top of a homogeneous half-space.

Earth's continental crust has many small layers and reflectors. These geological structures cause some of the energy to be scattered and arrive at a receiver later than the initial pulse. The scattered energy arrives from various directions and shows little or no preferred particle motion. Therefore, this scattered energy causes an arrival to have a *coda*, a trail of incoherent energy that decays over seconds or minutes.

Intraplate earthquakes occur within plate interiors, where they are not perfectly rigid. In some cases, for example in Hawaii, these earthquakes are associated with intraplate volcanism. These are studied, generally, to provide data about where and how the plate tectonic model does not fully describe tectonic processes.

For the Apollo Passive Seismic Experiments (APSE), a part of the Apollo missions, a network of four seismometers was installed on the surface of the Moon. Data were collected and transmitted back to the Earth from 1969 to 1977. The passive seismic experiment package sensor unit contained three long-period seismometers with resonant frequencies of 1/15 Hz, aligned orthogonally to measure surface motion in the vertical and two horizontal directions, plus a single-axis, short-period seismometer sensitive to vertical motion at higher frequencies with a resonant frequency of 1 Hz. During the eight years of network operation, more than 12,000 seismic events were detected.

Four distinct types of natural seismic sources have been identified: deep moonquakes, thermal moonquakes, meteoroid impacts, and shallow moonquakes.

Of those seismic events, deep moonquakes were the most abundant, with 7082 confirmed (and 317 unconfirmed) events. Most of their foci occurred within a clearly defined region between depths of 800 km to 1000 km. Their occurrences are strongly correlated with the tidal forces on the Moon due to the Earth and the Sun. These are low frequency, low magnitude events and are unlikely to pose a danger to a surface lunar structure. A body wave magnitude of 1.3 to 3.0 is estimated for such deep moonquakes.

Thermal moonquakes are also very unlikely to pose any threat to lunar structures, as these are very small seismic events caused by temperature variations at or near the lunar surface. Although their exact cause is unknown, it is believed they originate in young craters and large rocks.

A total of 1743 meteoroid impacts have been identified from the APSE data – while the actual number is much greater – the long period seismographs were capable of detecting mainly the signals from objects of mass 0.1 kg and higher. The largest meteoroid mass has been estimated at about 2000 kg, based on an assumed impact velocity of 22.5 km/s.

Shallow moonquakes, also called high-frequency teleseismic (HFT) events, were the most energetic and also the rarest seismic sources that have been observed on the Moon. A total of 28 events have been distinguished. The name HFT is used for these events, due to their unusually high frequency content and the great distance at which they were observed. Body wave magnitude estimates are in the range of about 5 to 5.5 for the largest shallow moonquake. The epicenters of all the detected HFT events were located outside the Apollo seismic array, and the number of recorded events was small, thus limiting our understanding of these events. The actual depths of the HFT events were undetermined, but are estimated at less than 200 km, and more likely below an approximate 55 km-thick crust, perhaps in the range of 60 km to 100 km. The cause of these events is as yet unknown.

While the terrestrial equivalent of the higher magnitude lunar quakes can cause moderate to severe damage to buildings, seismic effects on a lunar structure are not the same since the lunar geologic structure is very different than that of the Earth. The seismic wave train has been observed both on the Moon and the Earth. On the Moon, the seismic wave train is less attenuated and more intensely scattered than those of Earth. Thus, the lunar seismograms are very long, when compared to those of Earth. Although the lunar seismic wave trains are very long, the seismic energy is scattered over tens of minutes, likely reducing the potential damage to most, but not all, lunar structures.

Mathematical modeling

An elastic pulse propagates without change in shape in an ideal, infinite, homogeneous, non-dispersive, and elastic medium. However, in heterogeneous media, a single seismic pulse is rapidly converted into a long train of pulses. Therefore, the number of possible ray paths becomes quite large and the overall process takes a diffusive form. The nonstationary power spectral density of the ground velocity, $S_{vv}(\omega)$, is given by:

$$S_{vv}(R,\omega,t) = \frac{E_0(\omega)}{\rho(4\pi Dt)^{3/2}} \exp\left(\frac{-R^2}{4Dt} - \frac{\omega t}{Q}\right), \quad t > t_i, \tag{10.20}$$

where:

$$E_0(\omega) = \frac{10^{15.6}}{(2\pi)^{3/2}} \exp\left(-\frac{1}{2}\left(\frac{\omega - 6\pi}{2\pi}\right)^2\right), \tag{10.21}$$

and where t_i is the time after the start of the seismic event when the first seismic waves reach the structure, R is the distance from the focus, $E_0(\omega)$ is the total seismic energy generated by the seismic event within a unit frequency band around ω at $R=0$, ρ is the density of the regolith in the vicinity of the structure, Q is the non-dimensional intrinsic quality factor due to anelastic conversion of seismic energy to heat (does not include the loss by scattering), D is the diffusivity, and t is the time the seismic wave traveled.

The resulting power spectrum of the ground acceleration is found by multiplying Equation 10.20 by ω^2, and is depicted in Figure 10.7 for a normally distributed source, for $D=0.22$ km²/s, and $Q=8000$. For this analysis, the focus distance is chosen as $R=25$ km, the density $\rho=1740$ kg/m³, with Richter magnitude 7 (for conservative results), with the energy $E_0(\omega)$ normalized over the frequency range $0.5<\omega<10$ Hz. This normalization results in the factor 0.99379 in the denominator of Equation 10.21. The normally distributed source over the given frequency range has a mean value of 3 Hz, and a standard deviation of 1 Hz.

A realization of the seismic ground acceleration (at an instant of time) is shown in Figure 10.8, and a realization of a power spectral density of the ground acceleration is shown in Figure 10.9.

Even though we have much knowledge about the Earth, we still face many challenges in predicting earthquakes and their effects. Due to minimal data and limited knowledge of the Moon's interior, our understanding of moonquakes and their effects are, at best, very approximate. This seismic model is developed based on the best available data, and then applied to the proposed lunar structure shown in Figure 10.10.

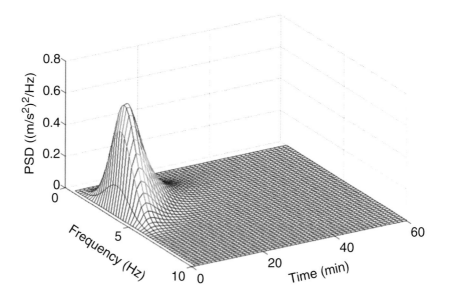

Figure 10.7. A nonstationary power spectral density for lunar ground acceleration. The non-stationarity can be seen in the changing form of the power spectral density with time.

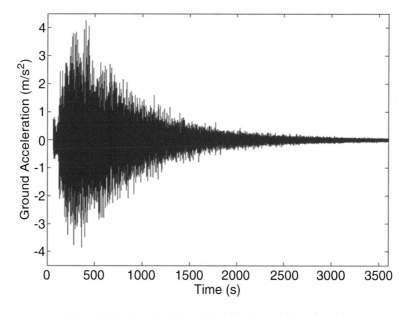

Figure 10.8. A realization of the seismic ground acceleration.

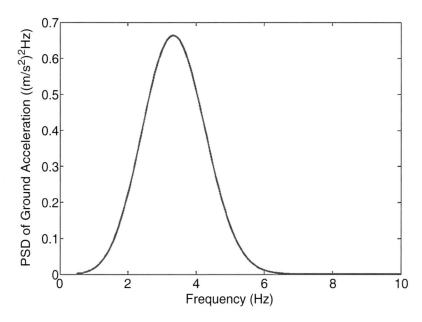

Figure 10.9. A realization of the power spectral density (PSD) of the seismic ground acceleration.

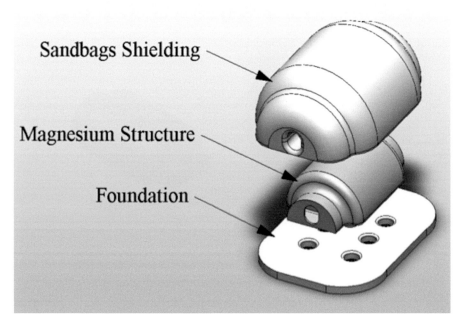

Figure 10.10. A representative image of the magnesium structure, its regolith shielding, and a mat foundation with holes for six piles.

Structural model

The lunar structure is a magnesium structure, covered with sandbagged regolith shielding, on a sintered regolith foundation that is envisioned to be built by autonomous minirobots using layered manufacturing technologies.[7] Although sintered regolith is the chosen material for the foundation, the technologies for sintering regolith are not ready to be used in the proposed construction method. When these technologies advance, the sintered regolith could have properties which, if not comparable to cast regolith, would be better than lunar concrete. Cast regolith possesses an ultimate strength that is an order of magnitude higher than lunar concrete.[8] Therefore, a lunar concrete foundation is chosen in this study as a conservative option. Six foundation piles are added to the magnesium structure that sits on a mat, which extends beyond the structure and supports the sandbags of regolith shielding. Engineering drawings of the magnesium structure are shown in Figure 10.11. The properties used in the structural analysis are listed in Table 10.1.

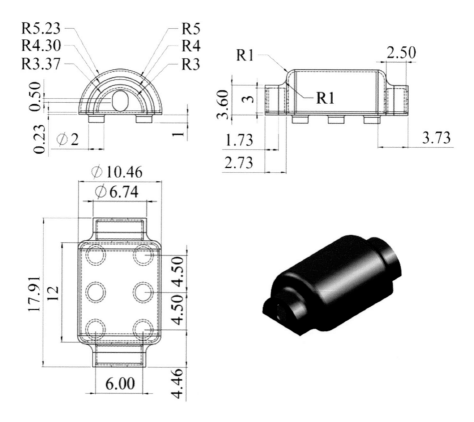

Figure 10.11. Engineering drawings for the surface structure showing the six piles.

Table 10.1.   Structural Material Properties

Property	MG	Foundation	Sandbags
Density [km/m³]	1738	2500	1650
Young's Modulus (E) [GPa]	40 Static 44 Dynamic	21.4	0.06
Poisson's Ratio [v]	0.35	0.21	0.3
Tensile Yield Strength [MPa]	21	5	
Compr. Yield Strength [MPa]	21	7.07	

Seismic response

A finite element analysis is used to model the structure computationally, with regolith shielding, and subject it to the random seismic loading. The nonstationary power spectrum is converted into the equivalent time history shown in Figure 10.8. The procedure is detailed and referenced in the original documents. In Figures 10.12 and 10.13, we see that the peak stresses are well below the 21 MPa yield stress for magnesium. Even so, the larger stresses in the Z direction in the arch midspan suggest that circumferential stiffeners are warranted, as they were in the static and thermal analyses. A more optimal design would build up the structure at locations of elevated stresses and deflections, and thin it out where stresses and deflections are very low. When 3D printing/autonomous technologies become viable, this fine tuning of the design would be simple to implement.

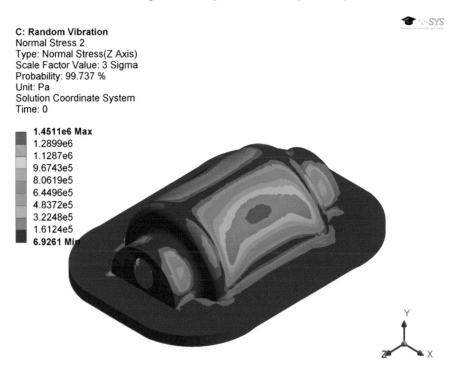

Figure 10.12.   Normal stress in the Z direction. Maximum stress equals 1.4511 x 10⁶ Pa.

The von Mises stress, shown in Figure 10.13, is often used to determine whether an isotropic and ductile metal will yield when subjected to a complex loading condition. The von Mises stress is compared to the material's von Mises yield condition. This comparison provides an indication of the stress state for a complex loading condition, regardless of the mix of normal and shear stresses.

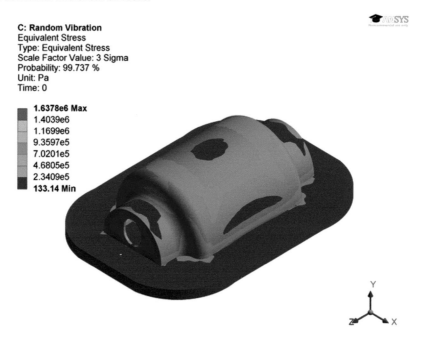

Figure 10.13.   Von Mises stresses for a normally distributed source.

Figure 10.14 shows the first six modes of vibration of the lunar habitat, including the regolith shielding, all coupled with the foundation. Each mode can be interpreted to be the displacement pattern of the structure for each natural frequency. The natural frequency of a vibrating system is a function of its mass and stiffness properties. In general, the vibratory characteristics of a complex structure such as this habitat are very intricate, which is why it is decomposed into modes of vibration. In principle, continuous structures such as this habitat have an infinite number of modes. Fortunately, for computational purposes, we need only a small number of modes to represent most of the energy of vibration accurately, in this case, six modes.

The structural response to a lunar seismic event reveals that the risk associated with designs that neglect seismic effects is low, since the designs use relatively high factors of safety. This low risk seems reasonable, especially since the lunar surface is covered with a layer of regolith and granular soil is proven to have excellent damping qualities. Moreover, it is very unlikely that a moonquake of such magnitude would occur at 25 km distance or less, as the estimated depth of these events are between 60 to 100 km.

It is estimated that a lunar structure, at a randomly chosen site, can experience a shallow moonquake of body wave magnitude greater than 4.5 within a range of 100 km about once in 400 years. This structure is highly damped due to the magnesium material property, and it is

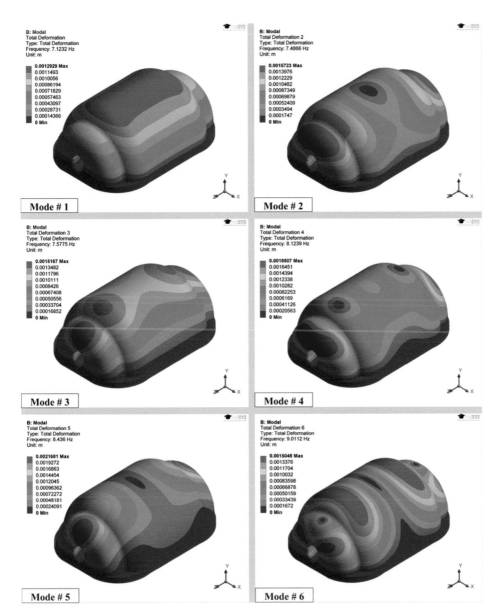

Figure 10.14. First six modes of vibration for the proposed lunar structure. These represent the deformation shapes at each natural frequency in free vibration. A continuous structure such as this has an infinite number of frequencies and modes, but for practical vibration analysis, a relatively small number of modes is representative of the structural behavior. The six modes shown here contain most of the vibration energy of this structure.

covered in a blanket of regolith, thus reducing the damage due to seismic events. The same statement might not be valid for tall or flexible structures. Also, seismic effects must be considered when designing the airlocks and nodes connecting different structures that require air seals, as these may be more sensitive to such excitation and to differential motions.

We expect second-generation lunar habitats to be sub-surface structures. Such structures require a more detailed seismic analysis, since the effects of regolith-structure coupling become more important. A buried structure in contact with the surrounding soil medium is more coupled to it, and directly experiences the stress waves moving through the volume. This is similar to the effects of stress waves on lava tubes.

Quotes

- "The Moon was the most spectacularly beautiful desert you could ever imagine. Unspoilt. Untouched. It had a vibrancy about it and the contrast between it and the black sky was so vivid, it just made this impression of excitement and wonder." Charles Duke.
- "The only time I had what you would call life-threatening fear was when I was on the Moon. Towards the end of our stay, we got excited and we were going to do the high jump, and I jumped and fell over backwards. That was a scary time, because if the backpack got broken, I would have had it. But everything held together." Charles Duke.
- "Perhaps the two greatest moments of my life were standing on the Moon and being outside of the room when my granddaughter was born! We tend not to remember the worst." Eugene Cernan.
- "Here I am at the turn of the millennium and I'm still the last man to have walked on the Moon, somewhat disappointing. It says more about what we have not done than about what we have done." Eugene Cernan.
- "Neil Armstrong was probably one of the most human guys I've ever known in my life." Eugene Cernan.

References

1. R. Kang, Q. Zhang, Z. Zeng, E. Zio, and X. Li: *Measuring reliability under epistemic uncertainty: Review on non-probabilistic reliability metrics*, Chinese Journal of Aeronautics, (2016) 29(3): pp.571–579.
2. **Probabilistic Models for Dynamical Systems**, Second Edition, H. Benaroya, S.M. Han, and M. Nagurka, CRC Press, Boca Raton 2013.
3. Y. Nakamura: *New identification of deep moonquakes in the Apollo lunar seismic data*, Physics of the Earth and Planetary Interiors, Vol.139, 2003, pp.197–205.
4. A.M. Jablonski, and K.A. Ogden: *Technical Requirements for Lunar Structures*, Journal of Aerospace Engineering, Vol.21, No.2, April 2008, pp.72–90.
5. S. Mottaghi: *Design of a Lunar Surface Structure*, thesis submitted in partial fulfillment of the M.S. Degree, Rutgers University, October 2013.
6. S. Mottaghi and H. Benaroya: *Design of a Lunar Structure. II: Seismic Structural Analysis*, Journal of Aerospace Engineering, Vol.28, Issue 1, January 2015.
7. H. Benaroya, S. Indyk, and S. Mottaghi (2012): *Advance system concept for autonomous construction and self-repair of lunar surface ISRU structures*. **Moon: Prospective, Energy, and Material Resources**, V. Badescu, Editor, Springer, Berlin Heidelberg, pp.641–660.
8. J.A. Happel (1993): *Indigenous materials for lunar construction*. Applied Mechanics Reviews, 46(6), pp.313–325.

11

Reliability and damage

"We need to make sure it survives for a while."

In this chapter, we introduce the issues of design life, a background to structural reliability and some possible tools that can be useful when we go from concept to reality in our designs for habitats on the Moon. Reliability can be interpreted as the probability that the designed object will function within the design guidelines for the *design life*. A reliability design often includes an estimate of a rate of failure, perhaps as a function of time. A corollary of reliability is *durability*, which is a quantitative assessment of how an object ages over time and is still able to operate within the design guidelines. For mechanical components, aging occurs due to high-cycle fatigue, corrosion, other forms of wear, design stress exceedances, impacts, and friction effects, for example.

We introduce ideas from reliability theory, in its simplest forms. Given our structural perspectives, our designs focus on strength and survival in the extreme lunar environment for a prescribed period of time. Design life is a time period that the designer uses as the basis for making many decisions about the structure, its configuration, what it is made of, and how it is constructed. Implied is a knowledge of the structural environment, and what load types and magnitudes the structure will experience over its design life.

For a lunar structure and its site, where our knowledge is both deep as well as lacking, reliability and probabilistic mathematical tools give us the ability to integrate our lack of knowledge into the analysis. The challenge is to balance risk and cost, and these do not exactly lead to a zero-sum game. It is possible to minimize risk without increasing cost by using a clever design. Similarly, it is possible to increase the risk, and the cost, by way of a poorly thought through design. Part of the challenge is to be able to envision the complete life of the structure. After all, the risks and costs begin at the conceptual level, long before the structure is erected on the lunar surface. Then there are transportation costs, decisions about the manufacture and construction of the structure, materials choices, and then there eventually comes a time to take down and recycle the structure. All these aspects, and many more, affect the complete risk and cost.

© Springer International Publishing AG 2018
H. Benaroya, *Building Habitats on the Moon*, Springer Praxis Books,
https://doi.org/10.1007/978-3-319-68244-0_11

11.1   RISK

This section is based on the lucid introduction to, and discussion of, risk by Kaplan and Garrik.[1] They draw a careful distinction between 'probability' and 'frequency'. And there is the connection to risk – a term that is applied to business, social systems, economics, safety of engineered systems, investment strategies, military operations, and political machinations, for example. The first notion that is clarified is that uncertainty is connected to risk by the threat of damage. Thus:

$$\text{risk} = \text{uncertainty} + \text{damage}.$$

We can be uncertain, but unless that uncertainty is connected to potential damage, there is no risk. Damage results from a source of danger, or a hazard. "Risk is the 'possibility of loss or injury' and the 'degree of probability of such loss.' Hazard, therefore, simply exists as a source. Risk includes the likelihood of conversion of that source into actual delivery of loss, injury, or some form of damage."

To avoid damage, we use devices to safeguard us against the hazard, and thus to reduce the risk. Qualitatively:

$$\text{risk} = \frac{\text{hazard}}{\text{safeguards}}.$$

This relation implies that, regardless of the number of safeguards, the risk cannot be eliminated. We can reduce the risk by increasing the safeguards, but never bring it to zero. Awareness is a safeguard that also reduces risk. "So, qualitatively, risk depends on what you do and what you know and what you do not know.

"In analyzing risk, we are attempting to envision how the future will turn out if we undertake a certain course of action (or inaction). Fundamentally, therefore, a risk analysis consists of an answer to the following three questions:

1. What can happen? (i.e., What can go wrong?)
2. How likely is it that that will happen?
3. If it does happen, what are the consequences?"

The first question leads to a list of scenarios, s_i; the second question leads to the probability of each scenario, p_i; and the third question links the outcome or scenario to a measure of damage, x_i. This 'triplet' is the risk:

$$R = \left\{ s_i, p_i, x_i \right\}, \quad i = 1, 2, \ldots N,$$

where N is the number of scenarios that we can enumerate. Since there may be scenarios that we are not aware of, we can 'add all the others' into scenario $N + 1$. This definition for risk replaces the qualitative 'uncertainty + damage' with 'probability + consequence'.

The group of scenarios can be plotted as a 'risk curve', or as multiple 'risk curves'. This implies that a single number is not sufficient to define risk; rather, a whole curve – a function – is needed to communicate the concept of risk. "Now the truth is that a curve is not a big enough concept either. It takes a whole family of curves to fully communicate the idea of risk."

We mentioned earlier that, typically, a frequency is interpreted as a probability. But there are two ways to interpret uncertainties:

"People have been arguing about the meaning of probability for at least 200 years, since the time of Laplace and Bayes. The major polarization of the argument is between the 'objectivist' or 'frequentist' school, who view probability as something external, the result of repetitive experiments, and the 'subjectivists', who view probability as an expression of an internal state – a state of knowledge or state of confidence.

"… both schools are right; they are just talking about two different ideas …

"What the objectivists are talking about we shall call 'frequency'. What the subjectivists are talking about we shall call 'probability'. Thus, 'probability', as we shall use it, is a numerical measure of a state of knowledge, a degree of belief, a state of confidence. 'Frequency', on the other hand, refers to the outcome of an experiment of some kind involving repeated trials. Thus, frequency is a 'hard' measurable number. This is so even if the experiment is only a thought experiment or an experiment to be done in the future. At least in concept then, a frequency is a well-defined, objective, measurable number.

"Probability, on the other hand, at first glance is a notion of a different kind. Defined, essentially, as a number used to communicate a state of mind, it thus seems 'soft' and changeable, subjective – not measurable, at least not in the usual way. … [W]e may define or 'calibrate' the entire probability scale using frequency as a standard of reference."

Thus, by attaching a frequency to the state of confidence we have in our uncertainty, "we calibrate the probability scale." The distinction between probability and statistics is "that statistics, as a subject, is the study of frequency-type information. That is, it is the science of handling data. On the other hand, probability, as a subject, we might say is the science of handling the lack of data. … When one has insufficient data, there is nothing else one can do *but* use probability."

We perform risk analyses in order to assess whether the risk is acceptable, and we compare various designs in order to compare their relative risks. This may be misleading. "There are two difficulties with the notion of acceptable risk: one major and one minor. The minor difficulty is that it implies that risk is linearly comparable. It implies that one can say that the risk of course of action A or design A is greater or less than that of design B." The difficulty with this is evident when we imagine two risk curves that cross each other somewhere in the middle. For a portion of the curve, one design is less 'risky', but for the other portion of the curve, the relative risks interchange. The risks are clearly different, yet we cannot readily say that one is bigger than the other. The key is that the risk curves are nonlinear and nonlinearly related to each other. They are not comparable. We can force a comparison by averaging the risk curves, and then compare the two averages, but this can be misleading and not satisfactory.

The conclusion is "that risk cannot be spoken of as acceptable or not in isolation, but only in combination with the costs and benefits that are attendant to that risk. Considered in isolation, no risk is acceptable!" Risk is taken for the possible benefits that come along with it. If the same benefits can be accrued with less risk, then that will be the chosen course of action, to minimize the risk. Or more benefits for the same risk. The analysis compares the costs, benefits, and risks of all the options. The 'best' option is chosen.

This is directly applicable to a comparison of lunar habitat designs. These designs are complex multidimensional entities. Even though we sometimes compare designs by using one or several numbers that ostensibly represent the designs, this can be misrepresentative. Since the risk is a family of curves, then so is the design risk we are interested in estimating. Comparing two designs requires a comparison of two families of design curves. This can be a daunting task if done right.

It is unclear that we can yet create sets of families of curves for comparisons of different lunar structural concepts.

11.2 RISK AT A LUNAR BASE

A key issue of concern that affects everything we do in space is the assessment of risks. All of our designs are explicitly the result of risk assessments. The key concerns of a reliability-based design method for lunar structures have been known for a long time, and have evolved.[2] In particular: What failure rate is acceptable? What factors of safety and levels of redundancy are necessary to assure this failure rate? What is the point of diminishing returns for added redundancies? Shortridge *et al.* summarized approaches to modeling hazards with deep uncertainties, alternatives to the standard probabilistic risk assessment methodologies.[3]

We are still at the point where every loss of life in a space venture brings space exploration to a halt. While all such losses are tragic, they result in a disproportionate disruption to progress. Because space is a new environment for people to inhabit, the novelty and high risk have perhaps led to over-caution. Terrible accidents with trains and airplanes do not bring these modes of transportation to a halt, though they are investigated, of course.

The development of aircraft has cost many lives. Test pilots are lost in that line of work, but rather than stop all test flights for years, as was done in the case of the Space Shuttle, lessons were learned and efforts were multiplied, assuring that these losses were not in vain.

The difficulty with space is that it is a hostile, unforgiving environment for which we have little data upon which to base risk estimates. Uncertain events can only be predicted on average, assuming we know all the possible outcomes and the likelihood of each outcome. For example, tossing a fair die results in a number from one to six – which number is unknown in advance. A single toss yields, say, the number two. Our knowledge of the structure of the die – knowing that it has six possible sides with all equally likely to come up in a fair die – leads us to the estimate that each face can appear with a likelihood, or probability, of 1/6. That number is the probability of a particular face ending up on top when the die is thrown. Similarly, a fair coin toss leads us to a probability of heads or tails of 1/2, because we know that a coin has two sides.

But what if we do not know the structure of the die or the coin? What if we do not know that a die has six sides or a coin two sides? And what if all the sides do not have the same probability of showing up? If our die has 27 sides, and each side has a different probability of showing up, a long series of tests are needed in order to figure out the probabilities of each possible outcome.[1] This is the difficulty for assessing risk in situations where there is much that we do not know, and at least some things which we don't know that we don't know. Space is that die with n sides, where we do not know the numerical value of n. We have improved our knowledge and reduced the range of uncertainties over the past hundred years, but there are still many challenges to our knowledge. And, of course, there are many aspects of space that we know are problematic, and for which we do not have solutions.

We do not yet have enough information or data in the early 21st century to estimate these risks, but even with this level of uncertainty, we will soon have to design structures for the lunar surface. Even with more mainstream technology like the Shuttle we lost two, along with many lives. Both events were a shock for the space program because each loss indicated a significant lack of knowledge and understanding about the Shuttle and its operations. We thought the Shuttle was a 6-sided die, but we found out it was really a 10-sided die after losing many astronauts.[2] There seems no way around this dilemma. We either accept the losses, or we stay in our caves.

Putting ourselves in the minds of the engineers who are considering the design of the first lunar bases, we have some questions to consider. What failure rate is acceptable for such a design? Since it is generally accepted that we cannot economically (or in reality) design for zero risk, the next logical consideration is the level of acceptable risk. One way to begin to answer such a question is to study the sources of natural risks to a system in its intended environment. In particular, we can examine all the natural phenomena and determine the risk exposure of the system to each phenomenon. Some, such as meteorites of a certain size, can destroy a facility, but occur infrequently, and therefore need not, and cannot, be designed against. Each of these risks defines a time limit or design life (in the probabilistic sense) for a structure or system; these may be independent or correlated. Therefore, the probability of occurrence of a catastrophic meteorite hit is a small risk, perhaps the smallest encountered risk. It may therefore be viewed as the *base risk* against which other risks can be weighed. Other natural risks may be ascertained, as best as possible – compared to the base risk – and then considered within an overall reliability analysis.

Next, the probabilities of man-made risks need to be assessed. Examples are: the probability of an explosion of a liquid oxygen tank; the likelihood of projectiles piercing a critical structural component due to an accident; the likelihood of thermal cycle fatigue; and the probability of various human errors. These may be estimated and compared to the base risk, or other known risks. Data on such risks can also be estimated from our Earth-based experiences. All of these 'component' risk factors need to be assessed. Using engineering judgment and calculations, weighted somewhat by political considerations, acceptable risk can then be decided upon. To this day, we recall the financial and psychological costs to the space program due to the Shuttle *Challenger* disaster. We need to develop methodologies for assessing risk for such untested, and in some cases untestable, situations.

---

[1] The Theorem of Total Probability encapsulates the requirement that the complete sample space, and the respective probabilities, be known so that probabilities of events can be calculated.

[2] Some at the time apparently knew that the Shuttle was not a 6-sided die but seemingly chose to ignore it, hoping for the best.

Suppose that we define $R_m = \Pr\{\text{meteorite}\}$ as the probability that a destructive meteorite will strike a site on the Moon during a year's time. Also, $R_f = \Pr\{\text{thermal fatigue}\}$ is defined as the probability that a certain number of thermal cycles in one year will result in material failure. Each such probability can be estimated independently, any correlation established, and then we can define the minimum prescribed design risk, $R_{min} = \min\{R_m, R_f\}$, as a measure of the smallest acceptable risk. This risk may be too small to be economically acceptable, but it is a starting point for an analysis. When we further consider that structures are designed to be compartmentalized and modular, accessible and repairable, then it begins to appear possible to increase the value of the acceptable design risk, R_{min}, to be used in the preliminary designs.

As engineers, we try to design the structure so that an impending failure becomes obvious to the inhabitants, so that they will have time to exit. Therefore, we cannot accept a first-excursion failure. We design for progressive failures. First-excursion failure occurs the first time a design is exceeded. There is no warning. Designers attempt to build a robustness into their structures to avoid first-excursion failures. For example, if a ceiling deflects in a way so that occupants know of an impending failure, then they have enough time to leave before the collapse. (Some of these terms are defined later in this chapter.)

What factors of safety[3] and levels of redundancy are necessary to assure this failure rate? Given an agreed upon acceptable level of risk, it becomes necessary, as a practical matter, to establish a design philosophy. For example, what factors of safety can lunar structural engineers of the 21st century build into lunar design codes? Since the lunar site provides designers with the most uncertainties of any engineering project, with few opportunities to obtain experience or data, one design philosophy demands higher-than-Earth factors of safety. However, it is possible to approach this question from another design perspective. We know that the site is inherently high-risk and – just as we accept high risks for test pilots – can accept a high-risk approach to a lunar outpost design concept. Both approaches can be justified.

In addition, the above risks can be compared to statistical data from occupational and normal population groups. Such data allow us to back out mission requirements and rates, such as the allowable (acceptable) individual death by illness during the mission, allowable death by injury, and allowable death from all causes, including spacecraft failure. Based on such an analysis, it is possible to set up reliability objectives for each mission scenario, and to make sure that the mission safety objectives are met.[4],[5]

Redundancy is a separate and double-edged question. Once a basis has been set for acceptable risk and safety factors, the designer must be ingenuous in the conceptual design, optimizing the design so that the overall risk is as close as possible to the acceptable level. Risk should also be distributed throughout the lunar habitat site in accordance with the criticality of the various parts to the overall mission. This is a difficult problem requiring the study of competing structural and system concepts. By adding redundancy,

---

[3]A factor of safety is an engineering parameter that is an indicator of the level of knowledge about the system that is being designed. If everything is exactly known, and there are no uncertainties regarding the structure or its intended environment, then the factor is 1. If there is some uncertainty, then the engineer will design a stronger structure so that there is some overdesign. If the structure is designed to be 30 percent stronger in order to account for the uncertainties, then the factor of safety is 1.3. Generally, the range of safety factors for well-known structures and systems is between 1.1–1.3. If there is too much uncertainty, then testing is necessary.

we significantly increase system complexity, adding layers of systems, each of which has its own reliability and risk.[6] There is a point of diminishing returns with added redundancy.

How does logistics interplay with considerations of risk and reliability? The links are close. Generally, one has two options when a component or system fails: replace or repair. During the early days of the return to the Moon, there will be no inventories on site to replace damaged or destroyed components. Self-repair systems are a long way off from application.

Reliability and safety are linked to the maximum amount of payload that can be brought to the lunar facility from the Earth in the minimum amount of time. This minimum-time-to-maximum-payload (MTMP) defines the absolutely necessary self-sufficiency time for the lunar inhabitants for survival. During this time, local replacements and/or repairs are mandatory in order to recover from, and survive, significant failures. Logistic requirements are therefore an important consideration at an early stage of the design-development cycle. (There is no possibility of rescue from Earth for a habitat on Mars.)

Eventually, an ISRU infrastructure on the Moon will permit the manufacture of any component needed for the safe operation of the settlements. Designs must allow for ease of repair and reconditioning. Commonality of parts is a strategic goal, and therefore a design constraint.

Here is one approach to a design framework. A large-scale lunar outpost, if designed for low risk to inhabitants, would be a complex and expensive undertaking, primarily because humans are very delicate. Instead, lunar settlements can be designed to higher risk tolerances with significant cost savings, but to ensure an overall high level of human safety, a number of smaller, much safer, emergency facilities can be placed throughout a settlement at easily accessible locations. These smaller facilities are designed to support the population for a significant amount of time – the time needed for rescue missions to arrive from the other settlements on the Moon, or Earth, depending on the time frame under consideration. The added mass cost for the emergency facilities is likely much less than the cost to bring the whole lunar settlement to those same high standards. A fleet of pressurized rovers can be a part of this safety net strategy.

With time, we can expect to evolve beyond frontier standards for safety. Smart structures that are completely sensor-rigged and monitored can have self-maintaining and self-repairing capabilities. Air leakages can then seal automatically, excess regolith dust on machines can be tracked and cleaned automatically, fracture and fatigue in structural components can be identified and repaired autonomously, for example.

A related discussion by Mendell and Heydorn (albeit about Mars) made some relevant points that affect lunar habitats:[7]

"In the case of the human Mars expedition, the Program Manager will be confronted with several major new systems, each incorporating new technologies. For a human program, the goal is a very small and manageable risk.

"The expedition itself will consist of a string of functions, each of which must be very reliable: Launch, Interplanetary flight, Mars landing, Mars surface operations, Mars ascent, and Earth return. We wish to discuss how to test and validate those aspects of the mission that take place far from Earth. Assurance that the systems function according to design is important, but also important is confidence that the system functions reliably, i.e., repeatedly without failure."

These functions challenge the lunar program manager as well. There will be new technologies, many 'rediscovered' technologies from the Apollo era updated for today's materials, and intelligent systems. Reliability testing and verification of systems designed to operate in micro- and low gravity, in a vacuum, under intense thermal and radiation conditions, is a serious challenge. Additionally, and as mentioned elsewhere in this book, assessing the reliabilities of individual systems is well understood, but then the reliability of a system of systems is at least an order of magnitude more complex to evaluate. Interacting components add dimensions that are difficult to predict, with new dynamics:

> "The quasi-permanence of the facilities raises new kinds of issues regarding the long-term performance of systems used by humans. Here, we emphasize the use of the word systems rather than equipment. Individual pieces of equipment can be tested for mean time between failure, but integrated systems can exhibit subtle malfunctions in addition to component failures. Unfortunately, at some scale, integrated systems become difficult to test with appropriate fidelity. They must actually be used (field tested) to understand their reliability."

In fact, the coupling between systems can lead to new failure modes. This can be an instance of the whole being less than the sum of its parts.

As we have observed in the habitat designs, in order to assure crew safety, escape options and safe havens are needed. If failure is to occur, it should proceed slowly enough that the crew can escape. These kinds of assurances are linked to the logistics of rescue and/or resupply, as identified above, and also offers hints to the designers of the first lunar colony:

> "Crew safety would depend upon options for crew escape, or at least safe haven. These may include concentration on fault isolation, reconfiguration, and finally redundancy design issues. For mission success, maintenance would play a major role. Designs that fail gracefully would be a prime consideration. These would allow time to recover from hardware failures that may accommodate logistic problems related to resupply and repair. The overall design would therefore involve a process flow that must allow for hardware/software and human reliability contingencies. The demonstration of such a process should be developed in stages to allow for the unanticipated. Judging from the proximity of the Moon, the inclusion of a lunar phase makes sense. ...
>
> "In fact, lunar exploration provides an ideal context within which to advance our technology to the level required for planning Mars exploration. Of course, properly conceived lunar missions would also produce important scientific findings and discoveries. ... Experience gained on the lunar surface is necessary for designers and engineers who will be two generations removed from the last human excursion out of Earth orbit. A lunar program will provide managers a benchmark for assessing the risks of Mars expeditions and will also increase public confidence in the institutions entrusted with the program. ... [Lunar] missions would serve not only to reduce the technical risk for expeditions to Mars but also to demonstrate to the nation that NASA has the expertise and management skill to embark successfully on the larger enterprise."

The lunar site has much to offer our nascent spacefaring attempts. It is a site both for testing our skills at building things that can survive the rigors of space for long periods of time, and for making sure that we first know how to survive and then thrive in that environment, again for long periods of time. Until we are able to do this on the Moon, any human forays beyond the Moon can only be brief visits, if they succeed at all.

A cable structure example

Cable structures, sometimes called cable nets, have been used with remarkable effectiveness and efficiency for terrestrial structures.[8] Considering the extraterrestrial site, some of the advantages of such structures are especially important.[9] In particular:

- Cable systems are easy to transport due to their light weight and compactness during transport. They are less prone to damage and therefore more reliable.
- They are inherently prefabricated, with generally simpler joint details than conventional systems, resulting in a faster construction process. This provides more access to pre-testing and weeding out of unacceptable components.
- They carry loads via tensile stresses, meaning that they are more efficient and permit longer spans than conventional structures, and are potentially more stable. Simpler structures will be possible, permitting easier maintenance and repair.

Along with the advantages of cable structures come potential difficulties:

- These types of structures are not typical, and engineers generally do not have an instinct regarding their behavior. Thus, an experience base must be developed.
- Erection of these structures poses different challenges due to the fact that their stiffness is engaged at the very end of the construction cycle, at which time they are stressed. As such, robotic or telerobotic construction will prove attractive in the (not so near) future.
- Failure modes are difficult to characterize generally. But specifically, failure will result due to the loss of a critical bar or cable, and the degree of structural redundancy will determine how critical each component is.

Pre-tensioning permits tension members, such as cables, to remain in tension under a broad range of loading conditions, and even reversals. The price for this ability is an increased sensitivity to fatigue and creep, and a generally shorter life.

In addition, due to the nature of the lunar regolith, as well as the difficulties expected to be inherent in lunar construction, surface structures should be designed to minimize the need for excessive or complex foundations. Thus, anchor and deep foundations should be avoided. Structural concepts that are self-contained, essentially 'sitting' on the surface, are preferable.

Cable-based structures are self-balancing, where the net loads are resisted by structural elements themselves, rather than being transmitted to a foundation.

Here, we seek to identify in more detail the key technical issues that must be addressed in order for tensile structures to be a viable option for the lunar base designer. These may be grouped as follows:

Selection of optimal configurations for specific applications:
It is not at all obvious what configurations or topologies are best for any particular application. The designer must address conflicting demands of economy, transportability,

and utility, while assuring that the final design is stable, robust, and reliable. Given such considerations, it is then necessary to map out the possible behavior modes of the chosen configuration.

Connections – cable, rod, and base:
The issues regarding structural connections can be very difficult in general. The cable-rod structure is connected to a fabric or membrane that contains the internal pressurized atmosphere, and pressure loads transfer between the two. The joint design is particularly challenging in the following respects: there are complex sets of cables and bars intersecting through the joints that need to be adjustable for a variety of topologies; and joints should be robust under the action of suspended fines, if so exposed. Connections and discontinuities are crucial aspects of analysis and design, and there is less tolerance for errors in the design than for more standard structures.

Reliability and safety issues

Key design considerations are the pre-tensioning mechanism, expandability of the facility, overall redundancy and reliability, system stability, ease of construction, and deployability:

Pre-tensioning mechanism:
This aspect of the design is related to the connections. Pre-tensioning is the essence of the structural concept considered here for the lunar structure. It is how the cable-rod structure attains its stiffness and shape. The design of the pre-tensioning mechanism is an integral part of the overall structural conceptualization and design, and reliability is a primary ingredient.

Expandability of facility:
One key criterion of a successful lunar structure is the ability to 'add on' more structures and create connections or paths between the facilities; openings are also required for light and mechanical access. This is difficult to achieve with a pressurized structure, in particular when structural integrity depends on continued pressurization, such as with an inflatable structure.

Overall redundancy and reliability:
Redundancy is a way in which the designer builds safety into the structure. This is especially important for a tensile structure, which is very integrated and acts almost as a single component. Secondary systems permit failures to transfer loads through other paths, providing time to make necessary repairs. To ensure safety, as with all structures designed for unforgiving environments, a monitoring system needs to be an integral part of the structural system. The structure needs to be equipped with diagnostic equipment that continually monitors its condition. This means being able to read stresses, strains and deflections for all members and cables. Such information is crucial to the scheduling of maintenance, repair and replacement of components. This is expensive and complicated, and can make reliability calculations ambiguous.

System stability:
Stability suggests the need for a 'uniform' distribution of pre-stressing tension in the cables. In addition, bars that are under compression must be analyzed for possible buckling.

Ease of construction:
Certainly, any viable design must be one that can be placed and built on the Moon using existing, or soon to be available tools. The design must be performed with a realization of construction capabilities. For example, if regolith is to be used as shielding, a containment system is required on the outside of the structure to hold the regolith in place. Designing a containment framework is challenging.

Deployability:
Tensegrity structures, due to the kinematic possibilities of their topology, have been considered for use in severe environments where deployment rather than construction is a more desirable approach, albeit more difficult on the design.[10]

Each of these technical issues has analytical, design, and construction aspects. Currently, one cannot expect to use off-the-shelf structural analysis computational tools, or existing design codes, to create these structures. While only the cable-rod structure has been discussed here, there are many layers of structure to be considered. Just as in a skyscraper, there is much more than the steel frame.

Framework for safety
Primary considerations for a cable-based lunar structural design are listed next and placed in a reliability design framework. Emphasis has been placed on factors that the designer will consider for the first time in a lunar reliability design. These are:

Load definition:
Regolith Q_1, internal pressure Q_2, thermal gradients Q_3, radiation Q_4, meteoroids Q_5, accidental impacts Q_6, operational vibrations Q_7, and explosions Q_8

Geometric aspects:
Cable-net configuration L_1, cable dimensions L_2, regolith shielding topology L_3, and thickness L_4

Unique environmental effects:
Suspended fines on joints/connections E_1, airlocks E_2, mechanisms E_3, and machines E_4

Material properties/strength parameters:
Vacuum R_1, radiation effects R_2, and material properties R_3

Design criteria:
Membrane *m*/cable *c* stresses S_m, S_c, deflections d, stability, and reliability, and

Humans:
Criteria related to human physiology and psychology. (These will be omitted below.)

There is scant or no information for some of the items listed in the above framework. An understanding of the system is required both with and without internal pressurization – the unpressurized case during construction, and in the event of a catastrophic decompression.

Our structural model has many parameters, z_i, that need to be modeled as random variables due to the inherent uncertainties of the lunar site and system.[11] For example, for a cable net, key parameters include: strand material, geometry (diameter, length) and strand configuration (spacing, intersection angle). One important design criterion, the deflection of the cable net, d, is a function of these parameters:

$$d = g(z_i),$$

and, along with the allowable deflection, $d_a(z_i)$, a failure function can be defined:

$$f(z_i) = d_a - d \equiv \Delta_d.$$

The safety of the system can be defined by:

$$\Delta_d \begin{cases} > 0 & \text{safe state} \\ = 0 & \text{limit state (onset failure)} \\ < 0 & \text{failure state.} \end{cases}$$

Thus, the safety margin of the cable network is a function of the random parameters. Similar expressions can be written for other responses of the system, for example cable strain. In principle, there are expressions that relate the correlations between the parameters.

Given estimators, such as the mean values and variances, for the distributions of the basic variables z_i, estimators for Δ_d can be calculated, as well as estimates for component and system reliabilities. For example, one can stipulate the allowable probability for failure due to excess deflection as:

$$\Pr(F) = \Pr(\Delta_d < 0) < 10^{-4},$$

which can be used in a simulation of the design space for the optimal set of design parameters that satisfy the inequality.

When discussing lunar, or any, structural reliability, two types of limit states are defined: ultimate limit states, and serviceability limit states. The ultimate limit state corresponds to the limit of the load-carrying capacity of a member, or the structure as a whole. Examples are the formation of a plastic yield mechanism, brittle fracture, fatigue fracture, instability, buckling, or cable pullout from a support. Serviceability limit states imply deformations in excess of tolerance without exceeding the load-carrying capacity. For example, cracks, permanent deflections, vibration, accidental limit states, or progressive limit states, and an accidentally damaged structure due to an explosion or fire. For a lunar structure, it can sometimes be difficult to differentiate between the two types of limit states. For instance, a serviceability limit state may lead to an ultimate limit state, and do so quickly. A micrometeorite impact initially results in making one part of the base unusable, but can also lead to a general weakening of the surrounding structure, making it too weak to survive other loads.

On the Moon, it is crucial to design for escape and rescue. This means that any potentially dangerous situation must be taken into account, so that base inhabitants have time to escape to a safe zone. While this may be the case for certain Earth structures, it must be the rule for lunar habitats, since there is no tolerance for operating in a damaged lunar structure. Further, rescue may be weeks away.

These dangerous possibilities (the design sample space) lead to the definition of limit states that can be mathematically modeled in the form of a *failure or limit state function*, much as the safety margin was above. The limit state function g of the relevant loading parameters Q_i, strength parameters R_j, nominal geometric parameters L_k, and environmental effects E_l, defining the structure, are chosen so that:

$$g\left(Q_i, R_j, L_k, E_l\right) = 0$$

characterizes the limit state, with failure indicated by $g < 0$. The design values for these basic random variables are denoted by q_{id}, r_{jd}, l_{kd}, e_{ld}, respectively, and may be written as:

$$
\begin{aligned}
q_{id} &= \gamma_i q_{ic} \\
r_{jd} &= \phi_j r_{jc} \\
l_{kd} &= \zeta_k l_{kc} \\
e_{ld} &= \eta_i e_{lc},
\end{aligned}
$$

where q_{ic}, r_{jc}, l_{kc}, e_{lc}, are characteristic values, γ_i are load factors, ϕ_i are resistance factors, ζ_i are geometrical factors, and η_i are environmental factors. Substitute these into the limit state function to ascertain that a safe state exists:

$$g\left(q_{id}, r_{jd}, l_{kd}, e_{ld}\right) \geq 0.$$

The characteristic values for an Earth-based structure are often the mean value for dead load, 98 percent fractile in the distribution of the annual maxima for movable loads, the 2-5 percent fractile for strength parameters, and the mean value for geometric parameters. The load factors are > 1, except when the load has a stabilizing effect. The resistance factors are usually < 1, and the geometric factors are usually $= 1$. How we would choose such factors for lunar structures is still an open question. Quite naturally, load and resistance factors may be further above or below 1, where there is greater uncertainty and less tolerance for risk. Geometric factors can be most easily controlled and therefore utilizing their average values in an analysis is reasonable. Note that the limit state function can be defined in two other ways, by defining ultimate g_u and serviceability g_s limit state functions. Limit states are further discussed in the next section.

The process of reliability design is an iterative one, requiring the designer to iterate between acceptable risk and economic costs. For a certain cost, there is an upper limit of possible safety. In addition, there is a principle of uncertainty, which rarely operates on the Earth, governing the lunar site. That is, only a limited amount of information can be gathered before operations must begin. The lead time required before construction can begin on the lunar surface is lengthy enough that we should expect some backtracking once the project is underway and new information becomes available.

The selection of characteristic values are likely based on optimization. Load, resistance, and geometric factors can be further decomposed into basic sources of uncertainties. For example, the loading parameter Q_i can be decomposed as $Q_i = Q_1 Q_2 \ldots Q_s$, where each Q_j represents particular loading uncertainties.

There are numerous measures of reliability. Such reliability indices are useful for comparisons of various designs. One such index due to Cornell is:

$$\beta_C = \frac{E[M]}{D[M]}, \tag{11.1}$$

where M is the safety margin, and β_C is a dimensionless coefficient. The idea here is that $E[M]$, the location measure from the failure set, is a good measure of the reliability. This number is non-dimensionalized using the dispersion $D[M]$. Such a measure is still valid for lunar applications since the safety margin M reflects all the choices made in a design. Equation 11.1 is the inverse of the coefficient of variation, Equation 10.9. A larger value for β_C implies a more reliable system.

To extend applicability to systems of components, which consist of many elements that may have different failure modes, the concept of *structure function* is introduced. This function is representative of how the failure of a component in the system affects the reliability of the system. The system may be modeled in series, where the system only operates if all the elements are functioning, or it may be modeled as parallel, where the system operates if at least one element is functioning. A hybrid model, *k-out-of-n*, is one that operates if at least k out of its n elements are functioning. Examples are a chain link for a series system, a multi-wire rope for a parallel system, and a suspension bridge for the hybrid system. For a lunar structure, which includes life-support, radiation management, and micrometeoroid absorbing systems, we also face a system (of systems) with complex time-dependent characteristics. Such systems are likely to age quicker in the lunar environment.

How materials are used in a design has a significant effect on system reliability and safety. Materials can be classified according to their failure characteristics. An ideal brittle material fails if a single particle within fails. An ideal plastic material yields if a single particle yields, but will carry the yield force. If a single particle fails, a fiber bundle material will redistribute forces, with the remainder of particles carrying the load. The Daniels system is an example. For second- and third-phase lunar surface structures constructed with local materials, research is required to understand how such components and structures will behave and can be fabricated. It is critical to know how they fail, and whether sufficient warning is built into the design in case of impending failure. Of course, this is true for the first lunar structures as well.

11.3   RELIABILITY ENGINEERING

Reliability engineering focuses on minimizing the number of failures and maximizing the design life, identifying potential sources of failure and minimizing the collateral damage due to failure, and gathering and assessing data from earlier correlated studies.

Objects can fail for a variety of reasons. The designs may have overlooked a critical loading or behavioral characteristic. For example, material stresses may have been underestimated in the design, fatigue calculations may have ignored some source of high-frequency oscillations, or resonant vibrations may have been induced by nearby sources.

Variability may have not been properly understood or quantified. Variables and environmental loads are inherently uncertain. The realized values may have been outside the range of the design. Testing of components is a critical aspect of understanding such variabilities. Mechanical, electrical and electronic components generally have low probabilities of failure due to manufacturing, transport, construction, or maintenance errors. They may fail due to human operator error. Further complicating analysis and design is that variabilities can change with time. This is due in part to human interaction with the component and in part to normal wear. The part changes with time due to usage and due to its interaction with the environment.

In the space and lunar environment, repairable and replaceable parts are problematic. Except for the most critical components, spare parts will be minimal. Reparability requires skills and resources.

The ability to design components and systems for reliability implicitly includes the ability to define failure, in all its manifestations, and to define the operational environment (loads, radiation, regolith dust, low gravity, impacts, dynamics, vibration, for example). Not all failures are of equal consequence. Some are of minimal impact and can be worked around. Others can terminate operations.

In engineering design, distinction is typically made between different categories of design criteria. These are frequently also referred to as *limit states*. The three most common categories are the Serviceability Limit State (SLS), the Ultimate Limit State (ULS), and the Fatigue Limit State (FLS). Many design documents also introduce the so-called Accidental Limit State and the Progressive Limit State, in order to take care of unlikely but serious structural conditions.

Engineering design rules are generally classified according to the depth of information available to the designer. *Level I* reliability design procedures apply point values (mean values) for the various design parameters and also introduce specific codified safety factors that are intended to reflect the inherent statistical scatter associated with the parameters. Level I methods are equivalent to standard deterministic procedures, where it is assumed that all parameter values are known exactly.

At the next level, second-order statistical information (information on variances and correlation properties, in addition to mean values) can be applied if such are available. The resulting reliability measure and analysis are then referred to as a *Level II* reliability method. Examples of Level II methods are first-order, second-moment (FOSM) methods, and reliability index methods.

For *Level III*, it is assumed that a complete set of probabilistic information is available. These are joint probability density and distribution functions that include correlations between all the design variables. Such information is very difficult, if not impossible, to obtain. Also included in Level III are numerical integration methods (because the functions of many variables are too complex for analysis), and approximate analytical methods (where the functions are simplified using order of magnitude arguments). FOSM methods are more widely used. We expect that it will take a significant presence on the Moon in order to obtain Level II and Level III data.

Refer to Figure 10.3, repeated here as Figure 11.1 for easy reference. Suppose that a structure is subjected to a load. X is the structural strength (in units of stress) and Y is the stress that the structure experiences due to the load. If $X - Y > 0$, then the structure can withstand the load and survives; it does not fail. If strength and stress are known exactly, then this is a deterministic problem, and we can state whether failure or survival has occurred with 100 percent confidence. However, if X and Y are not exactly known, then for a Level III analysis, X and Y are defined by their probability density functions, or, more precisely, they are jointly defined by their joint probability density function, $f_{XY}(x, y)$, which relates not only their probable realizations, but also how the two variables are correlated. In Figure 11.1, we show each marginal density function (the respective density function for each variable that is derivable from the joint density function). We can see the range of values for each parameter. Of course, we want the strength to be greater than the stress all the time. Usually this is not possible, and this reality is represented by the hatched region. Within that area, the stress is greater than the strength, and the structure fails to resist the load. The probability of such an occurrence is related to the area of the hatched region.

We define a function that represents the difference:

$$Z = X - Y,$$
(11.2)

where failure occurs for $X - Y < 0$. The boundary between failure and survival is given by the surface $X = Y$. The probability of failure, p_f, is then given by:

$$p_f = \Pr(Z = X - Y \leq 0)$$
$$= \iint_{X \leq Y} f_{XY}(x,y)dxdy,$$

where the integration is carried out over the failure domain (the hatched region).

In some instances, it is possible to assume the variables are statistically independent. In this case:

$$f_{XY}(x,y) = f_X(x)f_Y(y).$$

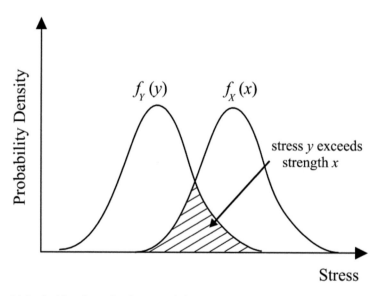

Figure 11.1.   In this schematic, the strength is X and the stress is Y. Here, they are modeled using normal probability density functions. When the stress exceeds the strength, the system has failed. A measure of the reliability is given by the overlapped hatched area.

Then:

$$P_f = \iint_{X \leq Y} f_X(x) f_Y(y) dx dy.$$

It is possible to integrate the strength variable X, and then:

$$P_f = \int_{-\infty}^{\infty} F_X(y) f_Y(y) dx dy,$$

where:

$$F_X(y) = \Pr(X \leq y) = \int_{-\infty}^{y} f_X(x) dx.$$

Figure 11.1 implies that X and Y are Gaussian. They can be any other probability density functions as well, depending on empirical evidence.

For the simplified case, where both X and Y are Gaussian random variables, with mean values μ_X and μ_Y and variances σ_X^2 and σ_Y^2, and where the two variables are assumed to be uncorrelated, the quantity $Z = X - Y$ is then also Gaussian, with the mean value and variance given by:

$$\mu_Z = \mu_X - \mu_Y$$
$$\sigma_Z^2 = \sigma_X^2 + \sigma_Y^2.$$

The probability of failure can then be written as:

$$P_f = \Phi\left(\frac{0 - \mu_Z}{\sigma_Z}\right) = \Phi\left(\frac{-\mu_Z}{\sigma_Z}\right),$$

where Φ is the standard normal distribution function that corresponds to a normalized Gaussian variable, with zero mean and a standard deviation equal to 1. Performing the above substitutions, we find:

$$P_f = \Phi\left(-\frac{\mu_X - \mu_Y}{\sqrt{\sigma_X^2 + \sigma_Y^2}}\right),$$

where:

$$\beta = -\frac{\mu_X - \mu_Y}{\sqrt{\sigma_X^2 + \sigma_Y^2}}$$

is called the *safety* or *reliability index*, and corresponds to Equation 11.1. By defining an acceptable failure probability, $p_f = p_A$, we can find the corresponding value of $\beta = \beta_A$ that represents an acceptable lower bound on β. A lower value of β results in a higher failure probability. This value can be used to determine, in a probabilistic sense, whether the strength X is within an acceptable range as compared to the load Y.

For problems with n random variables defining the failure surface or limit state, one of the challenges is performing the computation for the n-dimensional volume. Of course, the reliability index defined above is particular to the variable Z. More complex problems and limit states will have more complex reliability indices, and in many instances the indices are complex and cannot be written in a simple form, or as an equation. In many instances, functions with many variables are expanded in Taylor series, and the series truncated after two terms. These approximate representations are used to perform a probabilistic analysis that requires only the mean values and variances of the parameters. Their probability density functions are not needed. These are Level II methods because they only rely on mean values and variances.

11.4   FAILURE TYPES

The *reliability* of a component or a system is the probability of operating it according to design specifications for a prescribed time period. When considering manufactured products, the term *quality* is used as a measure of the variance of the part.[12]

These components are designed so that they function properly for a certain period of time, for a broad range of loadings. The time at which a part fails, T_f, is a random variable and as such varies from realization to realization. Empirical evidence clearly shows that T_f is not deterministic. 'Identical components' subjected to 'identical loads' will fail at different times, and the time to failure can only be described probabilistically. Of course, at some micro-scale level the components and the loads are not identical, but such small differences are not quantifiable.

Two types of failures are discussed here. The first is called the *first-excursion failure* or *first passage failure*. In this case, a structure may fail when one aspect of structural behavior exceeds a certain threshold for the first time, as shown in Figure 11.2. For example, a part may fail when the displacement or acceleration exceeds a threshold value, Z. If a part is designed to stay in the elastic range, it may be considered to fail when the stress at a critical location exceeds its yield strength. A rod that is in compression fails when the axial load exceeds the critical for buckling. A crack in a plate that is subjected to a tensile force will grow to failure when the stress intensity factor exceeds its critical value. First-passage failure is extreme, and unlikely to be designed in most systems. Preferably, systems will accumulate damage visibly, in time for repair. Certainly, surprises will not be welcome on the lunar surface, and neither will first-excursion failures.

The other type of failure is *fatigue failure*. When a part is subjected to cyclic loads, damage is done by repeated exposures or cycles. The part eventually fails when the accumulated damage reaches the total damage that a part can absorb. It can fail undergoing many oscillations, even at stress levels that are lower than the yield strength.

All these types of problems are nondeterministic, requiring probabilistic methodologies for analysis and design.

The dangers of first-excursion failures

Suppose that a component is subjected to stationary random forces and it is determined that, if the component fails, it does so when the component behavior $X(t)$ exceeds a critical, pre-determined value Z, for the first time. The behavior may be the displacement, acceleration, stress at a critical location, compressive load, or the stress intensity factor at a crack opening. We assume that $X(t)$ is a weakly stationary random process. If the critical value is also uncertain, and defined by a probability density function, then the first-excursion failure may occur earlier or later than expected. The uncertainty is two dimensional – one is the behavior, and the other is the cutoff for unsafe behavior.

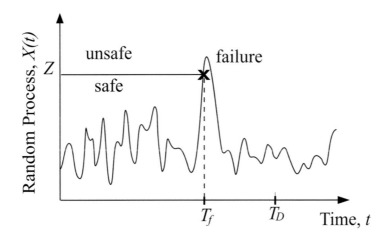

Figure 11.2.   First-excursion failure occurs when $X(t)$ exceeds Z for the first time. Here, Z is taken to be a specific deterministic value, but it may also be a random value.

Figure 11.2 shows a sample realization of $X(t)$. Failure is defined as the event when $X(t)$ exceeds the maximum allowable value Z. The goal of an engineering design is to produce a part that will function properly during its design lifetime, T_D, which is a deterministic quantity specified by the designers. If the time to failure, T_f, is greater than T_D, then the part has operated successfully. Figure 11.2 shows a first-excursion failure, where $T_f < T_D$, indicating that the design was unsuccessful and failed prematurely. A lunar habitat design must be designed to avoid first-excursion failures. This overdesign makes the unsafe cutoff Z very high, and assures that $T_D \leq T_f$.

Since T_f is a random variable, whether or not the part functions properly throughout its design lifetime can only be expressed as a probability, called the *reliability,* and denoted as p_r:

$$p_r = \Pr\left(T_f > T_D\right).$$

The probability of failure, denoted as p_f, is then given by:

$$
\begin{aligned}
p_f &= 1 - p_r \\
&= \Pr\left(T_f \leq T_D\right).
\end{aligned}
$$

T_D is prescribed by the engineer and is a deterministic quantity.

If the probability density function of T_f is known and denoted as $f_{T_f}(t_f)$, then the reliability, and the probability of failure, can be written, respectively, as:

$$p_r(T_D) = \int_{T_D}^{\infty} f_{T_f}(t_f)\,dt_f$$

$$p_f(T_D) = \int_{0}^{T_D} f_{T_f}(t_f)\,dt_f.$$

Therefore, the probability of failure is also the cumulative distribution function:

$$p_f(T_D) = F_{T_f}(T_D).$$

The reliability and the probability of time to failure are the areas under the probability density function of T_f, as shown in Figure 11.3.

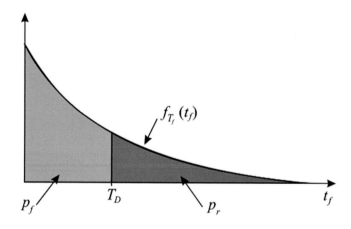

Figure 11.3. Reliability p_r and probability of time to failure p_f.

The general problem of obtaining a probability density function for the first arrival time, $f_{T_f}(t_f)$, is called the *first passage time problem*. Numerous probability densities can be used to model $f_{T_f}(t_f)$, depending on the assumptions made and the data. Examples include the exponential, gamma, normal, and Weibull failure laws. Will the environmental conditions on the lunar surface conform with these failure laws? Or do we expect other empirical probability densities for damaging conditions, such as dust abrasion, radiation aging of materials, dust damage to machinery and seals, as well as the probability densities that model biological risk factors?

The exponential failure law

Many components are governed by the exponential failure law, which is applicable when:

1. the parameter that indicates failure (stress, strain, acceleration) is weakly stationary, and therefore the failure rate is constant,
2. the disturbances are independent of each other so that the number of disturbances (forces or accelerations) are modeled as Poisson processes, and
3. the disturbances arrive at a constant rate.

The exponential failure law is given by:

$$ f_{T_f}\left(t_f\right) = v e^{-v t_f}, \quad t_f \geq 0, $$

where v is the *constant failure rate* or failures per unit time, often measured in hours. The mean value and the variance of time to failure are given, respectively, by:

$$ E\{T_f\} = \frac{1}{v}\,\mathrm{hr} $$

$$ \mathrm{Var}\{T_f\} = \frac{1}{v^2}\,\mathrm{hr}^2. $$

A disturbance rate of $v = 0.01$ failures/hr, for example, implies that the expected time to failure for the particular component is $E\{T_f\} = 1/v = 100$ hr, and the variance $\mathrm{Var}\{T_f\} = 1/(v)^2 = 10{,}000$ hr^2. The standard deviation is then 100 hr. Intuitively, we also know that the failure rate of $v = 0.01$ failures/hr means that we can expect an average of one failure out of 100 components every hour. Similarly, we can expect two failures out of 100 components in two hours, and so on. However, we should note that the probability of actually having one failure per 100 components per hour is not 100 percent, but relatively low.

The variance for the time to failure for an exponential law is relatively large, and the standard deviation equals the mean value, which indicates that the time to failure is spread over a large range. The constant failure rate v is interpreted to mean that after the item has been in use for some time, its reliability has not changed. That is, there is no *wearing-out* effect and the probability is *memoryless*. Thus, a steel beam that has not buckled is assumed to be as good as new.

The failure rate is an experimentally determined quantity. One way to estimate the failure rate is to record the times at failure for many components and find the average value $E\{T_f\}$, from which v is calculated using the above result, $E\{T_f\} = 1/v$. The failure rate can also be calculated from the joint probability density function, $f_{X\dot{X}}\left(x,\dot{x}\right)$, without having to test many physical components, where we must know the values of the random process $X(t)$ and its derivative in time, $\dot{X}\left(t\right)$. Whatever the mathematical model for the failure rate for components and systems on the lunar surface, the challenge is obtaining sufficient data to characterize the process. Figure 11.5 shows an example response for a Gaussian narrow band process, and its time of first-excursion failure.

Figure 11.4 shows a sample realization where $X(t)$ may exceed a desirable range N times, where N is a random variable with number of realizations n. Each exceedance of an arbitrary level Z is recorded as a *disturbance*. The cause of the disturbance may be the landing or take-off of a spacecraft, a micrometeoroid impact, a machine turning off or on, or large radiation gradients due to solar cycle activity. Exceeding from desirable into

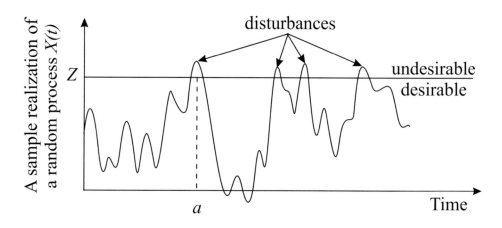

Figure 11.4. A sample realization of $X(t)$ as it wanders into undesirable behavior.

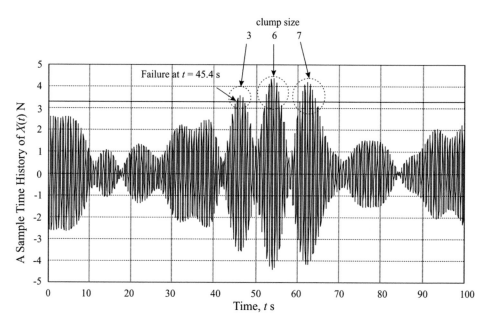

Figure 11.5. A sample response for a narrow band Gaussian random process $X(t)$, showing a first-excursion failure at $t = 45.4$ s when $X(45.4) = 3.3$.

undesirable behavior accumulates damage to the system, but does not necessarily result in immediate failure, though it does bring failure closer. Such damage accumulations are discussed in the next section.

11.5 FATIGUE LIFE PREDICTION

When a machine or a structural member is subjected to repeated dynamic stresses, even below their yield strengths, they eventually exhibit diminished strength and ductility. Fatigue is unavoidable even under the most benign conditions. When the cyclic stresses are continued, cracks in the material start to propagate and eventually *fracture* ensues. This phenomenon is called *fatigue* and the number of stress cycles prior to fracture is called the *fatigue life*. Before failure, a fatigue crack spreads from a location with high stress concentration, which occurs due to imperfections in material, surface roughness, and sudden changes in structural geometry.[13] For the lunar habitat, acoustic noise and machine vibration can result in local oscillations that can lead to local fatigue.

It is generally difficult to predict fatigue life due to the inherent scatter in experimental data and difficulties in translating laboratory data to real-life situations. The goal of this section is to introduce concepts in damage prediction and fatigue life using Miner's rule. Although it is unsatisfactory in some cases, Miner's rule is the simplest and most widely used stress-based fatigue failure rule. It assumes that:

- the mean stress equals zero
- the stress levels or amplitudes are discrete and can be specified, and
- the order in which the stress levels are applied has no effect.

In Miner's rule, the damage done to a component is expressed as a summation of ratios of the number of cycles applied, n_i, to the number of cycles N_i to failure at a given stress level. That is, the total damage done is given by:

$$D = \sum_i \frac{n_i}{N_i}.$$ (11.3)

The component is stipulated to fail when the damage D reaches 1. The total number of cycles to failure at a given stress level is obtained experimentally. The number of cycles applied can be obtained by counting the number of peaks at a certain stress level. Mathematically, the stress level s is a local maximum, or a peak, in a given cycle. The form of Miner's rule in Equation 11.3 is applicable when a discrete number of stress levels is applied.

However, for a continuous random process, any stress level in a range is possible. For instance, the number of cycles between stress levels s_1 and s_2 is given by:

$$n = \int_{s_1}^{s_2} n(s)\,ds,$$

where $n(s)$ is the number of cycles per unit stress range, and is indicative of how the number of applied cycles is distributed over the stress level. The continuous version of Miner's rule is then given by:

$$D = \int_s \frac{n(s)}{N(s)}\,ds.$$

$N(s)$ is the number of cycles to failure at the stress level s, and is obtained from the so-called failure curve, Wölher curve, or $S-N$ curve, the latter introduced in the next section. In practice, such curves are determined by experiments for various materials and geometries.

The term $n(s)$ can be normalized by the total number of peaks, n_t, resulting in the probability density function of the applied stress levels:

$$f_s(s) = \frac{n(s)}{n_t}.$$

Suppose the random process $X(t)$ is the material stress. The stress levels s are the peaks of $X(t)$. Therefore, the probability density function of peak heights, $f_H(h)$, is also the probability density function of the stress levels, $f_S(s)$. The total damage predicted by Miner's rule is then:

$$D = n_t \int_s \frac{f_s(s)}{N(s)} ds. \tag{11.4}$$

For fatigue failure problems, we first obtain the probability density of peak heights $f_H(h)$, or equivalently $f_S(s)$. Using $f_S(s)$, and an empirical law ($S-N$ curve) obtained from experimental data, the total damage is estimated. Unlike for first-excursion failure problems, it is often difficult to obtain analytical expressions for the probability density function of the time to failure.

Lunar structural oscillations that lead to an accumulation of damage are proportional to the system stiffness and inversely proportional to the system mass – actually $\sqrt{k/m}$. But system damping properties also affect the frequencies and amplitudes of oscillations. Damping also dissipates the oscillation energy, eventually drawing it down to zero, and the more damping, the quicker the oscillations decay. Regolith shielding that blankets the surface structure can dissipate the vibrational energy. If the structure is composed of a material such as magnesium, as suggested elsewhere in this book, then additional and significant damping is there to dissipate oscillations. All these effects can be calculated as part of a fatigue life estimate for the surface structure.

It is reasonable to expect that empirical damage models based on the Miner rule, and more sophisticated models, can help designers estimate damage due to abrasion, meteoroid impacts, and radiation aging, that is, $D_{abrasion}$, D_{meteor}, and $D_{radiation}$ respectively. All these damage components then need to be combined in a way that represents our experience on the Moon. Tests such as those depicted in Figure 11.6 can be carried out for abrasion, meteoroid impacts and radiation aging, in order to build the data set.

Failure curves

Fatigue behavior of structural components are often graphically described by $S-N$ curves. These curves are generated by classical uniaxial fatigue tests, as shown in Figure 11.6. A specimen is subjected to harmonic loading resulting in constant amplitude stresses, S, and the number of cycles to failure, N, for each stress level is recorded and plotted. Each dot in

Figure 11.7 corresponds to a single such experiment. The result is plotted in logarithmic scale for both S and N. There is a stress level below which the specimen will not fail regardless of the number of load cycles. This stress level is called the *fatigue limit* or *endurance limit strength*, denoted by S_e.

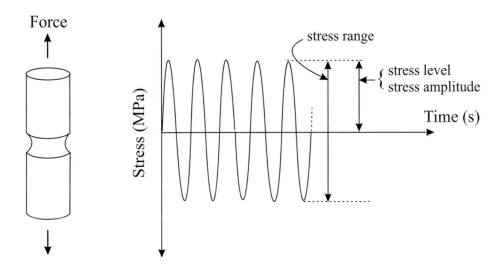

Figure 11.6. Uniaxial fatigue test to generate *S–N* curves.

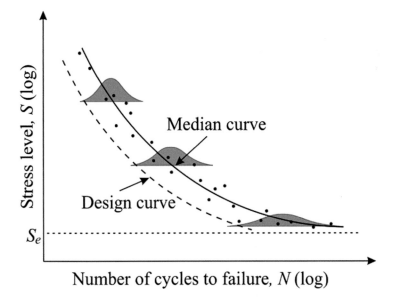

Figure 11.7. Results of fatigue tests and resulting *S–N* curves.

In most cases, the number of cycles to failure, N, corresponding to a certain stress level varies considerably, even if specimens with the same material and same dimensions are subjected to the same loading patterns. That is, for a given stress level, N is a random variable with a probability density function called the *life distribution function*. The life distribution function varies with the stress level. As the stress level approaches the endurance limit strength, the scatter in the data increases. When a large enough number of specimens is tested, the Gaussian probability density function provides an adequate data fit. The Gaussian distribution can be a good approximation to the distribution of log N.

$S-N$ curves such as the one shown in Figure 11.8 are used for design purposes. They are determined in a way to give the lower bound design. If we assume that log N is distributed normally,[4] and define the design curve to be mean-minus-two-standard-deviations, then the $S-N$ curves are associated with a 97.7 percent probability of survival.

In principle, Miner's rule applies to zero mean processes. However, in practice, the applied load will not have a zero mean. When the applied load has a nonzero mean, modifications are often made in the $S-N$ curve. One of the most important consequences of nonzero mean is the change in the fatigue life. Models that predict the fatigue life for nonzero mean processes are the modified Goodman, Gerber, and Soderberg models.

In Figure 11.8, the design $S-N$ curve may be used as a conservative representation of the data. We assume that the curve is definite, and can be written generally in the form:

$$NS^m = A, \tag{11.5}$$

where m and A are determined from experiments. We take the logarithm of both sides to find:

$$\log S = -\frac{1}{m} \log N + \frac{1}{m} \log A.$$

When log S is plotted against log N, a line called the linear model S-N curve with slope $-1/m$ is obtained.

Damage Equation 11.4 can be written as:

$$D = \frac{n_t}{A} \int_s s^m f_s(s)\, ds$$

$$= \frac{n_t}{A} E\{S^m\}. \tag{11.6}$$

---

[4]The Gaussian density is not the only, nor the best, approximation for life distribution. The Weibull probability density function has been proposed to give a better fit. On the Moon, we may expect other empirically-derived densities to govern.

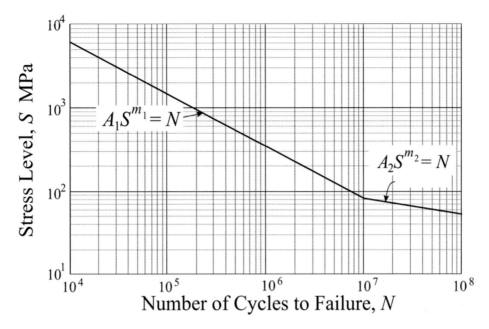

Figure 11.8.   *S–N* Curve Fit to Data.

11.6   MICROMETEOROIDS

"They're going to hit us, so let's prepare."

The nomenclature used to name the rocks that fly around Earth needs clarification since there is often mistaken usage. Based on the monograph by Yeomans, we begin with some terminology.[14] An *asteroid* is an inactive small rocky body orbiting the Sun. If the asteroid is partially composed of ices that vaporize in sunlight, then it is a *comet*. A one-meter or less particle that is ejected from a comet or asteroid is called a *meteoroid*. The meteoroid that enters the atmosphere, creating a light show as it vaporizes, is a *meteor*. If the meteor only partially vaporizes and a piece survives to land on Earth, it is a *meteorite*. If the particle is < 0.05 mm (0.002 in) in diameter, then it is called a *micrometeoroid*.

A near Earth object (NEO) is a comet or asteroid that can approach the Sun to within 1.3 astronomical units (AU) and has an orbital period of less than 200 years. It is considered hazardous if it can approach to within 0.05 AU of Earth's orbit (7.5 million km) and is large enough to cause impact damage.

Yeomans' monograph on NEOs is packed with data and history. There is an overview of where we are likely to find the asteroids, and how they were created. The dynamics between the asteroids and the planets is fascinating. It appears likely that the Moon was created about 50 million years after the Earth's formation, by an impact of a Mars-sized rock that ejected material into Earth's orbit, which then accreted into the Moon.

It also appears that some of the critical building blocks of life on Earth were delivered here in meteorites. Interestingly, there was no free oxygen in Earth's atmosphere until about 2.4 billion years ago.

Micrometeoroids are a serious threat in space and on the lunar surface due to their speed, concurrent with the lack of atmospheric shielding. These speeds are on the order of 10 km/s (22,369 mph). For Earth, most of these dust-size particles burn up in the atmosphere. The larger rocks that hit the ground can cause great damage, but these are rare. On the Moon, where there is no atmosphere, lunar surface structures need to be shielded from this sort of attack. For certain types of structures, such as depots and telescopes, shielding is only practical when they are not in use. New failure modes due to high-velocity micrometeoroid impacts need analysis with probabilities of impact.

Understanding the meteoroid environment has been a critical issue since the early stages of spaceflight, and of interest scientifically for much longer than that. An effort to model the meteoroid environment was in the NASA report by Cour-Palais.[15] Observations (photographic and radar) and direct measurements (using detectors on spacecraft and rockets) were used to derive the probability distributions of meteoroid velocities. The mass range tabulated was in the range 10^{-9} to 10^{-6} g, and by observations were in the range $> 10^{-6}$ g. Average velocities were 20 km/s. At an average speed of 10 km/s (2200 mph), a 1 cm fleck is capable of inflicting the same damage as a 550 lb. object traveling at 60 mph on Earth. A 10 cm projectile is comparable to 7 kg of TNT.

Micrometeorite fluxes were approximated in Tamponnet, and are provided in Table 11.1.[16] We have added the third column of probability estimates of craters of a certain size by using this data, and using a frequency interpretation of probability, where the number of impacts, as fractions of the total number, are used to estimate the probabilities.

Table 11.1. Micrometeorite Fluxes

Crater Diameter (mm)	Crater Flux (1/m²yr)	Probability (hb)
>0.1	30000.000	0.9524
>1.0	1200.000	0.0381
>10	300.000	0.0095
>100	0.600	1.905×10^{-5}
>1000	0.001	3.175×10^{-8}

Smith provided a full review of the lunar environment and discussed fracture and fatigue design criteria, as well as the phases of lunar design.[17] Impact loads were also discussed. In addition to the traditional consideration of direct hits by micrometeoroids, near-miss scenarios were suggested as having potential serious effects. A near miss may load the structure with cratering ejecta impacts, and may also cause significant vibration of the structure. Even low-level vibration can have implications for fatigue life.

A model by McNamara et al., known as the *Meteoroid Engineering Model*, identified sporadic radiants of meteoroids such as comets.[18] It was a physics-based approach that yielded accurate fluxes and directionality for interplanetary spacecraft anywhere from 0.2 to 2.0 AU, and obtained theoretically-derived velocity distributions that were validated against observations for a mass range of 10^{-6} g to 10 g. Oberst et al.:[19]

"… provide a summary of a workshop that brought together the different fields of meteor astronomy, hypervelocity impact physics, dust exosphere modeling, cratering, planetary interior modeling, and space instrumentation under the umbrella of the impact process. It reviews the current state-of-the-art on the study, both observational and theoretical, of the different effects of the impact phenomenon; highlights areas of investigation, such as the observations of impact flashes and the counting of new small craters on the Moon, where a significant volume of new results is becoming available; and quantifies the mutual consistency of inferences made on the current size distribution of impactors and their resulting craters."

As they noted:

"… as the meteoroids impact the lunar surface, their kinetic energy is partitioned into the excavation of craters, the production of plumes associated with a flash of light, and the generation of seismic waves that propagate through the lunar interior. Different observational techniques are used to examine different effects of the impact process, but the inferences made on the source parameters (e.g., impactor population, energy conversion efficiencies) are constrained by the fact that the source process is one and the same."

Such modeling efforts are important to the design of surface structures, for impact force data and frequency, local seismic consequences, and potential local ejecta risks. The question of how to account for the rare but catastrophic event is open. It is similar to the question in earthquake engineering, where it is not possible to design for the worst possible seismic event, but is also different. Modern skyscrapers are designed to withstand earthquake magnitudes of eight and higher. A building collapse is a tragic event, but civilization on Earth would not end. On the Moon, there will be very few settlements for a long time. Initially, there will be only one. An event that ends that settlement(s) also ends our manned presence on the Moon. But rather than worry about predicting the rate of occurrences of catastrophic events that cannot be defended against, we may be better off designing our structures and systems to survive the remaining smaller impact events, with warning systems to get us out of the way of the indefensible event. Clever designs are possible; for example, distributed structures, buried structures, or advanced impact absorbers, working in conjunction with the regolith shielding.

Empirical penetration and damage models

As structural designers, we care about the rate and geographical distribution of meteoroid impacts, and their size and velocity ranges. These are all probabilistic quantities for which data availability is relatively sparse, but improving. Impact can be on a structure or facility, or nearby. Direct impacts require an understanding of how meteoroid kinetic energy is converted into structural damage. That understanding informs design decisions regarding the necessary quantity of regolith, or other shielding, to minimize structural damage. It is necessary to understand the limits of such shielding and it clearly will not be possible to shield surface structures against all threats. Underground habitats and lava tubes offer possibilities for much improved shielding.

Nearby impacts offer different challenges to the designer. Impacts on the lunar surface transfer meteoroid kinetic energy into the ground as seismic energy, and propel ejecta into ballistic trajectories. There is also energy loss as light and heat. The risk associated with seismic effects are a function of the impactor mass, velocity, and proximity to the facilities. The risk associated with the ejecta can affect the entire facility, given how far these can travel. Nearby astronauts can be at risk, as can be more delicate components such as solar panels and optical devices that are not shielded, for example. Here, we focus on the surface structure.

We suggest applying the concept of the sacrificial plate, used in space and elsewhere, at the lunar surface. A set of such plates, strategically placed around the lunar site, and around unshielded equipment and instruments, can safeguard against at least shallow trajectory ejecta. Regolith-filled bags can also assist in such a strategy. The placement of these shields can be done optimally, depending on the layout of the habitat. They can also be placed around the habitat site at greater distances, providing some shielding for the astronauts who are performing tasks outside the habitat.

There have been numerous theoretical and experimental studies on impacts. Much of our understanding of impacts at all speeds, and on many different materials, comes from military studies. Efforts have focused on understanding how the kinetic energy of impactors are absorbed by structures. At very high speeds, due to the vaporization of both, computational fluid mechanics codes (hydrocodes) are used to model this behavior. We do not discuss these here, and only provide a brief set of examples from the literature in order to hint at the modeling and equations that result.

Hayashida and Robinson compared five single plate penetration equations for accuracy and effectiveness: *Fish-Summers*, *Schmidt-Holsapple*, two equations developed for the Apollo project (Rockwell and Johnson Space Center (JSC)), and the JSC *Cour-Palais*.[20] They were derived from test results, with velocities ranging up to 8 km/s, and developed for spacecraft subjected to orbital debris. We do not repeat the equations here, but they were all empirical, relating penetration depth to target thickness, projectile diameter, projectile density, target density, Brinnell hardness for the target, Young's modulus for the target, impact velocity, and speed of sound for the target. Even though the equations were based on test results for projectile velocities up to 8 km/s, predictions using these equations have been made for velocities up to 15 km/s. Parametric studies were carried out for various velocities, projectile masses densities, and targets of several materials and thicknesses. The penetration resistances (ballistic limit) of a single-wall structure for the five penetration equations were calculated. The ballistic limit is the velocity required for a particular projectile to penetrate a particular piece of material reliably (at least 50 percent of the time). Such statistics are costly to obtain, and are rarely acquired with high confidence.

The conclusions of this study were that these equations were good predictors of actual behavior. However, tests would be required to supplement the predictions of all the equations. "It is recommended that designed shields should be tested with the actual configuration, and realistic velocities and materials for projectiles, to prove the design will act as predicted when impacted by a meteoroid or debris particle. Only then can the design be declared acceptable for orbital operations." When applied to lunar surface structures, we are challenged by the inability to test various structural configurations, materials, and shielding options before placement. Overdesign, initially, seems to be the best course of action. "One can save much weight by choosing the most unconservative equation, which underestimates the value for the penetrating projectile diameter. Therefore, the cost can be reduced significantly, but this choice could result in unforeseen critical damage to the spacecraft or loss of the entire spacecraft. On the other hand, one can design a conservative, well-protected spacecraft, but the spacecraft could become very heavy. Thus, the launch and material costs would be increased." This summarizes the trade-offs and the challenges for many aspects of lunar structural design.

Steinberg and Bulleit studied the safety and reliability of lunar surface structures to meteoroid loading.[21] A homogeneous, filtered Poisson process was used to model the mass distribution of the meteoroids. This assumes that meteoroids impact the lunar surface statistically independently, with constant intensity, λ, and have exponentially distributed interarrival times. The intensity of strikes of masses less than or equal to m, $M \leq m$, is Poisson with:

$$\lambda_m = \lambda\left(1 - F_M(m)\right),$$

where $F_M(m)$ is the cumulative distribution function of the random variable M, λ_m is the rate of occurrence (intensity) of meteoroid strikes with $M \geq m$, and λ is the intensity of the base Poisson process.

Hypervelocity impacts lead to pulverization of the target and projectile. This velocity varies with the type of materials being considered, but, in this study, 5 to 6 km/s was adequate for typical structural materials:

> "The stresses caused by the impact [are] so great that the energy is dissipated equally in all directions creating a hemispherical crater. The crater depth is related to the 2/3 power of the velocity. Failure functions were developed from impact equations on semi-infinite targets. Complete penetration of the structure by the meteoroid was assumed to occur when the crater depth reached 2/3 the depth of the target thickness. This is known as the ballistic limit condition, which occurs when a spall plane propagating from the inner surface meets the crater.
>
> "Experimental procedures for hypervelocity impacts are limited to velocities of approximately 6 km/s with masses of approximately 3 g. Though these limits are far exceeded by the meteoroid impact problem, the equations developed from the experiments are the only ones available to analyze such a problem. In addition, the materials used for the projectiles as well as the targets in most experiments were not the same materials considered in this study."

Two sets of hypervelocity penetration equations were developed. The first set was for hypervelocity penetrations:

1. of semi-infinite targets, of materials such as concrete, sand, aluminum, and steel, for velocities of less than 6 km/s (used for the derivation of Equation 11.8)
2. of hemispherical craters, of materials such as aluminum, zinc, tin, steel, cadmium, copper, and lead, for velocities of less than 6 km/s, and masses less than 2.7 g
3. of targets and projectiles made of copper and lead, for projectiles with velocities up to 3.5 km/s and masses less than 2.8 g, and
4. of rock, sandstone, gypsite, permafrost, ice, desert alluvium, silt, sand and clay (used for the derivation of Equation 11.7).

The first set of (four) penetration equations above were used to develop a second set of (four) equations for the complete penetration of the regolith cover and the structure. In developing the second set, it was assumed that the energy remaining to penetrate the structure equals the total energy of the meteoroid less the energy required to penetrate the cover. This assumption does not include any interaction effects at the interfaces between the materials, where energy is dissipated. The relevant set of equations is shown next.[5]

Penetration of the regolith cover, by an assumed spherical meteoroid, is given by:

$$44.118h_s - \rho_P^{1/3} m^{1/6} (3280.8v - 100) > 0, \tag{11.7}$$

where h_s is the regolith depth in mm, ρ_P is the projectile density in g/cm^3, m is the meteoroid mass in g, and v is the meteoroid velocity in km/s. Note that such empirical equations are written for specific units (which don't have to be consistent). Equation 11.7 can be solved for the penetration depth of the regolith.

The equation used to calculate the penetration of the structure is:

$$hE^{1/3} - 384.79(mv^2)^{1/3} E^{0.09} > 0, \tag{11.8}$$

where h is the (structural) crater depth in mm, and E is the target modulus of elasticity in kPa. Using the above two equations, the equation for the complete penetration of the regolith cover and the structure is found:

$$hE^{0.243} - 384.79\left[mv^2 - 0.013466h_s vm^{5/6} \rho_P^{-1/3} - 0.3048mv \right]^{1/3} > 0. \tag{11.9}$$

This equation reflects the two separate calculations for the penetration of the regolith and then the structure.

For these failure functions, a positive result of the failure function defines a safe state, and the failure state is defined otherwise. The reliability index β is commonly used as a measure of the safety of the structure. High β values correspond to low probabilities of failure since they imply that the difference between structural strength and loading is larger.[22]

Without going through the details of the calculations in this study, we summarize the key ideas and conclusions. Equation 11.8 is solved for h, which is the damage, in the form of a crater, done to the structure by a single strike. While it may be unlikely that a single meteoroid will take out the structure, meteoroids of low mass are more abundant and are therefore more likely to strike the structure. Over time, these small mass strikes will lead to an accumulation of damage, and may cause safety problems. Therefore, the total damage, d, is the sum of the individual damages:

$$d = \sum_{i=1}^{N} h_i,$$

---

[5] It is worth mentioning that it is quite likely that classified sources have much data on high-velocity impacts.

where there are N strikes. This equation assumes that the individual damages h_i are independent of each other, even though repeated strikes will cause damage, and subsequent
strikes will be on a weakened structure, implying a coupling between sequential strikes
and respective damages. The mean values and variances of h and N are known, and so the
mean value and variance of d can be found. Then the amount of damage for certain levels
of β can be found. A first-order second-moment analysis is carried out. A similar procedure can be used for a regolith-covered structure, using Equation 11.9.

The use of regolith to shield against radiation, thermal effects, as well as meteoroids
was studied by Lindsey.[23] The *Fish-Summers* penetration equation was used:

$$t_s = k_s M_m^{0.352} V_m^{0.875} \rho_m^{1/6},$$

where t_s is the target or surface thickness in cm, k_s is a constant depending on the target
(0.57 for aluminum), M_m is the projectile mass in g, V_m is the impact velocity in km/s, and
ρ_m is the projectile density in g/cm³. In this study, to stop a meteoroid of size 7 cm, mass
89.75 g, and density 0.5 g/cm³ traveling at 20 km/s, we require an aluminum shield of
density 2.7 g/cm³ that is 34 cm thick, or equivalently, a regolith shield of density 2 g/cm³
and of thickness 45.9 cm.

The Smooth Particle Hydrodynamics impact simulation code (SPHC) was used to
assess the effects of high-speed and hypervelocity impacts, and to assist in the design of
surface lunar structures through the understanding of possible failure modes.[24] "The
potential failure modes caused by hypervelocity impacts are thermal and radiation protection degradation; structural degradation by partial or complete wall penetration or by internal spallation; and injury to personnel." This study focused on the verification of the code
through its use to predict experimental results for a variety of tests using concrete structures. Once verified, the code can be used to assess lunar surface structure designs to
hypervelocity impacts. In one such computation, a hemispherical concrete-walled dome
structure, with a 30 ft diameter and a wall thickness of 2 ft, was modeled by the SPHC. The
structure was internally pressurized at atmospheric pressure at Earth sea level, and gravity
was set to lunar gravity. The impactor was a porous stony meteoroid, with a diameter of 20
cm, mass 2.1 kg, and with a strike speed of 20 km/s. It is noted that historically, initial
design requirements for NASA missions have been set at a 95 percent Probability of No
Critical Failure (PNCF) for impacts, but for concrete or concrete-like walls, a penetration
of 1/3 of the total wall thickness is considered a fatal penetration since at that point extensive damage is expected, with secondary damage to the remaining wall thickness. We note
that a variety of failure criteria exist.

In this study, two damage models were used, one with ejecta, and the other without.
Ejecta damage can occur for lower velocities. The crater depths of the models were well
represented ($R^2 = 0.990$) by the following fit to the data:

$$D = 4.34d,$$

where D is the crater depth in cm, and d is the particle diameter in cm.

In order to include the ejecta damage, additional models were run at velocities of 100
m/s to 20 km/s, for several particle densities. The following fit represents the full data set
($R^2 = 0.997$):

$$D = 1.545\rho^{0.33}V^{0.67}r,$$

where D is the crater depth in cm, ρ is the projectile density in g/cm^3, V is the projectile velocity in km/s, and r is the projectile radius in cm. R^2 is a statistical measure of how close the data are to the fitted regression line. A value of 1 implies a perfect fit (a straight line) and a value of 0 implies no fit, or predictive value.

This study had a number of conclusions. One was that even if a single structure has an acceptable, low probability of serious impact damage, if there are numerous such structures then the probability of serious damage is multiplied by the number of structures, since the risk area is also multiplied. Also, certain parts of a surface structure may be more vulnerable to strikes or ejecta damage. Identifying such locations and improving their designs can significantly reduce impact damage risks:

"Most meteor impacts will be from micrometeoroids. Although not a direct threat to the walls of a structure, vital communications equipment, external sensors, solar power generators, thermal radiators, etc., will be more sensitive to small impacts. Over time, erosion of tougher structures will occur. Impact models can be used to evaluate these types of effects.

"Another consideration is the effect of a large meteor impact on or near the lunar base. Large impacts do occur every day on the Moon, and even though the probability of a nearby strike is very small, it could happen. Some warning might be available if a large impactor is dangerously near; in any case, a plan for evacuation to shelter, erection of a temporary shield, or relocation of the crew to a safer location should be developed in case of a catastrophic impact. A detailed analysis of the damage caused by such an event could help constrain the size, shape, and placement of individual structures in the base to minimize detrimental effects."

Schonberg, *et al.* discussed some approaches to shielding lunar habitats, focusing on shielding that is intended primarily to provide protection against meteoroid impacts, and on shielding approaches that could use resources mined or extracted from the Moon.[25] A number of shielding concepts were presented, as well as a testing regimen to assess the effectiveness of these concepts:

"Based on the flux model [of Cour-Palais 1969], a lunar habitat structure with an exposed surface area of 1000 m^2 can expect approximately 1.35 impacts per year by a 1 mm meteoroid particle and approximately 0.0065 impacts per year or 3.25 impacts every 500 years by a 5 mm meteoroid particle."

The exponential probability density function was used to estimate the probability of $N=n$ particle impacts per year. The exponential density function is given by the equation:

$$f_N(n) = \alpha e^{-\alpha n}, \quad n > 0,$$

and equals zero elsewhere, where $\alpha > 0$, and $1/\alpha$ is the mean value of N. The cumulative distribution function for N is given by the equation:

$$F_N(n) = \Pr(N \le n) = 1 - e^{-\alpha n}, \quad n \ge 0,$$

and equals zero elsewhere. Therefore, $\Pr(N>n) = \exp(-\alpha n)$.

Another approach to approximate the dynamic response of a surface structure was based on the use of an equivalent beam model for the structure, and to use an FAA airplane-debris impact code.[26] Based on the FAA code, a 1.2 kg projectile, traveling at a speed of 240 m/s, was assumed to strike a beam at mid-span, resulting in a sinusoidal force of half cycle, with a maximum force of 24.9 kN. This impact force was used as input to a finite element code. There is an extensive discussion of the results.

11.7 CONCLUDING COMMENTS

This chapter has provided an overview of reliability design that is so important for the design of structures, especially for a site as isolated and unforgiving as the lunar surface. The kinds of data and information that we require to design structures for certain design lives have been outlined here. The challenge, as repeated often, is the lack of the data needed to build the design curves, and the probability density functions, needed for the calculations. On Earth, these design curves have been developed for almost all possible engineering applications, based on long histories of usage and improvements learned from failures. For the Moon, there is no such data. We have to resort to Earth-based data in advance of being on the lunar surface, where we can develop local data. There appears to be no alternative to this dilemma, except to send structures to the Moon years in advance of astronauts, to gather data *a priori*. This is not likely.

Redundancy is often used to improve reliability. It is also used to prove high reliability without necessarily resorting to expensive and time-consuming testing. In the next chapter, we consider the fine line that is redundancy, and examine what can be learned from commercial airplanes, which are extremely reliable and complex structures.

Quotes

- "I'm substantially concerned about the policy directions of the space agency. We have a situation in the U.S. where the White House and Congress are at odds over what the future direction should be. They're sort of playing a game and NASA is the shuttlecock that they're hitting back and forth." Neil Armstrong.
- "Yeah, I wasn't chosen to be first. I was just chosen to command that flight. Circumstance put me in that particular role. That wasn't planned by anyone." Neil Armstrong.
- "I think the Space Shuttle is worth one billion dollars a launch. I think that it is worth two billion dollars for what it does. I think the Shuttle is worth it for the work it does." Pete Conrad.
- "The Space Shuttle has been a fantastic vehicle. It is unlike any other thing that we've ever built. Its capabilities have carried several hundred people into space." Robert Crippen.
- "For whatever reason, I didn't succumb to the stereotype that science wasn't for girls. I got encouragement from my parents. I never ran into a teacher or a counselor who told me that science was for boys. A lot of my friends did." Sally Ride.

References

1. S. Kaplan, and B.J. Garrik: *On the Quantitative Definition of Risk*, Risk Analysis, Vol.1, No.1, 1981.
2. H. Benaroya: *Reliability of Structures for the Moon*, Structural Safety, Vol.15, 1994, pp.67–84.
3. J. Shortridge, T. Aven, and S. Guikema: *Risk assessment under deep uncertainty: A methodological comparison*, Reliability and System Safety 159 (2017) pp.12-23.
4. G. Hormek, R. Facius, M. Reichert, P. Rettberg, W. Seboldt, D. Manzey, B. Comet, A. Maillet, H. Preiss, L. Schauser, C.G. Dussap, L. Poughon, A. Belyavin, G. Reitz, C. Baumstark-Khan, and R. Gerzer: *HUMEX, A Study on the Survivability and Adaptation of Humans to Long-Duration Exploratory Missions, Part 1: Lunar Missions*, Advances in Space Research, Vol.31, No.11, 2003, pp.2389-2401.
5. G. Horneck, and B. Comet: *General Human Health Issues for Moon and Mars Missions: Results from the HUMEX Study*, Advances in Space Research, Vol.37, 2006, pp.100–108.
6. J. Downer: *When Failure is an Option: Redundancy, Reliability and Regulation in Complex Technical Systems*, Discussion Paper No.53, Centre for Analysis of Risk and Regulation, London School of Economics and Political Science, May 2009.
7. W.W. Mendell, and R.P. Heydorn: *Lunar precursor missions for human exploration of Mars - III: Studies of system reliability and maintenance*, Acta Astronautica 55 (2004) pp.773-780.
8. **Tension Structures: Behavior and Analysis**, W.J. Leonard, McGraw-Hill, New York, 1988; **An Introduction to Cable Roof Structures**, H.A. Buchhold, Cambridge University Press, Cambridge, 1985; **Tensile Structures**, Frei Otto, MIT Press, Cambridge, MA, 1973.
9. H. Benaroya: *Tensile-integrity structures for the Moon*, Applied Mechanics of a Lunar Base, H. Benaroya, Editor, Applied Mechanics Reviews, Vol.46, No.6, June 1993; Also, H. Benaroya: *Reliability of Structures for the Moon*, cited earlier.
10. H. Furuya: *Concept of deployable tensegrity structures in space application*, International Journal of Space Structures, 7(2) (1992); A.S.K. Kwan, and S. Pellegrino: *A cable-rigidized 3d pantograph*, Proceedings of the 4th Euro Symposium on Space Mechanisms and Tribology, Cannes, 1989.
11. We have drawn on the book by Madsen *et al.*, for our subsequent discussion on structural safety: **Methods of Structural Safety**, H.O. Madsen, S. Krenk, N.C. Lind, Prentice-Hall, 1986.
12. J. Jin and Y. Chen: *Quality and Reliability Information Integration for Design Evaluation of Fixture System Reliability*, Quality and Reliability Engineering International, Vol.17, 2001, pp.355-372.
13. This is by no means the complete description of fatigue failure. The interested reader should refer to **Metal Fatigue,** N.E. Frost, K.J. Marsh, and L.P. Pook, Oxford University Press, 1974, reprinted by Dover Publications, 1999.
14. **Near-Earth Objects: Finding Them Before They Find Us**, D.H. Yeomans, Princeton University Press, 2012.

15. B.G. Cour-Palais: *Meteoroid Environment Model - 1969 - Near Earth to Lunar Surface*, NASA SP-8013, March 1969.
16. C. Tamponnet: *Life Support Systems for Lunar Missions*, <u>Advances in Space Research</u>, Vol.18, No.11, 1996, pp.(11)103-(11)110.
17. A. Smith: *Mechanics of materials in lunar base design*, in **Applied Mechanics of a Lunar Base**, H. Benaroya, Editor, ASME <u>Applied Mechanics Reviews</u>, Vol. 46, No 6, June 1993.
18. H. McNamara, R. Suggs, B. Kauffman, J. Jones, W. Cooke, and S. Smith: *Meteoroid Engineering Model (ME): A Meteoroid Model for the Inner Solar System*, <u>Earth, Moon, and Planets</u> (2004) 95, pp.123-139.
19. J. Oberst, A. Christou, R. Suggs, D. Moser, I.J. Daubar, A.S. McEwen, M. Burchell, T. Kawamura, H. Hiesinger, K. Wünnemann, R. Wagner, and M.S. Robinson: *The present-day flux of large meteoroids on the lunar surface – A synthesis of models and observational techniques*, <u>Planetary and Space Science</u> 74 (2012) pp. 179-194.
20. K.B. Hayashida, and J.H. Robinson: *Single Wall Penetration Equations*, NASA Technical Memorandum TM-103565, December 1991.
21. E.P. Steinberg, and W. Bulleit: *Reliability analyses of meteoroid loading on lunar structures*, <u>Structural Safety</u> 15 (1994) pp.51-66.
22. **Methods of Structural Safety**, H.O. Madsen, S. Krenk, and N.C. Lind, Prentice-Hall, 1986.
23. N.J. Lindsey: *Lunar Station Protection: Lunar Regolith Shielding*, International Lunar Conference 2003, Hawaii.
24. S.W. Evans, R. Stallworth, J. Robinson, R. Stellingwerf, and E. Engler: *Meteoroid Risk Assessment of Lunar Habitat Concepts*, NASA Marshall Space Flight Center, 2006.
25. W.P. Schonberg, F. Schäfer, and R. Putzar: *Some Comments on the Protection of Lunar Habitats Against Damage from Meteoroid Impacts*, <u>Journal of Aerospace Engineering</u>, Vol.23, No.1, January 1, 2010, pp.90-97.
26. R.B. Malla, and T.G. Gionet: *Dynamic Response of a Pressurized Frame-Membrane Lunar Structure with Regolith Cover Subjected to Impact Load*, <u>Journal of Aerospace Engineering</u>, Vol.26, No.4, October 1, 2013.

12

Airplanes, redundancy and lunar habitats

"Are two safer than one?"

As engineers, we work under the assumption that we can analyze and design anything with an acceptable level of reliability, given enough time and resources. We also extrapolate our design experiences into new domains, as much as possible, trying to learn lessons from successful designs and applying these to new, different, and perhaps more complex projects. In many ways, we can learn from the technological evolution of airplanes. Over the past hundred years, airplanes have developed from simple wooden structures built by one or two people using standard materials, to becoming one of the most complex and successful – reliable – structures humans have conceived and built. Perhaps the most complex and successful, given that tens of thousands of people put themselves into these structures daily and survive. The airplane passenger is safer than the automobile passenger.

This chapter considers why airplanes are so reliable, how they are regulated by the Federal Aviation Authority (FAA), and the role that redundancy plays in both their design and their regulation. We refer primarily to the work of Downer, coming from the science and technology studies literature, who aims to understand the limitations of engineering design, specifically in high-reliability industries, as pertaining to policy and government regulation and certification. How are airplanes so reliable, and are there lessons to learn as we move into space and the Moon with the purpose of settlement?

When considering airplanes, the foundations of uncertainty are at many scales, and are due to numerous environmental sources. The airplane is a complex system of tens of thousands of components – structural, electrical, electronic, fluidic, control and actuation, algorithmic – that are combined into one entity by humans and machines. The tens of thousands of components are made all over the world and put together at numerous sites, eventually creating the airplane at the last stage. Each component is individually tested, and its characteristics and reliability understood. The behavior of the components, when integrated, is generally understood, but the whole being more than its parts leads to uncertainties about how the parts will interact under rare or unforeseen events, and the need for repeated tests of prototype aircraft before commercial use is allowed.

© Springer International Publishing AG 2018
H. Benaroya, *Building Habitats on the Moon*, Springer Praxis Books,
https://doi.org/10.1007/978-3-319-68244-0_12

We can perhaps also make the argument that the whole is *less* than the sum of its parts, because the completed airplane now has characteristics that are not simply the linear sum of those parts. The characteristics are a nonlinear sum of its parts and as we know, nonlinearities imply large deviations over small distances and short times. This nonlinearity of the whole as a function of its parts stipulates the need for tests of prototype aircraft, even those that are only slight variations of earlier models. Computational models alone cannot certify a new airplane for commercial use.

A key problem in older aircraft is fatigue life, an inherently nondeterministic property that is a function of many parameters that define the whole system. It is only recently – in the last few decades – that the coupling between material properties at all scales and the behavior of the structure is beginning to be understood. However, even with greater understanding, the process of damage accumulation and the resulting fatigue life prediction is inherently uncertain. Part of that uncertainty stems from the structure of matter and the random distribution of dislocations and imperfections in the material. Another component of uncertainty is due to the complexity of the structural topology, or configuration – the location of components with respect to each other and how they are connected to each other. This is not fully or exactly definable, even with the most sophisticated computers and software.

The most sophisticated engineering mathematical models include assumptions and approximations in order to become tractable (solvable). There are also variabilities that cannot be defined in the manufacturing process: quality control across hundreds of facilities that manufacture components, and where the components are integrated into one airplane. Once the airplane is put into operation, it is flown by numerous pilots following equally numerous flight plans, each in a unique weather system, and is subjected to a wide variety of inspections, maintenance and repair regimen. Each pilot, each flight plan and weather system, each cycle of inspection, maintenance and repair regimen, is unique. There are variabilities in each aspect, and they are not quantifiable. Designing, building, operating, and then maintaining the airplane involves unique paths that challenge our analytical abilities. Damage to the airplane is a function of all of these aspects of its life. Quantitative models used by owners and operators to predict damage, and to schedule inspections, attempt to account for uncertainties. Certification agencies such as the FAA must also accommodate the same uncertainties as part of their regulatory function.

However, airplanes are ubiquitous in our society and they function remarkably well, giving us comfort that they are very well designed and maintained. We are interested to understand the basis for this success, as reliable workhorses of modern society, and to transfer the essence of that understanding to those who will design and eventually construct structures and habitats for the lunar surface.

12.1 AIRCRAFT CERTIFICATION

As a critical technology, aircraft are certified by the FAA to fly and carry passengers. The FAA also manages the maintenance of these aircraft by way of rules and advisories that commercial carriers must follow. Two National Research Council Reports (National Academy of Sciences) are valuable assessments of the FAA Airworthiness Certification procedures.[1] The recommendations of these reports have been incorporated into the

FAA's procedures for certification, and the effectiveness of certification and rule-making is incontrovertible given the very low rates of death and injury in commercial aviation. We mention a few findings from these reports that appeared 18 years apart.

Key elements of a safe fleet include data collection, database management, risk analysis, risk management/action, and monitoring effectiveness:

"Industry should collect, organize, and analyze safety data and take appropriate corrective action to protect the safety of the fleet. The FAA should not independently collect, organize, or analyze safety data for large transport aircraft. Instead, the FAA should oversee the entire process, providing direction, assessing the accuracy and objectivity of industry's risk analyses, and mandating corrective action, as appropriate. The overall objective is to produce a more effective safety management process that routinely monitors operations and maintenance, uses data on incidents and other abnormalities to identify potential hazards proactively, and takes corrective action before hazards cause an accident." [1980 report]

Incident data need to be databased in a way to inform the operators and the manufacturers effectively. A major recommendation is that [1998]:

"The committee believes that safety would be enhanced if the FAA focused its design approval process on determining that applicants' design organizations are technically qualified and have internal review processes that ensure compliance with the applicable airworthiness standards, rather than continuing to rely on its own ability to determine compliance through spot checks of the applicant's analyses and tests. The FAA should examine the technical qualifications and integrity of design organizations, including their understanding of regulations and policies and their ability to properly implement them. Qualified organizations should then be certificated as ADOs (Approved Design Organizations), allowing them to make detailed findings of compliance in accordance with published policies. FAA audits would verify continued compliance, in part by ensuring that ADOs' level of involvement in specific projects is appropriate in light of the technical issues involved."

We now proceed to consider how the framework of redundancy impacts design, verification, and certification.

12.2   REDUNDANCY

We recall that redundancy is a design philosophy aimed at increasing reliability and minimizing risk, by adding a backup capability to a system that can engage in case the primary system fails. A thrust of Downer's analysis, cited in the following discussion, is that it is not possible to calculate the reliability of a complex system to the levels demanded of safety-critical technologies using engineering tests and models. Manufacturers are able to fabricate, and the FAA is able to manage, a very large fleet of commercial airliners by using service data from earlier models that show reliable performance, and then building new models that will be as reliable by making only small changes in the designs.

Another key issue is that: "Complex, dangerous technologies … demand 'ultra-high' levels of reliability, and their assessments must be prospective, as much rides on them being known to be reliable before they are deployed.[2] Yet reliability at such levels cannot simply be 'inspected' in a laboratory setting, where billions of hours of observation would be required to achieve statistically meaningful measures …"

But this is not the only thesis put forward. Another of great interest is that redundancy sometimes *increases* the risks of failure. This may appear counterintuitive, but it has also been argued by Boeing (for the 777) in suggesting that a two-engine airplane is more reliable than one with four engines, because the redundancy adds a complexity that cannot be modeled, and is therefore ignored in the engineering models. Some ideas regarding redundancy are presented that clarify the stereotype:[3]

> "Although first established as a design tool, engineers soon realized that redundancy's practical virtues were coupled with an epistemological advantage: redundancy not only allowed them to design for high reliability, it also allowed them to quantitatively demonstrate reliability – something that had previously proved very difficult. Redundancy could do this because it offered a powerful, straightforward, and convincing rubric with which engineers could mathematically establish reliability levels much higher than they could derive from lab testing."

The essential challenge is that redundant systems require extra elements in the design to determine if a failure has occurred, and to engage/turn-on systems that can replace the failed systems. The failure of an engine during flight, for example, requires a number of actions by the automatic control systems. The failure must first be sensed, then the fuel cut off from the failed engine, and the remaining engines must be given more thrust while the rudder is adjusted to compensate for the now asymmetrical thrust. The pilot is sometimes used as a redundant system as well, mediating competing information in order to make decisions in an emergency. Redundancy, as highlighted above, requires additional elements, and these can diminish the overall reliability of the airplane.

An additional assumption regarding the calculation of reliability for redundant systems is that the extra elements, the so-called mediating elements that establish the need for the redundant system to activate in the event of a failure of a primary system, must themselves be completely reliable. Such mediating elements do not have their own redundant systems. "Despite being critical to the reliability of redundant systems, however, mediating systems cannot be redundant themselves, as then they would need mediating, leading to an infinite regress. 'The daunting truth,' to quote a 1993 report to the FAA, 'is that some of the core [mediating] mechanisms in fault-tolerant systems are single points of failure: they just have to work correctly'."

The assumption of independence of elements in the system, whether for purposes of redundancy or as part of the system model, can also be a cause of failure. Interdependencies (correlations) exist in complex systems at the least because they are operating in, and are driven by, the same environment. Also, "seemingly redundant and isolated elements frequently link to each other at the level of the people who operate and maintain them. … Technological failures open a window for human error."

If we want to double the reliability, or survival probability, of a lunar population, we might consider building two habitats, say, ten miles apart. This would be close enough to run back and forth by rover and exchange equipment in case of failures, for example. In order to simplify the design and fabrication, we may consider having both habitats designed to be almost identical. The question is whether we have actually doubled the reliability. In general, the answer is no. Even though both habitats are far apart, they are still coupled to each other in significant ways. The first way is by the environment. It is true that Habitat One can survive in the event that Habitat Two is struck by a large meteoroid, but they are both subject to the same radiation environment, and any design flaws, or repair issues, can affect both in the same way. If both structures are maintained by the same people, then human errors can propagate to both habitats. Habitat operations are heavily dependent on software-based systems. It is likely that both habitats will have essentially identical hardware and software systems, so any glitch can affect both habitats. The inhabitants of both habitats are trained identically, have been selected for the Moon mission using identical criteria, and are likely to be temperamentally similar. All of these couplings are real, even though the two habitats are geographically apart, and the presumed redundancies, and independence, may be ephemeral.

A fundamental issue is that it is impossible to verify the high levels of reliability that are demanded of airplanes experimentally. Redundancies are attractive, since they mathematically demonstrate very large increases in reliabilities without resorting to expensive and lengthy testing regimens, which often cannot be carried out because of the very low failure rates required by the certifying agencies. The added elements of the redundant system are generally simplified, or ignored, in order to satisfy the assumptions of independence between these and the larger system. Otherwise, it would be necessary to estimate a quantitative measure of the interdependencies (correlations) and this is generally not possible.

Failures also propagate, and the path of the failure cannot be predicted due to the inherent uncertainties in the system and its working environment.

> "Aviation manufacturers could never demonstrate the levels of reliability required of their products unless regulators were willing to overlook the uncertainties inherent in the frame it provides (and regulators would be unable to demand the kinds of proof that the public and policy makers request)."

12.3   ACCIDENTS

Downer discussed the differences between "Normal Accidents" and "Epistemic Accidents."[4] The differences are summarized:

> "A normal accident is one that occurs due to a series of seemingly trivial events and non-critical failures sometimes interacting in unexpected ways that thwart the very best engineering designs and cause catastrophic system-wide failures. ... No complex system can function without trivial errors: that is, with absolutely no deviation from platonically perfect operation. They must tolerate minor irregularities and small component failures: spilt milk, small sparks, blown fuses, misplaced luggage, stuck valves, obscured warning lights and suchlike. Engineers might not like these events, but eliminating them entirely would be entirely impracticable, and they are considered irrelevant to system safety. [Downer continues:]
>
> 1. Normal accidents are unpredictable and unavoidable because they are a result of a confluence of failures on innocuous components, each of which alone would not lead to an accident.

2. Normal accidents are more likely in tightly coupled and complex systems. Tightly coupled implies rapid and automatic interactions between its components. Complex systems are those with many interacting elements.
3. Normal accidents are unlikely to reoccur because the bundle of minor failures occurring together is a random occurrence.
4. Normal accidents rarely challenge established knowledge, again because the set of precipitating failures are not unexpected individually.
5. Normal accidents are not heuristic[1] since, again, they do not challenge the engineer's understanding of the world.

"An example of a normal accident is the Three Mile Island nuclear reactor meltdown. Unavoidable accidents that do not fall into the above Normal case are called Epistemic Accidents. While there are similarities between the two kinds of accidents, [those] similarities have different causes.

1. Epistemic accidents are unpredictable and unavoidable because engineers build technology around fallible theories, judgments and assumptions.
2. Epistemic accidents are more likely in highly innovative systems since these are new and do not have a history of practice.
3. Epistemic accidents are likely to reoccur unless the underlying causes become understood.
4. Epistemic accidents challenge design paradigms since they reveal a lack of fundamental understanding.
5. Epistemic accidents are heuristic since they provide a path to learning by engineers."

An example of an epistemic accident is the 1988 Aloha Airlines Flight 243, where the top of the fuselage tore off due to multi-site fatigue damage. This failure points to the divergence between the aircraft's performance in the laboratory and its performance in operation. "Engineers cannot exactly reproduce the real world in their laboratories so they must reduce its variables and make subjective judgments about which are relevant." Aloha 243 led to revolutionary changes in our understanding of fatigue. Beyond cycles as the basis for fatigue calculations, Flight 243 was a very old plane, manufactured in 1969 and its operations in the warm, saltwater environment of Hawaii had accelerated fatigue and corrosion damage. Finally, imperfect bonding in the plane's fuselage, allowing saltwater to permeate the metal sheets, also accelerated damage and corrosion. These aspects were not, and perhaps could not, be replicated in a laboratory.

The important point to note from the Aloha failure is that while laboratory testing is crucial in the development of complex technologies, real world testing is as important. Even so, unique situations can still result in failures because of our lack of fundamental understanding of something critical. There are *known knowns*, *known unknowns*, and *unknown unknowns*. The last category can lead to accidents that are essentially blameless because these accidents are resident in a part of the domain that we do not know exists. Perhaps a distinction can be made between the unexpected (because it is an unknown unknown) and the unpredictable (because it is an event of very small probability, with almost no data to define its rate of occurrence).

---

[1] A heuristic (adjective) pertains to the process of gaining knowledge, enabling a person to discover or learn something.

We referred to this idea earlier when noting that unless we know how many faces a die has, we cannot calculate any probabilities. We can only calculate probabilities if we have the complete sample space. While uncertainties result in events that we cannot predict, it can also be due to a lack of knowledge about a process, even if that process is a deterministic one. The underlying process does not have to be probabilistic, it only has to be unknown to us.

The connection of this to the design of lunar habitats is profound. The lunar site is full of unknown unknowns, for engineered systems and for biological systems. The unknowns of the remote lunar site can only be partially uncovered from afar. On-site presence is required for a fuller understanding. We see that, even with aircraft in an environment to which we have complete access, there are serious unknowns that are missed until full operations have been in place for many years.[5] Various parts of the aircraft that are not coupled within the structure become coupled because of the environment. The operational environment couples subsystems in the structure from the outside. That outside coupling can be the source of the accident. Similar situations can be expected for a structure on the lunar surface. We can design the structure on Earth, test it here under almost identical circumstances – except for the lunar gravity – and yet, on the lunar surface, the real environment kicks in and couples the structural subsystems in ways that we could not duplicate in our test facilities. On the lunar surface, our structure is transformed from being a die with six faces, to one with n faces. Our predictive models are transformed into non-predictive models. New dimensions have been added to the structure's sample space.

There are accidents other than the unavoidable Normal or Epistemic. There is a larger pool of potentially avoidable failures, many of which result from organizational errors. Continuing from Downer's discussion on the different kinds of accidents:

> "Simply put, the designs of all complex systems rely on elaborate knowledge claims. These contain unavoidable uncertainties and ambiguities, which inevitably give rise to fundamentally unresolvable doubts and disagreements. Accounts of accidents that directly or indirectly explain them by reference to unresolved disagreements or ignored doubts, therefore, fundamentally misconstrue the nature of engineering knowledge. To work with real-life systems is, by necessity, to make judgments with imperfect information. To say that we should not fly airplanes unless all disagreements have been resolved and all facts are known is, in essence, to say that we should never fly airplanes."

In a similar vein, the challenges introduced here do not imply that we should not go to the Moon and settle. It means that we need to try to avoid some of the pitfalls that lead us to believe that our systems (engineered and biological) are safer than they really are. The two Space Shuttle accidents are examples where we believed that our systems were safe, but the reality proved otherwise.

12.4   REGULATION AND CERTIFICATION

Downer examined the FAA's type certification process.[4] This is the way that it evaluates civil aircraft and certifies that they are sufficiently reliable for public use. The essential thesis is that "regulators of high technologies face an inevitable epistemic barrier when making technological assessments, which forces them to delegate technical questions to people with more tacit knowledge, and hence to 'regulate' at a distance by evaluating 'trust' rather than 'technology'."

Because the regulators do not and cannot have the same level of technical knowledge as those who manufacture and repair the airplanes, the regulators make judgments based on their assessments of the people who build and repair airplanes, in lieu of assessing the actual airplanes. This is accomplished to some extent by using Designated Engineering Representatives (DERs), who are deputized engineers at the facilities that act for the regulator. "They give the FAA access to a reservoir of tacit 'hands-on' knowledge, based on a level of involvement not practical for FAA personnel."

Conflicts of interest are possible, also known as regulatory capture, where over time, powerful industries tend to dominate the agencies that regulate them. In this context:[2]

"Perhaps more consequential than the direct effects of censure on institutional cultures, moreover, are its indirect effects. Several studies have observed that 'institutional risks' – such as the threat of legal, vocational or reputational damage – create incentives for managers to adopt an extremely 'defensive' organizational posture. They further suggest that this can conflict with the management of the 'societal risks' (such as passenger safety) that should be the primary concern of organizations like the FAA."

The two Space Shuttle accidents can be argued to be examples of failed institutions. Refer again to the Nimrod accident review, cited earlier. Taleb also mentioned the effectiveness of the airline industry:[6]

"Every plane crash brings us closer to safety, improves the system, and makes the next flight safer ... these systems learn because they are antifragile and set up to exploit small errors; the same cannot be said of economic crashes, since the economic system is not antifragile the way it is presently built. Why? There are hundreds of thousands of plane flights every year, and a crash in one plane does not involve others, so errors remain confined and highly epistemic – whereas globalized economic systems operate as one: errors spread and compound."

Taleb considered the commercial airline industry to be a 'good' system since it learns from (small) mistakes, unlike the 'bad' economic system that cannot learn, and therefore each mistake leads to a bigger one later. Taleb emphasized the point that an antifragile system is different to a robust system. Antifragile systems learn from mistakes and become stronger. Robust systems, at best, survive mistakes, but remain as they were, no better than before. They do not benefit from the survival of a mistake.

Now that there are non-governmental organizations designing and building rockets, and then launching them into space and to the space station, the FAA's role has been expanded. The Office of Commercial Space Transportation (AST - Associate Administrator for Commercial Space Transportation) was established in 1984 as part of the Office of the Secretary of Transportation within the Department of Transportation. In November 1995, it was transferred to the Federal Aviation Administration as the FAA's only space-related line of business. AST was established to:

- regulate the U.S. commercial space transportation industry, to ensure compliance with international obligations of the United States, and to protect the public health and safety, safety of property, and national security and foreign policy interests of the United States

- encourage, facilitate, and promote commercial space launches and reentries by the private sector
- recommend appropriate changes in Federal statutes, treaties, regulations, policies, plans and procedures, and
- facilitate the strengthening and expansion of the United States space transportation infrastructure.

AST manages its licensing and regulatory work, as well as a variety of programs and initiatives, to ensure the health and facilitate the growth of the U.S. commercial space transportation industry through the Office of the Associate Administrator, along with its five divisions: The Space Transportation Development Division, the Licensing and Evaluation Division, the Regulations and Analysis Division, the Safety Inspection Division, and the Operations Integration Division.

The Office of the Chief Engineer directs AST activities in the areas of technical oversight and research development for commercial space transportation licensing requirements. This office also conducts program management for the FAA Center of Excellence for Commercial Space Transportation and the Safety Management System.

The Regulations and Analysis Division, formerly the Systems Engineering and Training Division, develops, manages and executes the AST Rulemaking Program, and the AST Tools and Analysis Program. Under the Tools and Analysis Program, the Regulations and Analysis Division conducts flight safety analyses, system safety analyses, and specific types of hazard and risk analyses.

The AST safety inspection function, previously implemented under the former Licensing and Safety Division, was established as a separate division in March 2011 for the continued purpose of executing AST's Compliance and Enforcement Program. Safety inspection involves monitoring all FAA licensed, permitted and otherwise regulated commercial space transportation activities, to ensure compliance with FAA regulations and to provide for the protection of public safety and the safety of public property. Because the FAA licenses and permits commercial space transportation operations and not vehicles, safety inspection correspondingly involves the monitoring of those public safety-related pre-operational, operational and post-operational activities.

While AST is still small, compared to the rest of the FAA that focuses on aviation, we can expect to see it grow as commercial space activities grow and begin to carry passengers. Will it, or another government entity, regulate activities, manned or otherwise, on the Moon? We have discussed how the space legal framework is incomplete and needs broadening to include commercial activities. Will nations take it upon themselves to oversee their own commercial activities in space? Likely, there will be national oversight, as well as organizations that coordinate between national entities.

12.5 LESSONS LEARNED

We have an interest in learning from the success of highly reliable commercial aircraft and applying these lessons to the high-risk endeavor of creating a permanent habitat on the Moon. Efforts to transfer the remarkable and successful efforts of the commercial airplane industry to other high-reliability industries, such as nuclear power plants, and high-reliability enterprises such as lunar habitats, have not been possible for a number of reasons.[7] The central challenge for the design of such systems, and to assess their reliabilities, is that new designs must be assessed prior to their use. The validity of designs, and an assessment of their reliabilities, cannot be tested empirically. There are very few nuclear reactors, with a limited number of hours of operations, and, of course, there are no lunar bases or habitats, as far as we know.

Commercial aircraft safety, however, is of exceptionally high reliability. "Jetliners are important in this context for a very different reason. This is that, almost uniquely among critical technologies, their reliability (and thus the validity of expert calculations that anticipate that reliability) can be assessed empirically after they enter service (because we build large numbers of near-identical jetliners), and, confoundingly, they appear to be as reliable as calculations predict." Their reliability assessment is not *a priori*. Their success is based on 'design stability' and 'recursive practice', and by "leveraging service experience rather than tests and models."

Reliability is unusual because it is a 'negative' property, defined by the absence of failure. Predicting reliability is very different from proving that something is reliable based on empirical data. We look forward to designing and building lunar structures, and need to assess safety and reliability before they are built. We need to be able to do this so that the venture can go forward. Our calculations are based on very limited testing of subsystems and data with many gaps. Yet it is still necessary to gain confidence in these calculations before moving forward. We do this to some extent by testing components individually, and a few as a group, and run simulations on these systems. If all these tests and simulations lead to the result that the components and systems are reliable, some confidence ensues.

> "The fundamental basis for these predictive calculations are bench tests performed in advance of actual service. This poses a problem, however, because critical systems demand reliabilities that are higher than tests can claim to demonstrate."

In order to demonstrate high reliabilities under such circumstances, reliability is 'assured' via redundant designs. But, as discussed above, such redundant designs are complex and not adequately modeled in engineering designs. This becomes critical when a new structure or system is put into operation without a long history of similar structures or systems having been operating reliably.

There is also the sensitivity of a design to the so-called 'relevance' of the mathematical model or the testing system. All models, whether mathematical or physical, are approximations of the real world they are meant to represent. Engineers attempt to capture the most important aspects of the real world using as simple models as possible. Mathematical predictions in the form of engineering designs are then verified by testing and prototyping. This is a recursive process requiring many design corrections and adjustments. We know that mathematical models and tests "struggle to capture the ambiguities of real performance." For critical technologies, small differences and ambiguities between model and reality can lead to erroneous reliability calculations.

Design stability, the process of improving new versions of the same product in incremental steps, is a conservative design procedure. In the case of commercial aircraft, design stability refers to the very slow improvement of designs, so that there are only small changes between models. In this way, a provably reliable system is modified only slightly, based on the confidence gained by a long history of reliability and data. This is not true for nuclear power plants, of which there are numerous different designs and many generations. Of course, for space structures and lunar habitats, the population is small and nonexistent, respectively. There is no experience base for a lunar structure reliability analysis. Essentially, space structures are each unique – except perhaps for rockets, which tend to be reliable.

Recursive practice is the process of extensively studying accidents and then bringing the lessons learned into new designs, as well as into retrofitting existing models. The FAA and the airlines meticulously oversee the implementation of any shortcomings found as a result of inspections and accidents into the fleet. Based on Downer, we summarize:

1. Design stability, where there is significant similarity between past and present designs, is predictive of the performance of future aircraft.
2. Design stability coupled with extensive service experience forms the foundation for recursive practice. Service experience is necessary for recursive practice because many operational hours are needed to witness rare and unusual events – the unknown unknowns. Design stability quickly allows engineers to identify what design changes might have led to the accident. Slowing down technological innovations helps to stabilize the systems.
3. Recursive practice is a way to keep high levels of reliability by quickly learning the causes of accidents and implementing improvements in current systems, and future systems that are not too different than the current ones. High reliability is not being predicted for future systems, but assured by past performance, and progress in small steps.

The above does not transfer to any other high-tech industry. The airplane industry is unique in this way, but there are lessons to be learned.

12.6   LUNAR STRUCTURES

We wonder how this new understanding about the high level of reliability of airplanes, and the subtle understanding of redundancy, translates to the engineering design of habitats for the lunar surface. There is no track record for designing, transporting and constructing habitats on the Moon. We have a limited dataset characterizing some, but not all, aspects of the lunar environment. There are serious gaps. There is no avoidance of the reality that the initial surface structure will be designed, and mostly built, on Earth, for transport and erection on the lunar surface. Of course, components will be tested exhaustively in conditions that are thought to be representative of the target environment, but we know that the lunar environment cannot be replicated on Earth or in orbit. Costs preclude sending more than one such structure (if even that one) to the lunar surface in advance of astronauts to test it in order to gather additional data with which to improve the design.

Different sources state that NASA accepts a reliability for its space missions of between 0.95 to 0.99. This is an averaged number. Some components will have very much higher reliability, other components will come in with less reliability. The overall reliability is not a useful number, since the failure of a critical component can effectively fail the mission. The FAA requires a failure rate of between 10^{-6} to 10^{-9} for commercial aircraft (which is unprovable, with verification based on approaches such as those discussed in this chapter).

A major difference between a lunar habitat and a commercial aircraft or a nuclear power plant, is that a lunar mission failure, while tragic in personal terms, does not directly affect the public at large. This appears to imply that a relatively low reliability is tolerable. However, if a lower reliability is tolerable, then so is the occasional loss of an astronaut or a facility. Such a loss should be viewed as par for the course, just as we accept the loss of test pilots and prototype aircraft. We accept and move on. We do not shut down the program for two years. However, given the time and cost required to set up that first lunar facility, we are not ready to risk it.

We need to converge on a design, and then proceed with it without falter. We cannot accurately predict the reliability of the complete system, even though we can estimate the reliability of the individual components. There are no shortcuts to achieving or assessing the reliability of complex technologies. The accuracy of reliability assessments is tied to a long history of testing and operations, with design improvements that are evolutionary, not revolutionary. Overly complex systems with many redundant elements may be less reliable than simpler systems that use standardized components which are simply repaired or replaced. Developing too many different technologies for doing the same thing is not beneficial if we eventually want to send habitats and astronauts to the Moon. We need to focus on designs and systems that are proven, with as few deviations as possible. To use NASA terminology, our systems and components should have very high Technology Readiness Levels (TRLs). The higher the number, the more standard is the technology, with a long history of reliable operations. A low TRL implies a technology that is still in the research or R&D stage, where only prototypes exist.

In principle, we need to have multiple copies of a lunar habitat, so that we can absorb the loss of a single unit. We need multiple launchers and numerous trained astronauts for the missions. Even with such a depth of personnel and resources, losses will occur, but the mission continues in the face of tragedies and we have to be prepared to accept the losses. The population, the astronaut corps, NASA and all corporate entities must realize that while losses are to be avoided as much as possible, they are expected.

We are either serious about placing humans on the Moon, or we are not.

Quotes
- "After Apollo 17, America stopped looking towards the next horizon. The United States had become a space-faring nation, but threw it away. We have sacrificed space exploration for space exploitation, which is interesting but scarcely visionary." Eugene Cernan.
- "For some time, I thought Apollo 13 was a failure. I was disappointed I didn't get to land on the Moon. But actually, it turned out to be the best thing that could have happened." Jim Lovell.
- "I think most astronauts are not risk takers. We take calculated risks for something that we think is worthwhile." Michael J. Massimino.
- "I believe every human has a finite number of heartbeats. I don't intend to waste any of mine." Neil Armstrong.
- "Well, I think we tried very hard not to be overconfident, because when you get overconfident, that's when something snaps up and bites you." Neil Armstrong.

References

1. Committee on FAA Airworthiness Certification Procedures, National Research Council, 1980, **Improving Aircraft Safety**, 118 pages; Committee on FAA Airworthiness Certification Procedures, National Research Council, 1998, **Improving the Continued Airworthiness of Civil Aircraft: A Strategy for the FAA's Aircraft Certification Service**, 86 pages.
2. J. Downer: *On audits and airplanes: Redundancy and reliability-assessment in high technologies*, Accounting, Organization and Society, Vol.36, 2011, pp.269–283.
3. J. Downer: *When Failure is an Option: Redundancy, reliability and regulation in complex technical systems*, Centre for Analysis of Risk and Regulation at the London School of Economics and Political Science, May 2009.
4. J. Downer: *Anatomy of a Disaster: Why Some Accidents are Unavoidable*, Centre for Analysis of Risk and Regulation at the London School of Economics and Political Science, March 2010.
5. There are many lessons to learn from this accident review: **The Nimrod Review**: An independent review into the broader issues surrounding the loss of the RAF Nimrod MR2 Aircraft XV230 in Afghanistan in 2006, C. Haddon-Cave, October 28, 2009.
6. **Anti-Fragile: Things That Gain From Disorder**, N.N. Taleb, Random House, 2012, 2016; See also: **The Black Swan - The Impact of the Highly Improbable**, N.N. Taleb, Random House 2007, 2016.
7. J. Downer: *The Aviation Paradox: Why We Can 'Know' Jetliners But Not Reactors*, Minerva (2017) 55, pp.229–248.

13

Advanced methodologies

"Other ways to deal with uncertainties."

In this Chapter, we suggest analytical and decision options that have not been applied to studies of lunar structures, but offer potentially powerful frameworks to take into account the broad set of issues that are part of the design process. In particular, we suggest the Dempster-Shafer Theory, and performance-based engineering. The published literature does not show the application of the Dempster-Shafer Theory to lunar habitat design. Performance-based engineering has been suggested as an approach, but is, as yet, in the early stages of research.[1] [2]

13.1 DEMPSTER-SHAFER THEORY

This is a general framework for reasoning with uncertainty based on the paper by Dempster.[3] It was later developed into a framework for modeling epistemic uncertainties by Shafer.[4] The Dempster-Shafer Theory is also known as the theory of belief functions, and is a generalization of the Bayesian theory of subjective probability. Bayesian theory requires expert, opinion-based, subjective probabilities for each question of interest. Belief functions base degrees of belief for one question on probabilities for a related question. These degrees of belief may, or may not, have the mathematical properties of probabilities. How closely the two questions are related affects how much the degrees of belief differ from probabilities.

According to Shafer:

"The Dempster-Shafer Theory is based on two ideas: the idea of obtaining degrees of belief for one question from subjective probabilities for a related question, and Dempster's rule for combining such degrees of belief when they are based on independent items of evidence. ... We obtain degrees of belief for one question from

H. Benaroya, *Building Habitats on the Moon*, Springer Praxis Books,
https://doi.org/10.1007/978-3-319-68244-0_13

probabilities for another question. Dempster's rule begins with the assumption that the questions for which we have probabilities are independent with respect to our subjective probability judgments, but this independence is only *a priori*; it disappears when conflict is discerned between the different items of evidence.

"Implementing the Dempster-Shafer Theory in a specific problem generally involves solving two related problems. First, we must sort the uncertainties in the problem into *a priori* independent items of evidence. Second, we must carry out Dempster's rule computationally. These two problems and their solutions are closely related. Sorting the uncertainties into independent items leads to a structure involving items of evidence that bear on different but related questions, and this structure can be used to make computations feasible."

The theory has been applied to a variety of fields.[5] It has also been used in fault tree analyses.[6] It is considered an alternative to probabilistic modeling if both a large amount of uncertainty and a conservative treatment of this uncertainty are necessary, as in early design stages. As per Limbourg *et al.*, it is a synthesis of probabilistic analysis and interval arithmetic, the latter case where parameters are known by their upper and lower bounds. Dempster-Shafer Theory has an advantage in the framework it provides for combining different sources of evidence. A variety of approaches to combining evidence has been proposed.

From the perspective of a lunar design, there are large gaps in the data that characterize potential lunar sites and the environment. There are unknowns about human psychological reactions to the close quarters of a lunar base. Similar concerns exist about the reactions of the human body to the low gravity and high-radiation fields. Since a lunar design is likely to go forward before these knowledge gaps are filled, a framework is needed in order to integrate the parts of the puzzle we know well with the remaining parts that are unknown and yet must be informed by expert judgements. Dempster-Shafer Theory is suggested as one such possible framework. Performance-based engineering is put forward as another such framework.

13.2 PERFORMANCE-BASED ENGINEERING

Performance-based engineering (PBE) is a probabilistic design approach that links the structural design to its end-user, where the model outputs are the decision variables for the end-user. Its methodology addresses system-level performance in terms of the risks of significant events, fatalities, repair costs, and post-event loss of function. PBE began as a methodology to assess the designs of nuclear power plants but the method has evolved to address seismic performance assessment, wind engineering, and hurricane engineering.[7-11] All these applications are mathematically modeled by engineers as random processes due to their complexities. There is much to learn from all these applications moving forward, to create a PBE framework for a lunar habitat design.

PBE is structured to meet specific performance expectations of a structure's occupants, an owner, and the public. Generally, the framework can be structured into four categories, or steps:

Hazard Analysis
results in frequencies of occurrence of key events. For a lunar structure or facility, examples of key events include meteoroid strikes and velocities, radiation types and levels, possible accidents, and structural aging/degradation. Data are needed for the probabilistic modeling of these events. Some of this data can be found from test results, but others require subjective assessments by experts.

Structural Analysis
is performed in terms of system and component uncertainties. A lunar structural analysis includes structural member strengths and geometric properties, regolith mechanics, internal pressurization, dynamic and impact loads, thermal and gradient loads, radiation fluctuations, and other material aspects. Some of these are deterministic, and others are probabilistic.

Damage Analysis
is based on the use of fragility curves, which are used to model damage probabilistically. The fragility curves are cumulative probability distribution functions. They relate parameter values to respective probabilities. The reliability of a system, a lunar habitat, is a complex nonlinear function of the reliabilities of all its components. It is complex because the components interact in ways that cannot be fully modeled mathematically, or tested in advance. Some of the interactions, or correlations and cross-correlations, can be nonlinear, for which common sense and extrapolations do not work. Examples of metrics include: wear rate of components, materials and shielding; failure rates – anticipated vs. actual; and a number of biological fragility curves that represent how the human body reacts to the low gravity, radiation dosage, and other psychosocial stressors.

Loss Analysis
is based on frequency and probability data, and is used to extract performance metrics that are meaningful to facility stakeholders; metrics such as upper bound economic loss during the owner-investor's planning period. Risk-management decisions can be made based on the loss analysis. The project manager for the design and fabrication of a lunar facility has concerns that go beyond investor metrics. The lunar facility is more than a project; rather, it is something that has an almost metaphysical hold on those who have devoted their lives (figuratively and literally) to its creation. Performance metrics for a lunar base need to be based on survivability and development, as well technical aspects of its operation. Examples of metrics include: psychological well-being; physical health; timely accomplishment of tasks and goals; and rates of component failures.

Some of these metrics are useful in multiple categories. Research is needed to create, and populate with data, the detailed structure of these categories and it would not be surprising to find additional categories for analysis. Since these are probabilistic models, a challenge is gathering the needed data to populate the density functions.

More formally, the probability equation:

$$\Pr(D) = \int \Pr(D|H)\, dH,$$

where Pr($D|H$) is the probability of damage D conditioned on the hazard H. This equation can be written in terms of probability density functions:

$$\Pr\left(d_0 < D < \infty | H = h\right) = \int_{d_0}^{\infty} f_{D|H}\left(d|h\right) dd,$$

where $f_{D|H}(d|h)$ is the conditional probability density function of damage D conditioned on the hazard value $H=h$. Note that rather than integrating with respect to H, we integrate with respect to damage d. The way to interpret this equation is as follows: For a particular hazard value, $H=h$, the probability that the damage is greater than the value d_0 is given by Pr($d_0<D<\infty|H=h$), which is evaluated by performing the integration on the right-hand side, and where:

$$f_{D|H}\left(d|h\right) = \frac{f_{DH}\left(d,h\right)}{f_H\left(h\right)},$$

where $f_{DH}(d,h)$ is the joint probability density function of the random variables D and H, and $f_H(h)$ is the marginal probability density of H, given by:

$$f_H\left(h\right) = \int f_{DH}\left(d,h\right) dd.$$

We can expand the above relations by introducing intermediate useful variables. For example, we can introduce the force F in the following way:

$$\Pr\left(D|H\right) = \Pr\left(D|F\right)\Pr\left(F|H\right).$$

Here, on the left-hand side of the equation, the damage D conditioned on the hazard H is expanded to include an intermediate variable, the force F. So the conditional probability of damage conditioned on the hazard, Pr($D|H$), is related to the conditional probability of damage conditioned on the force, Pr($D|F$), and the conditional probability of the force conditioned on the hazard, Pr($F|H$). Therefore, the damage probability becomes:

$$\Pr\left(D\right) = \iint \Pr\left(D|F\right)\Pr\left(F|H\right) dH dF.$$

And similarly, in terms of the conditional probability density functions, we have:

$$f_{D|H}\left(d|h\right) = f_{D|F}\left(d|f\right) f_{F|H}\left(f|h\right).$$

Therefore, the probability of damage for a particular hazard realization, $H=h$, is:

$$\Pr\left(d_0 < D < \infty | H = h\right) = \iint_{d_0}^{\infty} f_{D|F}\left(d|f\right) f_{F|H}\left(f|h\right) dd df.$$

We note that the single integral becomes a double integral, where we integrate over the range of damage, and over the range of forces. In this instance damage occurs for $D > d_0$.

We can generalize these ideas for any input I, and find the probability of damage λ, as:

$$\lambda = \Pr(D) = \iiint \Pr(D|S)\Pr(S|I)\Pr(I|H)\,dH\,dI\,dS.$$

Here, the intermediate parameter S has been introduced, where S is the structural response (can be stress, displacement), and λ can be interpreted as the decision variable. It is a probability, a number between zero and one. The larger its value, the higher the probability of damage.

Other possible risk parameters include: G = low gravity, R = radiation dosage, Q = small spaces, Is = isolation. Gravity and radiation have both engineering implications and physiological implications. Is and Q are reflected in the physiological and psychological wellbeing of the inhabitants.

For a particular lunar site, what are the Hazards? Forces? Inputs? A goal is to evaluate λ for different concepts, designs, and locations. Performance-based engineering can be a useful tool to evaluate and compare various concepts. Clearly, one challenge is to gather enough data to populate the respective probability functions. The other challenge is the numerical integration of the multiple integrals; that is, identifying the ranges and domains of each of the integration variables.

Possible models

Here is a look at a simple set of equations. Our generic case study is the following:

$$\lambda_y = \Pr(Y > y)$$
$$\lambda_y = \iint \Pr(Y > y|m,r) f_M(m) f_R(r)\,dm\,dr,$$

(13.1)

where:

$$\Pr(Y > y) = \left[1 - F_Y(y)\right] \equiv G_Y(y)$$
$$\Pr(Y > y|m,r) = \left[1 - F_{Y|M,R}(y|m,r)\right],$$

and M and R have been assumed to be statistically independent:

$$f_{MR}(m,r) = f_M(m) f_R(r).$$

The quantity $[1 - F_{Y|M,R}(y|m,r)]$ is called the *fragility*. It is the probability of exceeding the value of a parameter that indicates failure.

We can assume density functions for M and R, and we need expressions for $F_{Y|M,R}(y|m,r)$. One possibility is the two-parameter Weibull density:

$$f_{Y|M,R}(y|m,r) = \frac{m}{r^m} y^{m-1} e^{-(y/r)^m}, \quad 0 < y < \infty,$$
$$m > 0, \quad r > 0,$$

with cumulative distribution:

$$F_{Y|M,R}\left(y|m,r\right) = \Pr\left(Y \le y\right) = 1 - \exp\left\{-\left(\frac{y}{r}\right)^m\right\},$$

or:

$$\Pr\left(Y > y|m,r\right) = \exp\left\{-\left(\frac{y}{r}\right)^m\right\}.$$

The hazard rate can be found by differentiating the hazard function $R(y)$:

$$R\left(y\right) = -\text{In}\left[1 - F_{Y|M,R}\left(y|m,r\right)\right],$$

to find:

$$r\left(y\right) = \frac{m}{r^m}y^{m-1}.$$

If $m = 1$ in the Weibull, then we get the exponential density. Then M is a deterministic quantity and the double integral becomes a single integral, with the exponential being a function only of R. This may be a good test of the method, with R given by a uniform density over a range:

$$\lambda_y = \int_{r_1}^{r_2} \exp\left\{-\left(\frac{y}{r}\right)\right\}\frac{1}{r_2 - r_1}dr.$$

Although incomplete, we put forward these specialized equations to help begin to understand their meanings.

A lunar PBE model

The interesting aspect of the PBE framework for hurricane engineering by Barbato *et al.*, cited above, is that "while other PBE methodologies focus on single hazards, the landfall of a hurricane involves different hazard sources (wind, wind-borne debris, flood, and rain) that interact to generate the hazard scenario for a given structure and to determine its global risk. Thus, hurricanes can be viewed, and must be analyzed, as multi-hazard scenarios." When we consider the lunar surface environment and the many hazards that affect a structural design, clearly it is also a multi-hazard environment. In addition to the physical hazards, there are the biological and human factors hazards. Similar to the hurricane model, the lunar model can group hazards as:

1. Independent hazards, because no mutual interaction between the two hazards has the effect of modifying the intensity of the corresponding actions. These hazards can occur individually or simultaneously. An example is the meteoroid hazard or the radiation hazard.

2. Interacting hazards are interdependent. The radiation environment and the temperature environment are correlated. At some level, radiation can affect the behavior of the electrostatically charged regolith. We also know that low gravity and psychosocial stressors affect human physiology.
3. Hazard chains are those where the effects of some hazards can sequentially affect other hazards. For example, the shielding and structure can be damaged due to a meteoroid strike, leading to a weakened structure requiring repair or evacuation. Similarly, radiation on a structure can cause the release of secondary particles.

For the lunar PBE model, a hazard analysis for each hazard results in an equation that characterizes the relation between a specific damage state and the environmental causes. Since the structure needs to shield the astronauts, a hazard analysis that relates human physiology to the radiation, low gravity and dust environment is needed. These are likely coupled (interacting) equations that, it seems, may need to be coupled to human psychological behavior. Of course, it is unclear that these models can be developed from data available today. It may initially be necessary to use crude qualitative models to begin to understand these relationships. It is also expected that these relationships can have probabilistic components due to the lack of data; that is, due to our lack of knowledge.

Clearly, the PBE method needs extensive development for its application to the assessment of lunar habitat designs, while accounting for all the stages of the design, transportation, fabrication or construction, operations, possible losses, and the implications of failures at any stage. This is challenging, but worth the attempt. It is ongoing research.

References

1. H. Benaroya: *Performance-Based Engineering for Lunar Settlements*, p.33, Report of the Space Resources Roundtable VII: LEAG Conference on Lunar Exploration, October 25-28, 2005, League City, Texas.
2. H. Benaroya: *Performance-Based Framework for the Design of Lunar Structures*, 66th International Astronautical Congress, Jerusalem, Israel, 2015. IAC-15-E5.1.4.
3. A.P. Dempster: *A generalization of Bayesian inference*, Journal of the Royal Statistical Society, Series B 30, 1968, pp.205-247.
4. **A Mathematical Theory of Evidence**, G. Shafer, Princeton University Press, 1976.
5. L.A. Zadeh: *A Simple View of the Dempster-Shafer Theory of Evidence and its Implication for the Rule of Combination*, AI Magazine, Vol.7, No.2 (1986), pp.85-90.
6. P. Limbourg, R. Savić, J. Petersen, and H.-D. Kochs: *Fault tree analysis in an early design stage using the Dempster-Shafer theory of evidence*, **Risk, Reliability and Societal Safety**, Aven and Vinnem, Editors, 2007, Taylor & Francis Group.
7. C.A. Cornell, and H. Krawinkler: *Progress and Challenges in Seismic Performance Assessment*, PEER Center News, Vol.3, No.2, Spring 2000.
8. K.A. Porter: *An Overview of PEER's Performance-Based Earthquake Engineering Methodology*, in Proceedings of the 9th International Conference on Applications of Statistics and Probability in Civil Engineering, Rotterdam, A. Der Kiureghian, S. Madanat, and J.M. Pestana, Editors, Civil Engineering and Reliability Association, Millpress, 2003.

9. F. Petrini, M. Ciampoli, and G. Augusti: *A probabilistic framework for Performance-Based Wind Engineering*, EACWE 5, Florence, Italy, July 19-23, 2009.
10. M. Ciampoli, F. Petrini, and G. Augusti: *Performance-Based Wind Engineering: Towards a general procedure*, Structural Safety 33 (2011) pp.367-378.
11. M Barbato, F. Pertini, V.U. Unnikrishnan, and M. Ciampoli: *Performance-Based Hurricane Engineering Framework*, Structural Safety 45 (2013) pp.24-35.

14

Concluding thoughts

"Parting words."

We close this book with a short summary of its contents and with a few additional thoughts. The Moon is an ideal location for the settlement of people for the purposes of expanding our understanding of ourselves, science, human nature's demand for exploration and learning, and of our place in the universe. As by-products, but no less important, we will be creating another world of intellectual and economic opportunities for people, initially the pioneers, but soon afterwards the motivated along with their families.

If there is any takeaway from this book, it is that we need to understand how the reliability of engineered systems can be improved in the unforgiving space and lunar environment and, synergistically with reliability, how to ensure that humans and other living systems can survive and thrive physically and psychologically in that environment.

Uncertainty and high risk are two topics that we focused on heavily. We cannot overemphasize the notion that the success of a lunar facility, as defined above, rests on our abilities to understand and work with incomplete knowledge. Low-probability, high-impact events define our lives in space and on the Moon. Low-probability events cannot be predicted, and are almost impossible to quantify. Therefore, we must make ourselves and our systems hardened to these events. We have control over our actions in this way. We have experienced numerous low-probability, high-impact events and they fulfilled their moniker; no one predicted them in a way that we could either stop them from happening, or completely negate their effects. Their impacts were severe.

Here are the key issues:

Environmental Definition
is critical in order for a rational design of a habitat to go forward. The impact of the environment is on engineered systems as well as biological systems. Our engineered systems and our biological systems are coupled in numerous ways. The former need to be designed so that the latter survive in them, can perform operations easily and without much stress, and can thrive and live full lives in them.

© Springer International Publishing AG 2018
H. Benaroya, *Building Habitats on the Moon*, Springer Praxis Books,
https://doi.org/10.1007/978-3-319-68244-0_14

The latter need to be able to manage and operate the systems of the former. They need to be able to repair the former, and crucially, be able to anticipate critical situations before they become critical.

Complexity and Reliability
are considerations that play against each other. Some level of complexity is unavoidable given the monumental task of placing habitats on the Moon, but too much complexity, and the reliance on many layers of redundancies, can be catastrophic. The Moon should not become a testing ground for new technologies more than it has to be. Anything that can be done as well, or sufficiently well, with a simpler and older technology should be designed that way. Considering the near-term, where layered manufacturing and 3D printing, ISRU, and automation and robotics are still in their infancies, the likely first habitat will be a pressure vessel, or two of these combined. They will be delivered intact to the lunar surface, with astronauts not far behind to finish the task of shielding. We cannot *a priori* estimate the reliability of a lunar habitat, because we cannot duplicate the lunar environment on Earth where we will build the habitat. At best, we can guesstimate orders of magnitude. (Perhaps the use of slide rules should be considered again, since they do not provide more than 3 or 4 significant digits. Computers can deceive by providing tens of significant digits, regardless of the accuracies of the input parameter values.)

Redundancies
are needed for almost everything. The astronauts will need to have overlapping multiple skills. Our engineered systems cannot have a single point of failure, but systems that appear to have redundancies may be deceptively redundant. Couplings occur through unanticipated mediums. Two habitats at distant locations on the lunar surface are coupled by virtue of the environment, the manufacturers of the habitats, the software managing the habitats, and the personalities and the training of the astronauts. At some level, the engineering of the lunar habitat is not fully understood. Mathematical and computational models are all approximate, regardless of their sophistication and complexity. Tests using habitat prototypes on Earth cannot account for everything that will happen on the lunar surface. We need to make our systems redundant, but we also need to map out the operations domain with the possibility that redundancies may be coupled in the end.

Design Stability
implies that new generations of designs should not be revolutionary, but evolutionary, building upon successful and reliable designs that have withstood the test of time. On the Moon, there is no such history, so the first structures on the lunar surface should be of designs that have a long history on Earth; for example, pressure vessels. Inflatable structures are attractive, but their development and deployment on Earth is comparatively limited. It is reasonable to expect that once a first habitat is placed on the lunar surface, improvements will proceed rapidly. Recursive practice will ensure that any lessons learned are quickly implemented in subsequent structures.

The Moon
needs to be the site of our first extra-planetary habitats. Mars can only be considered as such decades after we survive on the Moon, can build larger habitable volumes, can develop a minimal infrastructure for mining and ISRU, and begin to understand how the lunar environment, in all its magnificent desolation, affects humans and plants.

Haym Benaroya
September 2017

Index

A

Abrasion, 75, 79–81, 120, 122, 126, 175, 176, 268, 272
Accidents, 152, 252, 290–293, 296, 301
Apollo, 12, 14, 16, 19, 21, 22, 32, 36, 37, 42, 44, 46, 50, 53–55, 58, 67, 69, 71–74, 77–79, 81, 89, 95, 96, 98–100, 134, 136, 137, 142, 147, 148, 159, 160, 166, 175, 184, 187, 217, 225, 238–240, 256, 278
Architecture, 32, 92–94, 98, 100, 102–105, 116, 130, 148, 150, 158, 184, 194
Astronomy, 9, 20, 55, 149, 172–174, 176, 277
Automation, 15, 87, 88, 90, 110, 179, 308

B

Bacteria, 167, 168
Base
    construction of, 79, 149, 153, 157
    design of, 73, 106, 149, 150, 257
    lava tube, 187–193
    lunar, xiii, 6–11, 15, 18–21, 31–33, 48, 62, 64, 73, 79, 89, 90, 101, 106, 109, 110, 112, 121, 122, 142, 145, 147–151, 153, 154, 157, 163, 180, 197, 213, 252–262, 282, 295, 300, 301
    Mars, 36, 43, 50, 64, 133
    mobile, 102, 110, 112, 149
    modular, 32, 112
    planetary, 25, 26, 110
Bigelow, 96, 103, 126, 128, 131

C

Chemical exposure, 43
Cohen, M., 93, 101, 103, 105, 112, 161

Complex, 4, 13, 20, 36, 39, 49, 53, 55, 68, 73, 77, 89, 91, 97, 115, 117, 128, 130, 134, 155, 158, 167, 183, 186, 191–193, 209, 224, 225, 235, 246, 252, 255–258, 262, 263, 266, 283, 286, 288–292, 295, 297, 301
Composite, 67, 91, 115, 116, 127–130, 148, 151, 154, 174, 178, 180, 201
Construction, 9, 15, 42, 71, 79, 88–91, 93, 94, 97, 100, 106, 121, 142–145, 148, 149, 151–155, 157, 162, 173, 174, 176, 179, 180, 183, 197–199, 201, 203, 204, 244, 249, 257–259, 261, 262, 305
Control, 6, 14, 16, 48, 49, 55, 56, 60, 62, 72, 80, 81, 87, 97, 98, 108, 117, 119, 127–129, 149, 159, 182, 193, 208, 209, 216, 218, 261, 286, 287, 289, 307
Cosmic rays, 10, 60, 73, 178
Cylinder, 62, 106, 108, 114, 132, 143

D

Damping, 73, 79, 182, 246, 272
David, C., 128, 130
Dempster-Shafer Theory, 299, 300
Deploy, 11, 53, 77, 90, 92, 95, 103, 110, 111, 114–117, 120, 122, 123, 125–132, 135, 143, 156, 174, 192, 193, 258, 259, 289, 308
Design
    codes, 47, 120, 152, 254, 259
    constraints, xiii, 68, 128, 150, 154, 174, 255
    life, 154, 249, 253, 262
    loads, 152
    materials, 116, 149, 180
    parameters, 80, 150, 260, 263
    philosophy, 66, 108, 109, 134, 135, 254, 288

© Springer International Publishing AG 2018
H. Benaroya, *Building Habitats on the Moon*, Springer Praxis Books,
https://doi.org/10.1007/978-3-319-68244-0

risk, 252, 254
  structural, xii, xiv, 36, 47, 66, 93, 97, 117, 130,
    134, 145, 150, 152–154, 159, 186,
    197–206, 213, 225, 235, 259, 279, 300, 304
Design stability, 295, 296
Dosage, 46, 47, 61, 62, 154, 301, 303
Dual-use, 23–27
Dust, 9, 36, 47, 52, 53, 71–73, 77, 79–81, 87, 97,
    98, 118, 119, 134, 135, 142, 150, 154, 174,
    175, 179, 186, 188, 190, 219, 255, 263,
    268, 276, 277, 305

E
Economics, 35, 99, 104, 108, 250
Ejecta, 19, 175, 276–278, 281, 282
Engineering
  design, 75, 93, 145, 174, 182, 263, 267, 286,
    290, 295, 296
  NASA, 17, 19, 52, 72, 94, 102, 256
  reliability, xiii, 99, 187, 262–266, 307
  system, 36, 53, 97, 98, 104, 135, 186, 250,
    292, 294, 307, 308
Environment
  artificial, 158
  extraterrestrial, 167
  hostile/extreme, 99, 109, 114, 131, 136, 158,
    163, 167, 178, 208, 249
  isolated, 25, 27, 36, 98, 135, 186
  low/reduced gravity, 14, 28, 43–45, 159, 160,
    263, 305
  lunar, xii–xiv, 12, 14, 15, 42–81, 93, 116, 148,
    149, 152, 153, 156, 167, 168, 175, 178,
    181, 216, 249, 262, 263, 276, 296,
    307–309
  plasma, 77
  physical, 42, 152, 307
  radiation, 60–70, 97, 101, 142, 145, 148, 229,
    263, 290, 305
  shielding from, 62
  thermal, 97, 208
  vacuum, 173, 201
Erection, 90, 152, 203, 204, 257, 282, 296
Ethics, 28

F
Fabrication, 69, 85, 90, 178, 179, 197, 289, 301,
    305
Fatigue, 42, 45, 87, 149, 152, 158, 178, 182, 213,
    249, 254, 255, 257, 260, 262, 263, 266,
    271–274, 276, 287, 291
First-excursion, 254, 266, 267, 269, 270, 272
Fold, 64, 68, 114–116, 122, 125, 126, 129

Force, 3, 8, 15, 43, 58, 68, 73, 90, 91, 98, 117,
    152, 153, 190, 202, 203, 212, 224, 226,
    229, 231, 233–236, 240, 251, 262, 266,
    267, 269, 277, 283, 292, 302, 303
Foundation, 21, 33, 34, 49, 55, 73, 77, 79, 90, 98,
    120, 122, 145, 153, 180, 197, 201, 203,
    210, 212, 216, 218, 221, 222, 243, 244,
    246, 257, 286, 296
Framework, 43, 93, 106, 108, 109, 122, 152–157,
    192, 226, 234, 255, 259, 288, 294, 299,
    300, 304

G
Gravity, 2, 14, 43–45, 47, 48, 51, 52, 57, 65, 66,
    71, 75, 78, 79, 88, 98, 108, 118, 127, 150,
    158, 159, 163, 165–168, 174, 182, 188,
    190, 215, 238, 256, 263, 281, 292, 300,
    301, 303, 305

H
Habitability, 117, 150, 158, 159, 161, 162
Habitat, 13, 85, 142, 143, 148–163, 166–168, 178,
    197, 199, 201, 202, 204, 208, 249,
    286–297, 299, 308
Hazard, 3, 6, 15, 44, 60, 62, 66, 71, 153, 192, 250,
    252, 275, 288, 294, 302–305
Heat flux, 209, 218, 219, 222
Heat pipes, 209
Heat transfer, 149, 208, 209, 216
Human
  environmental challenges to, 148, 151
  habitats/habitation, xi, 15, 159
  health, 42, 43, 98
  motivation, 158
  physiology, xi, 31, 43, 48, 159, 163–166,
    259, 305
  psychology, xi, 31, 159–163, 259
  risks/hazards to, 37, 62, 153, 304, 305
  settlement, 15, 33
  survival/safety, xi, 21, 37, 54, 91, 99, 136, 167,
    178, 187, 255
Human factors, 93, 94, 104, 213, 226, 304

I
Igloo, 179, 197, 206, 210
Impact, 17, 19, 22, 29, 31, 32, 38, 43, 67, 71–76,
    79, 81, 117, 128, 129, 132, 136, 152, 153,
    156, 160, 163, 174, 175, 179, 182, 184,
    188, 190, 210, 217, 219, 226, 233, 239,
    240, 249, 260, 263, 269, 272, 275–279,
    281–283, 288, 301, 307

Inflatable, 26, 86, 91, 103, 114–133, 135, 149,
    151, 155, 162, 179, 192, 193, 197, 258, 308
In-situ resource utilization (ISRU), 13, 35, 36, 49,
    94, 97, 102, 114, 127, 130, 134–136, 148,
    151, 154, 162, 178, 180, 181, 184, 185,
    192, 255, 308, 309

K
Kennedy, J.F., 1, 2, 14

L
Limit state, 260, 261, 263, 266
Livingston, D., 16, 28, 35
Load, 47, 69, 91, 103, 115, 119, 125, 127, 128,
    152, 153, 156, 165, 174, 178, 198, 199,
    203, 209, 212, 214–216, 222, 226, 249,
    257, 258, 260–262, 264, 266, 273, 274,
    276, 301
Logan, J., 48
Loss, 16, 80, 94, 100, 150, 161, 165, 217, 221,
    241, 250, 252, 253, 257, 278, 279, 297,
    300, 301
Lunar Exploration Systems for Apollo (LESA),
    106, 142, 143, 147

M
Magnesium, 13, 181–183, 210, 212, 214, 219,
    243, 244, 246, 272
Mars
    colonization, 52
    environment of, 36, 42, 62, 135, 186
    exploration, 31
    habitat, 27, 31, 66, 102
    human mission, 31, 37, 54, 99, 136, 162, 184,
        185, 187
    ISRU, 23, 133, 179, 184–187
    resources, 15, 27, 39, 133
    settlement of, 31, 161
Materials, 7, 13–15, 17, 28, 42, 44, 46, 49, 52, 60,
    64, 66, 68, 72, 79, 80, 86–88, 93, 95, 107,
    114–116, 127–129, 132, 133, 135, 145,
    148, 149, 152, 154, 155, 174, 178, 179,
    181, 183, 185, 201, 223, 231, 249, 256,
    262, 268, 272, 278, 279, 286, 301
Medical, 14, 15, 21, 48, 49, 52, 54, 56, 58, 98,
    142, 150
Meteoroid, 32, 43, 66, 71, 87, 88, 91, 152, 156,
    174, 182, 192, 210, 217, 226, 227, 233,
    239, 240, 272, 275–277, 279–282, 290,
    301, 304
Microgravity, 45, 47, 65, 98, 142, 152, 158, 163

Micrometeoroid, 42, 71, 75, 125, 126, 129, 135,
    145, 153, 160, 173, 174, 190, 262, 269,
    275–283
Miner rule, 272
Module, 55, 62, 66, 72, 74, 77, 81, 86, 92,
    97, 103, 105, 107, 108, 112, 116, 117,
    121, 128, 132, 147, 152, 155, 156,
    163, 204
Moonquake, 156, 182, 238, 240, 241, 246
Mottaghi, S., 210, 239
Multiplier, 17, 18

N
Nonlinear, 78, 79, 115, 116, 128, 251,
    287, 301

O
Objective, 19, 49, 60, 88, 106, 110, 154, 155, 172,
    208, 251, 254, 288

P
Package, 108, 117, 119, 126–128, 240
Performance-based engineering, 299–305
Physiological pressure, 93, 150, 160
Physiology, 48, 159, 163, 226, 259, 305
Plants, 48, 52, 65–67, 160, 162, 167, 168, 294,
    296, 300
Power spectrum, 235, 236, 241
Pressurization, 87, 133, 150, 155, 160, 199, 258,
    259, 301
Probability, 37, 43, 54, 61, 62, 66, 99, 129, 136,
    150, 156, 187, 224, 226–228, 230–237,
    239–241, 244–246, 249, 250, 252, 253,
    260, 264–269, 272, 274, 276, 281–283,
    289, 291, 299–303, 307
Project Horizon, 142, 144, 145
Psychological, 14, 31, 32, 52, 89, 93, 117, 150,
    158–162, 167, 253, 300, 301
Psychology, 32, 59, 91, 104, 159
Psychosocial, 150, 158, 161, 301, 305

R
Radiation, 8, 11, 14, 42, 44–47, 49–51, 53, 59–62,
    65–69, 73, 86, 91, 97, 98, 101, 102, 105,
    112, 115, 120, 125, 128, 129, 142, 145,
    148–150, 154, 158, 160, 167, 173, 174,
    178, 179, 182, 184, 187, 188, 190,
    208–210, 216, 217, 219, 229, 235, 236,
    256, 259, 262, 263, 268, 269, 272, 281,
    290, 300, 301, 303, 304

Random, 55, 64, 128, 156, 224–237, 239, 245, 259–261, 265–267, 269–272, 274, 279, 287, 291, 300, 302
Rapp, D., 66, 183
Recursive practice, 295, 296, 308
Redundancy, 110, 180, 252, 254, 256–258, 283, 286–297
Regolith
  composition of, 180, 181
  conductivity (thermal), 47, 149, 182, 183, 218, 219
  density, 64
  dust, 47, 72, 73, 80, 179, 219, 255, 263
  grains, 73, 75, 79
  lunar, 13, 14, 47, 53, 61, 66, 71, 72, 74–77, 79, 80, 112, 119, 148, 149, 168, 179–181, 219, 257
  mechanics, 74, 149, 301
  properties of, 73–77, 79, 80, 216
  refining, 180
  shielding, 42, 62, 67, 68, 95, 106, 108, 109, 114, 121, 151, 153, 163, 200, 205, 206, 210, 212, 215, 216, 219, 229, 243–246, 259, 272, 277
  simulant, 179
  sintered/fused, 179, 210, 218, 244
Reliability, 36, 37, 53, 88, 91, 93, 97–100, 110, 117, 122, 129, 134–136, 147, 153, 156, 161, 174, 180, 186, 187, 209, 224, 226, 229, 235, 249, 250, 252–269, 271–283, 286, 301, 307
Rigid, 85, 103, 105, 106, 108, 110, 112, 114–117, 121, 127, 128, 131, 132, 197, 240
Rigidization, 116, 128, 129, 131
Rigidize, 114, 115, 122, 123, 129, 131, 192
Risk, 14, 36, 37, 42, 43, 53, 59, 61, 62, 64–68, 81, 88, 98, 100, 122, 127, 132, 135, 136, 150, 154, 158, 159, 161, 178, 182, 186, 187, 214, 226, 246, 249–262, 268, 277, 278, 282, 288, 289, 293, 294, 297, 300, 301, 303, 304, 307
Robotics, 12, 13, 15, 21, 23, 25, 32, 33, 38, 51, 87, 88, 90, 91, 94, 101, 110, 127, 150, 151, 156, 162, 172–174, 179, 180, 257, 308
Rovers, 24, 26, 27, 64, 101, 106, 111, 118, 127, 186, 209, 255, 289
Ruess, F., 197–202

S
Science, 3–5, 12, 14, 16–18, 20, 21, 25, 27, 29, 31, 33, 34, 37–39, 51–53, 56–58, 65, 86, 87, 91, 94, 95, 98, 99, 104, 112, 115, 118, 134, 145, 152, 162, 172, 173, 176, 183, 184, 194, 251, 283, 286, 287, 307

Seismic, 43, 76, 77, 79, 148, 152, 172, 188, 190, 192, 224–226, 229, 233–248, 277, 278, 300
Self-healing, 36, 98, 116, 117, 125, 129, 135, 186
Settlements, 12, 14, 15, 18, 21, 22, 31–33, 36, 47, 49–51, 57, 65, 69, 71, 77, 89, 93, 96, 99, 101, 106, 136, 142, 146, 147, 149, 151, 153, 159–161, 163, 167, 168, 179, 180, 185–188, 204, 255, 277, 286, 307
Shell, 67, 86, 112–114, 117, 125, 126, 128, 212, 214
Shield, 9, 14, 42, 46, 47, 49, 50, 60–62, 64–69, 75, 88, 91, 95, 101, 102, 106, 108, 109, 111, 112, 118–121, 123, 125–127, 129, 149, 151, 153, 155, 163, 167, 172, 174, 178–180, 182, 188, 198–201, 204–206, 209, 210, 212, 215–217, 219, 229, 239, 243–246, 259, 272, 276–278, 281, 282, 301, 305, 308
Sinter, 155, 179, 210, 218, 244
Soil, 32, 53, 65, 67, 71, 73–80, 89, 90, 120, 123, 145, 148, 149, 167, 168, 179–181, 183, 191, 208, 209, 246, 248
Solar particle, 60, 64, 67
South Pole, 9, 72, 173, 216, 217, 220–222
Space
  architecture, 92–94, 100, 103–105
  budget/money/funding, 4, 16, 22, 166
  commerce, 34, 35, 39
  environment, 14, 22, 28, 43, 45, 52, 87, 97, 99, 167, 192
  exploration, 3, 12, 15, 16, 18–21, 29–33, 49, 57, 59, 133, 136, 137, 149, 172, 252, 297
  habitat/structure, 31, 97, 98, 103, 115, 116, 129–131, 135, 296
  industry, 4, 17, 104
  policies, 15, 283
  race, 3, 14, 49, 96, 145
  radiation, 61, 67, 68, 98
  resources, xiii, 12
  science, 3, 4, 29, 31, 58, 173
  settlement, 19, 21, 32, 36, 49, 65, 99, 136, 185, 186
  tourism, 12, 34, 35
  treaties, 29, 38
Spectral density, 234–237, 241–243
Structure
  composite, 115, 121, 128, 148, 151, 180
  cylindrical/cylinder, 91, 106, 108, 109, 128, 143

Structure (*cont.*)
  deployable, 90, 114–116, 128, 131, 135,
    192, 193
  design of, xii, xiii, 60, 69, 90, 128, 145, 152,
    204, 239, 258, 281, 283
  3D-printed, 35, 85, 90, 91, 94, 130, 133, 184
  engineered, xi–xiii, 14, 31, 93, 121, 131, 156,
    191, 192, 212, 244, 254
  inflatable, 26, 86, 91, 103, 114–117, 121–123,
    125–129, 131–133, 135, 149, 151, 155,
    162, 179, 192, 193, 197, 258, 308
  lightweight, 127
  lunar, xiii, 35, 36, 42, 47, 71, 79, 88–91, 93,
    94, 110, 115, 129, 130, 145, 148, 149, 151,
    153–155, 159, 162, 178, 181, 184, 185,
    201, 204, 209, 212, 213, 216, 220, 226,
    235, 237, 239–241, 244, 246, 247, 249,
    252, 254, 258–262, 272, 276, 278, 279,
    281, 295–297, 299, 301
  rigid, 85, 103, 105–117, 121–123,
    127–132
  shielding of, 47, 66, 68, 75, 149, 151, 180,
    200, 204, 205, 239
  surface, xiii, 42, 66, 68, 88, 149, 151, 159,
    162, 183, 204, 209, 210, 219, 225, 235,
    237, 239, 240, 244, 248, 249, 253, 257,
    262, 272, 276–279, 281–283, 292, 296
  underground/buried, 110, 142, 148, 197, 239,
    248, 277
Sulfur, 179

T
Technology transfer, 22
Telescope, 55, 172–175, 276
Teller, E., 6–11, 18
Temperature, 6–10, 27, 42, 43, 46, 49, 52, 77, 98,
    112, 116, 120, 127–129, 148, 149, 160,
    167, 174, 179, 182, 184, 187, 188, 191,
    198, 203, 208–210, 216–221, 240, 305
Thermal, 24, 42, 46, 49, 50, 66, 87, 88, 97, 115,
    116, 125, 127–129, 145, 149, 152, 156,
    157, 161, 174, 175, 178, 182, 190, 191,
    208–223, 240, 245, 253, 254, 256, 259,
    281, 282, 301
Thermal control, 24, 62, 80, 87, 97, 127, 149, 208,
    209, 216, 218
Toxicity, 43, 45, 72, 88
TransHab, 102, 103, 116, 125–128, 130

U
Uncertainty, 43, 61, 62, 68, 73, 90, 128, 152, 156,
    224–226, 229, 231, 233, 237, 239,
    250–254, 259, 261, 267, 286, 287, 290,
    292, 299–301, 307

V
Vibration, 79, 87, 132, 158, 161, 182, 226,
    236–239, 246, 247, 260, 262, 263, 271,
    272, 276